Nutritional Science and Technology

Scrivener Publishing
100 Cummings Center, Suite 541J
Beverly, MA 01915-6106

Bioprocessing in Food Science

Series Editor: Anil Panghal, PhD

Scope: Bioprocessing in Food Science a series of volumes covering the entirety of food science, unit operations in food processing, nutrition, food chemistry, microbiology, biotechnology, physics and engineering during harvesting, processing, packaging, food safety, and storage and supply chain of food. The main objectives of this series are to disseminate knowledge pertaining to recent technologies developed in the field of food science and food process engineering to students, researchers and industry people. This will enable them to make crucial decisions regarding adoption, implementation, economics and constraints of the different technologies. Bioprocessing has revolutionised the food industry by allowing for more efficient and sustainable production methods. This comprehensive series focused on microbial fermentation, enzyme technology, genetic engineering, microbial transformations, and bioreactor design. As we continue to face challenges such as population growth and climate change, bioprocessing will play an increasingly important role in ensuring a sustainable food supply for future generations.

Manufacturers are looking for new opportunities to take a significant position in a food market that is continually changing as demand for healthy food rises in the current global environment. In the current scenario, academia, researchers and food industries are working in a scattered manner and different technologies developed at each level are not implemented for the benefits of different stake holders. Compiled reports and knowledge on bioprocessing and food products is a must for industry people. However, the advancements in bioprocesses are required at all levels for betterment of food industries and consumers.

The volumes in this series are comprehensive compilations of all the research that has been carried out so far, their practical applications and the future scope of research and development in the food bioprocessing industry. The novel technologies employed for processing different types of foods, encompassing the background, principles, classification, applications, equipment, effect on foods, legislative issue, technology implementation, constraints, and food and human safety concerns are covered in this series in an orderly fashion. These volumes are comprehensively meeting the knowledge requirements for the curriculum of undergraduate, postgraduate and research students for learning the concepts of bioprocessing in food engineering. Undergraduate, post graduate students and academicians, researchers in academics and in the industry, large- and small-scale manufacturers, national research laboratories, all working in the field of food science, agri-processing and food biotechnology will be benefitted.

Publishers at Scrivener
Martin Scrivener (martin@scrivenerpublishing.com)
Phillip Carmical (pcarmical@scrivenerpublishing.com)

Nutritional Science and Technology

Concept to Application

Edited by

Tejpal Dhewa
Anil Kumar Puniya
and
Anil Panghal

Scrivener Publishing

WILEY

This edition first published 2023 by John Wiley & Sons, Inc., 111 River Street, Hoboken, NJ 07030, USA and Scrivener Publishing LLC, 100 Cummings Center, Suite 541J, Beverly, MA 01915, USA

For more information about Scrivener publications please visit www.scrivenerpublishing.com.

Wiley Global Headquarters
111 River Street, Hoboken, NJ 07030, USA

For details of our global editorial offices, customer services, and more information about Wiley products visit us at www.wiley.com.

Library of Congress Cataloging-in-Publication Data

ISBN 9781119808961

Cover images: Dermcidin, Molekuul | Dreamstime.com, Doctor with Microscope, Kurhan | Dreamstime.com, Probiotic Bacteria, Kateryna Kon | Dreamstime.com
Cover design by Kris Hackerott

Set in size of 11pt and Minion Pro by Manila Typesetting Company, Makati, Philippines

Printed in the USA

Contents

Preface

The book, *Nutritional Science and Technology: Concept to Application*, in the "Bioprocessing in Food Science" series is an up-to-date and comprehensive overview of this area. Written and edited by a team of experts in the field, this book covers topics such as human nutrition, technological processes, the health benefits of fermented foods, and food safety concerns. The book's primary goal is to provide extensive knowledge about recent technologies in nutritional science and technology to students, researchers, and industry professionals. It will enable them to make informed decisions regarding adopting appropriate processing technology, implementation, economics, and constraints of different technologies. As the demand for healthy, nutritious, and safe food increases, manufacturers are looking for new possibilities to occupy a growing share of the rapidly changing food market.

The book covers a range of emerging topics, from concepts to applications of human nutrition, supplements, gut microbiota, mechanistic aspects of dietary modulation, food metabolism, chronic diseases, health benefits of fermented dairy products, advances in nutritional biomarkers in the detection of metabolic disorders, food omics, their implications in nutritional sciences, nanotechnology in food systems, applications of nanomaterials in the food system, emerging techniques in food preservation, molecular nutrition, nutrient-gene interactions, food allergy and intolerance, and food safety and quality assurance in the food chain. The book also provides insights on advances in nutritional science and technology for healthy and safe nutrition with illustrations of how to ensure public health safety and adequate nutrition. The authors and editors discussed the need for innovative food products, contamination in the food chain, risk assessment, regulatory frameworks, and their challenges within the context of the nutritional sciences in the global food market.

Editors
Tejpal Dhewa
Anil Kumar Puniya
Anil Panghal

1

Human Nutrition and Supplements

Mahak Sharma*, **Divya Sanghi and Ankita Sharma**

Department of Nutrition and Dietetics, Faculty of Allied Health Sciences, Manav Rachna International Institute of Research and Studies, Faridabad, Haryana, India

Abstract

Dietary supplements are isolated or bound nutrients consumed by individuals to improve their nutritional status. They are usually consumed when an individual is deficient in a particular nutrient or wants to increase his/her muscle mass. The ingredients usually used as supplement are amino acid, herbs, minerals, vitamins and enzymes. Dietary supplements are used in the form of softgels, powder, capsules, liquid, and gelcaps. There are many health benefits of dietary supplements like Iron, Zinc, Protein supplements, Vitamin D, Calcium, Omega–3 fatty acids and many more. To maintain optimal growth and development while prevent deficiency, adequate intake is advisable but excess use or high dose cause the possibility of toxicity. This chapter emphasizes various health benefits of dietary supplements and their uses.

Keywords: Food supplements, sports supplements, nutrition, health

1.1 Dietary Supplements – Introduction

There is a broad variety of dietary supplements in the market. More than 56,000 distinct dietary supplements are available, according to the US Food and Drug Administration (FDA), with thousands of new products/supplements being introduced to customers each year [1].

Dietary supplements, by definition, are formula that can supplement our regular diet with nutrients and reduce the risk of deficiency diseases and health problems like Protein Energy Malnutrition (PEM) and osteoporosis. Dietary supplements can be used as remedy in case of nutrient deficiencies

**Corresponding author*: mahak.fas@mriu.edu.in

Tejpal Dhewa, Anil Kumar Puniya and Anil Panghal (eds.) Nutritional Science and Technology: Concept to Application, (1–20) © 2023 Scrivener Publishing LLC

either due to any chronic disease course or lower intake of nutrients but it cannot replace a normal/regular diet and daily nutrients of a meal, which are necessary for a healthy life. Many ingredients, such as herbs/medicinal plants and hormones, are used to make dietary supplements that have not been identified as nutrients or as having any nutritional functions [2]. It was explained that, when labeled correctly, these compounds should be called supplement ingredients rather than drug ingredients.

Along with natural supplements, many synthetic supplements are also in trend. Dietary supplements [3] are presumed healthy, when taken as directed dosages, they do not pose a serious or unreasonable risk of injury or illness. The FDA now regulates the dosage limit, marketing, advertising, and most importantly the labeling of these dietary supplements.

Dietary supplements come in many forms such as pills, capsules, powders, gels and extracts, although the most popular form of consumption is in powder form. They contain many nutrients which are difficult to get from daily normal diet like amino acids, vitamins, minerals, fiber and enzymes.

1.2 Global History of Dietary Supplements

In most countries, supplements are not regulated as a whole. Instead, they are covered under a collection of already existing laws and regulations [4, 5]. That is, they are "hanging from the hook" of pre-existing law. Previously, the dilemma was whether these goods should be classified as medications or foods; however, a third alternative is to classify them as biologics under existing rules. The Dietary Supplement Health and Education Act (DSHEA) of 1994 established supplement control in the United States. In the US, the FDA regulates dietary supplements as foods rather than drugs as a result of DSHEA; however, they are treated differently from traditional foods. Even if the ingredients used in supplement preparation can have nutritional benefits or occasionally adverse consequences, they are not tested for protection or effectiveness before being released into the market, unlike drugs. When a nutritional supplement is launched/available in the market, the FDA has the duty of demonstrating that it is having adverse consequences in order to limit its use or withdraw it from the market. The FDA bases its decision on a "MedWatch" programme in which healthcare professionals have reported supplement-related complications and side effects. Consumers, on the other hand, were expected to notify the FDA of any potential supplement-related side effects. Some nutritional supplements, particularly botanical formulations, and pharmaceuticals were regulated. In most of the countries only those supplements got permitted that met recommended criteria.

1.3 Usage of Dietary Supplements

The usage of dietary supplement is popular in all age groups to get the perfect body image or according to disease course to prevent chronic nutrient deficiencies. Usage of supplements is much more common in Western countries or developed countries in comparison to underdeveloped countries. Ethnographic studies revealed that cultural factors and family rather than individuals also have the impact on potential usage of supplements [6]. Many misconceptions exist in society about the use of iron tablets. Honduran women believe that using iron tablets will result in larger infants, and they are also concerned about gaining too much weight and creating deformities in their children. Due to the low cost of iron tablets, it has little social standing. Tiredness, fatigue, and dizziness were considered common symptoms of pregnancy in India, so women did not seek treatment [7].

There are several research studies that show that various forms of perceiving or responding to public health messages or initiatives may have a substantial effect on their success or failure.

There are also other factors which act as a barrier for usage and selection of different supplements.

1.4 Types of Dietary Supplements

Supplements can be classified in three categories according to ingredients selected for their preparation: 1. Natural Dietary Supplements, 2. Semi-Synthetic Supplements, 3. Synthetic Dietary Supplements.

1.4.1 Natural Dietary Supplements

Natural supplements are made up with original extract of the natural ingredients or botanicals (herbs, fruits, vegetables, etc). No added chemicals are used in preparation of these supplements. These supplements are more expensive in comparison to synthetic supplements.

1.4.2 Semi-Synthetic Supplements

These types of supplements contain some amount of natural ingredients but their nutrient density will depend on the synthetic chemicals also, which are easily prepared *in vitro* and easily added to natural ingredients to increase the nutrient density of the product. These are a cheaper source of nutrients and sometimes have low bioavailability.

1.4.3 Synthetic Supplements

These types of products are prepared by using all synthetic ingredients *in vitro*. No natural ingredients are required to prepare these products. Mostly these products contain predigested or simple monomers which can easily be digested and absorbed by the human body. Many differences were reported between certain natural and synthetic vitamins in regard to physiochemical properties, proven clinical advantages of natural vitamins. Recent studies also proved the fact that natural vitamins are nutritionally better than synthetic vitamins [8].

To prove the difference between natural and synthetic supplements, many randomized control trials were conducted. Meinrad Lindschinger [9] in 2019 conducted a trail for effect of B complex vitamins on Homocysteine, Oxidative Stress, and Antioxidant Level and found no statistically noticeable difference between both the groups but some favorable tendencies like more sustainability, and less oxidative stress and homocysteine levels were observed in natural vitamins, which have more sustainability in comparison to synthetic ones.

Apart from this, supplements can also be classified in another way by number of nutrients present in one entity such as single nutrient mineral/vitamin (high protein, caloric dense, etc.) and multiple micronutrients or having multiple nutrients in single tablet/capsules. Many supplements are available in the market which is targeting organ-specific disease such as kidney disease or liver disease.

The quality of supplement is a challenge; either it is botanical or synthetic, and many adulterants or misidentifications are related to this. Certain categories of supplements, like supplements related to sport and sexual performance, particularly use some unlabeled prohibited substances to deliver "spiking" and even sometimes the addition of active synthetic drugs.

1.5 Nutritional Adequacy and Dietary Diversity

Nutritional Adequacy is defined as a diet providing all the nutrients in balanced form, thus providing adequacy.

A diet is called a diverse diet when it includes various food groups (cereals, pulses, milk products, vegetables, fruits, meat product, etc.), which is important for attaining sufficient nutrient adequacy and sufficient development and body expansion [10, 11, 14].

Dietary diversity is defined as "the number of different foods or food groups consumed over a given reference period" [12, 14]. A diet covering all the food groups in balanced amount will provide adequate macronutrients and micronutrients, which could improve overall health especially in schoolchildren and decrease the prevalence of many chronic diseases such as cardiovascular diseases (CVDs), diabetes mellitus, metabolic syndrome and cancers, etc. [13, 14].

The best indicator used for assessing diet quality is Dietary diversity score (DDS), which is linked inversely with risk of chronic diseases [14].

Thus, it is important to focus on dietary diversity as it can help in improving the health status, especially in relation to abdominal obesity and its related diseases [15]. Generally, there is poor dietary diversity leading to poor nutritional adequacy, making it necessary to provide dietary supplements to treat some of the chronic diseases.

1.6 Calcium and Vitamin D Supplements

As the population ages, osteoporosis remains a major public health issue. Calcium and vitamin D have long been identified as essential nutrients for bone health and its maintenance. Adequate doses of calcium and vitamin D are important for patients with bone loss. Unfortunately, around 90% of women are not getting adequate calcium and more than 50% of women getting treatment for bone loss may have inadequate levels of vitamin D [16, 17].

The most important element in the body is Calcium, which is essential for many cell functions. Apart from bone health, Calcium plays an essential role in blood coagulation, neuromuscular activity and normal cardiac function apart from bone function.

It is an important part of the architecture of bone and is required for bone mineral deposition for the entire life. The majority of the calcium (99%) is in the bones and teeth plasma and extracellular fluid (ECF) also contains some amount of calcium. High or low levels of plasma calcium indicate positive or negative calcium balance. Bone resorption increases to restore plasma levels, if the plasma level decreases. Sufficient calcium intake is important to maintain the balance. With the help of vitamin D, calcium is adequately absorbed in small intestines [18, 19].

At around the age of 30, the peak bone mass is adequately achieved; therefore, sufficient physical activity and taking the recommended doses of vitamin D and calcium in adolescence and young adulthood helps in ensuring adequate bone mass development [19]. Therefore, to ensure adequate intake of calcium, a lot of calcium supplements are readily available for all age groups.

The Calcium Carbonate and Calcium Citrate are the two most popular calcium supplements. These supplements have been observed to be adequately utilized when taken with any food [19, 20].

Among the calcium salts, Calcium carbonate supplements provide maximum percentage of elemental calcium. Calcium carbonate has composition of 40% of elemental calcium in comparison to 21% of elemental calcium present in calcium citrate. Around 13% of it is found in calcium lactate, and around 9% is found in calcium gluconate [19, 21].

Vitamin D is another necessary nutrient required for maintaining bone health. The most important function of vitamin D is intestinal calcium absorption regulation along with the stimulation of bone resorption leading to the serum calcium concentration maintenance [19, 22]. The main sources of vitamin D are sunlight, diet, and supplements. Some studies have shown that around 90% of people between the ages of 50 to 70 years do not receive adequate vitamin D from the diet they consume [19, 23].

Calcium intake influences optimal intake of vitamin D. High intake of dietary calcium along with vitamin D is required for optimum bone health [19, 24]. Patients can be counseled regarding intake of calcium supplements, which also contain vitamin D, and through this adequate supplementation can be achieved.

Therefore, proper nutrition education about healthy lifestyle, with intake of adequate vitamin D and calcium along with proper physical activity can prevent bone loss and osteoporosis. This nutrition education can be given from time to time to all age groups since bone loss increases as the age progresses [19].

Proper bone health is not maintained by many people as they do not receive an adequate amount of nutrition, although there are a variety of dietary sources of the two nutrients. Also, a very limited amount of vitamin D is synthesized from the skin since people have been made more aware of the damaging effects of sunlight. Therefore, people have to depend on the supplements to meet the amounts necessary for bone health [19].

1.7 Omega-3 Fatty Acid and Health Benefits

Since the 1900s, cardiovascular diseases (CVDs) are the main reason behind morbidity along with mortality worldwide.

The most used dietary supplements are probiotics, which are the fastest-growing supplements along with fish oil (Omega-3 fatty acids), multivitamins, Vitamin D, and coenzyme Q10 (CoQ10), in that order.

Providing multi-vitamin supplementation helps in reducing cardiovascular events; occurrence of coronary artery disease, a trial fibrillation and heart failure is directly linked with Vitamin D levels; CoQ10 deficiency is linked to myocardial dysfunction and with statin myopathy. It is recommended to take probiotics for lowering blood pressure and circulating lipids. However, studies done on the effects of dietary supplementations are not adequately quoted, as people believe that dietary supplements have a beneficial effect in maintaining a healthier diet and lifestyle, but rare randomized studies performed could not prove this [25].

Consumption of two fatty acids per week, i.e., 0.3-0.5 grams of eicosapentaenoic acid and docosahexaenoic acid per day is recommended since omega-3 fatty acids intake is found beneficial in decreasing the levels of triglyceride, blood pressure, platelet aggregation, arrhythmia, and atherogenesis [26].

Recent studies have shown that increased intake of LCn3 can slightly reduce the risk of coronary heart disease mortality, and reduce serum triglycerides. Increasing ALA reduces the risk of cardiovascular events and arrhythmia slightly [27].

1.8 Zinc Supplementation and Health Benefits

The biological growth of the human body, immune function, cell proliferation, enzyme functioning, gene expressions, genome stability, reproductive functions, and stabilization of DNA are some of the vital functions of a trace element called Zinc. Zinc plays a very important role in maintaining the homeostasis of the human being. In categorical pattern, the role of this trace element is grouped into three major functions – Zinc acts like a catalyst, it has structural functions and regulatory functions [28]. The effective absorption is less of the Zinc present in solid food and varies depending on the diet composition; therefore Zinc supplement administered is done in aqueous solutions and has great absorption of around 60-70% [29].

To estimate the average physiological requirement of Zinc, a factorial method was used, keeping in mind the absorption of zinc and losses through both intestinal and non-intestinal pathways as represented in Table 1.1.

The "Zinc deficiency" is more prevalent in young children, as required for their growth and development, in adolescents for pubertal growth

Table 1.1 Zinc physiological requirement in regard to age group and gender.

Variables age	Who		Variables age	FNB		IZINCG	
	Reference wt. (kg)	Requirement (mg/day)		Reference wt. (kg)	Requirement (mg/day)	Reference wt. (kg)	Requirement (mg/day)
6-12 mo	9	0.84	6-12 mo	9	0.84	9	0.84
1-3 year	12	0.83	1-3 year	13	0.74	12	0.53
3-6 year	17	0.97	4-8 year	22	1.20	21	0.83
6-10 year	25	1.12					
10-12 year	35	1.40	8-13 year	40	2.12	38	1.53
12-15 year	48	1.82					
15-18 year M	64	1.97	14-18 year M	64	3.37	64	2.52
15-18 year F	55	1.54	14-18 year F	57	3.02	56	1.98
Pregnancy	-	2.27	Pregnancy*	-	4.1-5.0	-	2.68
Lactation	-	2.89	Lactation*	-	3.8-4.5	-	2.98

*Different stages of pregnancy/lactation. WHO=World Health Organization; FNB=Food and Nutrition Board; IZiNCG=International Zinc Nutrition Consultative Group
International Zinc Nutrition Consultative Group (IZiNCG) [30].

spurt, and during pregnancy and lactation period, as Zinc is essential for infant growth. The elderly population also take insufficient amount of Zinc required by the body. The statistics states that 13% of mortality and 4% of global child morbidity is prevalent because of the deficiency of Zinc.

Ota E. *et al.* (2015) conducted a review that includes data from 21 randomized controlled trials (RCTs) with a sample size of 17,000 women and their infants. The data revealed that the supplementation of Zinc during pregnancy leads to a relative reduction of 14% in birth of preterm infants with placebo, but no other beneficial evidence was discovered regarding intake of zinc supplementation during pregnancy [31].

The deficiency of Zinc can lead to growth retardation, hypo-gonadism, hepato-megaly, spleno-megaly, wrinkled and dried skin, and severe iron deficiency anemia [32]. Severe immune deficiency can also develop and can cause many fatal infections before 25 years of age [33]. Recent studies also indicated that a decrease in cell–mediated and humoral immune response and increased susceptibility to infections like diarrheic diseases is observed in Zinc deficient patients. Administration of zinc in acute diarrhea conditions was found to be a beneficial therapeutic remedy in children [34, 35]. A recent research [36] stated a positive impact of Zinc supplementation in conditions like pneumonia and acute diarrhea. The mortality rate in children due to diarrhea has been reduced by supplementation of Zinc in developing countries over a period of less than five years. Zinc supplementation also has a beneficial effect on conditions like "enteropathicacrodermatitis, Wilson's, chronic hepatitis C, shigellosis, leprosy, leishmaniasis" and the common cold [37, 38].

Recent studies indicated the increased risk of neuro-degenerative disorders, affecting "neurogenesis" and increasing "neuronal apoptosis" due to the deficiency of Zinc, which further can lead to difficulty in understanding and increasing forgetfulness. This further links zinc deficiency with cerebral-aging, depression, anxiety, Alzheimer's disease and Parkinson's disease [39–41].

1.9 Iron Supplementation and Health Benefits

As compared to zinc, an abundance of iron is present on earth [42, 43], and every living organism requires iron biologically for proper development, growth and functioning of the human body [44, 45]. Even if it is present in abundant amounts, Iron in the environment act like a growth-limiting [43]. Iron in the presence of oxygen forms oxides, becomes insoluble in nature, and is thus unavailable for uptake by living organisms [42]. In the human body, Hemoglobin contains iron, and is present in bound form. Iron is bound to a protein called hemoprotein as heme compounds [46]. Iron is essential for formation of hemo-globin and myo-globin, which carries oxygen and transport throughout the human body. Almost 2/3rds of the human body's Iron is used for circulating erythrocytes in the living organism and 15% of iron is bound to myoglobin in muscle tissue. The presence of Iron is required even for the synthesis of heme-enzymes and many other iron containing enzymes [46, 47]. Around 25% of Iron is present in mobilizable iron store [48].

Breast milk contains very little amount of Iron, which is sufficient in early infancy. 0.7 – 0.9 mg/day iron is required by the infant from 4-6 months after birth till the remaining part of the first year as represented in the Table 1.2. A double amount of Iron is required between the age of 1 and 6 years. During the period of growth spurts in adolescents, there is an increase in demand for Iron by the human body [49].

1 to 3 gms of iron is stored in the body of an average adult. The intake by diet and loss of iron needs to be balanced by the human body. The shedding of cells from the mucosal surface and skin leads to a loss of about 1 mg of iron [50]. Among premenopausal female adults, an average loss of 2 mg per day of Iron takes place due to menstruation [51]. The enhancement of human body mass during the neonatal stage and for growth and development in the childhood stage transiently boosts iron requirements [52].

When the supply of Iron is compromised, including the erythron, the medical condition is defined as Iron deficiency. Iron deficiency also takes place due to no mobilizable iron stores [53]. Iron deficiency can be present in human beings with or without anemia. The development of Anemia can cause many functional defects which include cerebral development, immune system, and work capacity [42, 54–56]. Deficiency of Iron during the different stages of pregnancy is correlated with negative outcomes for both maternal health and neonatal health, including high risk of infection, maternal-mortality, low birth weight infants and peri-natal mortality [57].

The depletion of Iron stores, and the body not being able to supply enough iron as compared to the metabolic demands to replenish the iron loss and to sustain development of the human being, become the major cause of iron deficiency [42]. Lack of consumption of bio-available iron and impaired absorption of Iron are also considered to be one of the primary causes of Iron deficiency [58, 59].

Iron supplementation is one of the major strategies which is used to treat Iron deficiency among populations with different age groups. "Ferrous

Table 1.2 Iron requirement by World Health Organization.

Age/sex	**mg/day**[b]
4-12 months	0.96
12-24 months	0.61
2-5 years	0.70
6-11 years	1.17
12-16 years (girls)	2.02
12-16 years (boys)	1.82
Adult males	
Pregnant women[c]	1.14
First trimester	0.8
Second and third trimester	6.3
Lactating women	1.31
Menstruating women	2.38
Postmenopausal women	0.96

Iron requirements of 97.5% of individuals in terms of absorbed iron[a], by age group and sex (World Health Organization, 1989).
[a]Absorbed iron is the fraction that passes from the gastrointestinal tract into the body for further use. [b]Calculated on the basis of median weight for age. [c]Requirements during pregnancy depend on the woman's iron status prior to pregnancy [49].

iron salts (ferrous sulfate and ferrous gluconate)" are used for oral iron supplementation, due to their high bioavailability and low cost [60]. The best way to take an iron supplement is to take it on an empty stomach as the iron absorption is higher during that stage but can cause nausea and epi-gastric pain due to the high iron dosage (usually 60 mg Fe/day) [61]. In developing countries, during pregnancy, Iron supplementation is advisable due to the low iron stores in the human body [62].

1.10 Dietary Supplements and Sports

Dietary supplements are isolated or bound nutrients consumed by the individuals to improve their nutritional status. They are usually consumed when an individual is deficient in a particular nutrient or wants to increase his/her muscle mass. The ingredients usually used as supplement are amino acid, herbs, minerals, vitamins and enzymes. Dietary supplements are used in the form of "softgels, powder, capsules, liquid, and gelcaps".

In today's era of sports competition and zeal, it is a common false belief of athletes that supplement consumption will magically improve their performance. But experts on sports nutrition still emphasize a well-balanced diet. The human body's requirement for vitamins and minerals can be met easily by consumption of fruits and vegetables (yellow orange, red, deep green, and blue). Moreover, they contribute to the body's antioxidant pool and reduce oxidative stress. Scientific studies have revealed that taking extra vitamins and minerals supplements has no role in increase of muscle mass, strength or VO_2 max. An athlete should consume five to nine servings of fruit and vegetables in a day. Egg, meat, milk and milk products are wholesome foods containing both vitamins and minerals in good amount. It is good nutrition and sound training techniques that yield improvements and better performance rather than supplements. Intense training and balanced nutrition and diet should be the first priority in an athlete's nutrition program.

As per the DSHEA, dietary supplements should be well labelled as per the guidelines and should be identified as "Dietary supplement and not drug". Among the various forms of supplements, proteins are the commonest choice for athletes. Here are three common forms of protein supplements:

1.10.1 Protein Concentrates

They are the product of protein extraction by heat and acid or enzymes. They have 60-80% protein; the other 20-40% is fats and carbohydrates.

1.10.2 Protein Isolates

Further filtration of concentrates separates large amount of fats and carbohydrates, and concentrates the amount of protein. Protein-isolate powders have around 90–95% protein.

1.10.3 Protein Hydrolysates

Further by the use of acid or enzyme and by heating of isolates detached the amino acids bonding which produces hydrolysates. They are absorbed more quickly by the body [63].

Protein supplements can also be classified on the basis of their source. Some of most available in the market are given in Table 1.3 below.

Table 1.3 Types of sports supplements.

Type of supplements	Benefits	Limitations
Milk Protein	• Contains both slow and fast proteins • Has 80% casein (slow) and 20% whey (fast) protein • Less expensive • Other nutrients are also there	• Contain fats and cholesterol • Shelf life is less • Not suitable for lactose intolerant person
Whey protein	• Contains fast proteins • They are rapidly digested • Highest amount of BCAA, increase muscle growth and repair • Less muscle fatigue and wasting	• Protein powder needs to be mixed with water
Soy Protein	• Comparable to animal protein • Also a fast protein • Some forms may contain isoflavones	• May not build muscles as that of whey and casein • Taste may be unfamiliar to athlete
Egg protein	• Perfect protein due to essential amino acid profile • Contains all 13 essential nutrients • Cholestrol and fat is not present in egg whites	• Half of the protein is present in egg yolk but its also high in cholesterol. Salmonella bacteria can develop easily in raw eggs. • May be allergic for some

1.11 Dietary Supplements and FDA

The formulation and regulation of dietary supplements is taken care of by FDA. The regulations are applied by FDA in different sets, covering "conventional" foods and "drug products". Under the Dietary Supplement Health and Education Act (DSHEA) of 1994:

1.11.1 Manufacturers and distributors of adulterated or misbranded dietary supplements and dietary ingredients – market products are prohibited. This means that the responsibility of a particular company for evaluating and testing the product for safety and correct product labelling has to be done before marketing to make sure that the product satisfies all the requirements of DSHEA and FDA regulations.

1.11.2 Any adulterated or misbranded dietary supplement product that reaches the market is the responsibility of the FDA to take action on it.

Dietary supplements includes both liquid and solid external substances. So before jumping to supplement types, it is important for sports nutrition students to understand fluid and its importance.

1.12 Dietary Supplements and Toxicity

In today's scenario, it has been observed that there is a growing trend of increased intake of multi-vitamin/multi-mineral supplements in developed countries [64, 65]. As reported by the study, 33% of US adults consume multi-vitamin and multi-mineral supplements [66] and intake is high among "male military personnel" (32–47%) and among "military women" 32–47%) [67]. About 64-81% of "long-term cancer survivors", consume large amounts of multi-vitamin and mineral supplements [66]. To maintain optimal growth and development and prevent deficiency, adequate intake is advisable but excess use or high dose cause the possibility of toxicity [68]. The dosage of pyridoxine (vitamin B6) higher than 500 mg/d reported the conditions of Photo-sensitivity and neuro-toxicity. Vitamin E doses of 800–1200 mg/d can cause bleeding which is further associated with anti-platelet action, and dosage more than 1200 mg/d give the symptoms of acute diarrhea, muscle cramping, cloudy vision, and gonadal dysfunction [69]. Excess intake of vitamin A is also linked with toxicity. In two large clinical trials, the "Retinol Efficacy Trial" [70] and the "ATBC study" [71], the increased risk of lung cancer is associated with male smokers consuming β-carotene supplements. The ABTC study reported the increased mortality and incidence of prostate cancer in male alcohol users consuming the β-carotene supplement. The consumption of "fish liver oils" having vitamin A and multi-vitamin supplements in them can cause hypervitaminosis A [72, 73].

References

1. A. Sarubin, *The Health Professional's Guide to Popular Dietary Supplements.* Chicago, IL: American Dietetic Association, 2000.
2. M.C. Nesheim, What is the research base for the use of dietary supplements? *Public Health Nutr,* 2:35–38, 1999.
3. L.S. Balluz, S.M. Kieszak, R.M. Philen, J. Mulinare, Vitamin and mineral supplement use in the United States: Results from the Third National Health and Nutrition Examination Survey, *Arch Fam Med,* 9,258–262, 2000.
4. A. Dickinson, History and overview of DSHEA. *Fitoterapia,* 82, 5–10, 2011.
5. C.L. Taylor, Regulatory frameworks for functional foods and dietary supplements. *Nutr. Rev,* 62, 55–59, 2004.
6. L. Williams, L. Semu, D. Behague, C. Sibale, C.France, A qualitative study of constraints to reducing iron deficiency and anaemia in women of reproductive age in Thyolo district, Malawi MotherCare. John Snow Inc. Arlington, VA, 1996.
7. M.B. Bentley, A. Parekh, Perceptions of anemia and health seeking behavior among women in four Indian states (Technical Working Paper #9) MotherCare, John Snow Inc. Arlington, VA, 1998.
8. R. J. Thiel, Natural vitamins may be superior to synthetic ones, *Med Hypotheses,* 55(6), pp. 461-9, 2000.
9. M. Lindschinger, F.Tatzber, W. Schimetta, I. Schmid, B. Lindschinger, G. Cvirn, O. Stanger, E. Lamont, W. Wonisch, A Randomized Pilot Trial to Evaluate the Bioavailability of Natural versus Synthetic Vitamin B Complexes in Healthy Humans and their Effects on Homocysteine, Oxidative Stress, and Antioxidant Levels, *Oxidative Medicine and Cellular Longevity,* 2019.
10. L.Azadbakht, A.Esmaillzadeh, Diet variety: a measure of nutritional adequacy and health, *J Qazvin Univ Med Sci,* 13, pp. 88–97, 2009.
11. M.Vakili, P.Abedi, M.Sharifi, M.Hosseini, Dietary diversity and its related factors among adolescents: a survey in Ahvaz-Iran, *Glob J Health Sci,* 5, pp. 181–186, 2013.
12. E.H. Pangaribowo, N. Gerber, M. Torero, Food and nutrition security indicators: a review. Bonn: Center for Development Research, 2013.

13. P. Mirmiran, L. Azadbakh, A.Esmaillzadeh, F.Azizi, Dietary diversity score in adolescents - a good indicator of the nutritional adequacy of diets: Tehran lipid and glucose study, *Asia Pac J Clin Nutr*,13, pp. 56–60, 2004.
14. K.F. Tiew, Y.M. Chan, M.S. Lye, S.C. Loke , Factors associated with dietary diversity score among individuals with type 2 diabetes mellitus, *J Health Popul Nutr*, 32, pp. 665–676, 2014.
15. S. M. Nachvak, H. Abdollahzad, R. Mostafai, S. Moradi, Y. Pasdar, M. Rezaei, and S. Esksndari, Dietary Diversity Score and Its Related Factors among Employees of Kermanshah University of Medical Sciences, *Clinical Nutrition Research*, 6(4), pp. 247–255, 2017.
16. (NDC) National Dairy Council, USDA's Continuing survey of food intakes by individuals, 1994–1996 (online), Accessed on March 22, 2008.
17. M.F. Holick, E.S. Siris, N. Binkley, M.K. Beard, A. Khan, J.T. Katzer, R.A. Petruschke, E. Chen, A.E. de Papp, Prevalence of Vitamin D inadequacy among postmenopausal North American women receiving osteoporosis therapy, J Clin Endocrinol Metab, Jun, 90(6), pp. 3215-24, 2005.
18. F.R. Bringhurst, M.B. Demay, *et al. Harrison's Principles of Internal Medicine*, 16th ed. II. New York: McGraw Medical Publishing Division, Bone and mineral metabolism in health and disease, pp. 2246–9, 2005.
19. J. A. Sunyecz, The use of calcium and vitamin D in the management of osteoporosis. Therapeutics and *Clinical Risk Management*, 4(4), pp. 827–836, 2008.
20. R.P. Heaney, S.D. Dowell, J. Bierman, *et al.*, Absorbability and cost effectiveness in calcium supplementation, *J Am Coll Nutr*, 20, pp. 239–46, 2001.
21. S.M. Weisman, Calcium connection to bone health across a woman's lifespan – A roundtable, *J Reprod Med*, 50(11 Suppl), pp. 879–84, 2005.
22. I.R.Reid, Clinical aspects of the use of vitamin D and its metabolites, In: E.S. Orwoll, M. Bliziotes, eds. *Osteoporosis: Pathophysiology and Clinical Management*, Totowa, NJ: Humana Press, pp. 293–307, 2003.
23. C. Moore, M.M. Murphy, D.R. Keast, *et al.*, Vitamin D intake in the United States, *J Am Diet Assoc*,104, pp. 9980–3, 2004.
24. C.M. Weaver, J.C. Fleet, Vitamin D requirements: current and future, *Am J ClinNutr*, 0(suppl), pp. 1735S–9S, 2004.
25. S. Bronzato,, A. Durante, Dietary Supplements and Cardiovascular Diseases, *Int J Prev Med*, Sep 17, 9:80, 2018.
26. A. Saremi, R. Arora, The utility of omega-3 fatty acids in cardiovascular disease, *Am J Ther*, Sep-Oct,16(5), pp. 421-36, 2009.
27. A.S. Abdelhamid, T.J. Brown, J.S. Brainard, P. Biswas, G.C. Thorpe, H.J. Moore, K.H. Deane, C.D. Summerbell, H.V. Worthington, F. Song, L. Hooper, Omega-3 fatty acids for the primary and secondary prevention of cardiovascular disease, *Cochrane Database Syst Rev*, Feb 29, 3(2), 2020.
28. R.J.Cousins, Zinc, *Present Knowledge in Nutrition,* 7th ed., Washington DC: International Life Science Institute Nutrition Foundation, pp. 293–306, 1996.

29. FAO/WHO, Expert Consultation on Human Vitamin and Mineral Requirements, Vitamin and mineral requirements in human nutrition: Report of joint FAO/WHO expert consolation, 2nd ed. Bangkok, Thailand, pp. 341, 2004.
30. K.H. Brown, J.A. Rivera, Z. Bhutta, R.S. Gibson, J.C. King, B. Lönnerdal, M.T. Ruel, B. Sandtröm, E. Wasantwisut, C. Hotz, International Zinc Nutrition Consultative Group *(IZiNCG), Food Nutr Bull,* 25(1 Suppl 2), pp. S99-203, 2004.
31. E. Ota, R. Mori, P. Middleton, R. Tobe-Gai, K. Mahomed. Zinc supplementation for improving pregnancy and infant outcome. Cochrane Database Syst Rev, 2015 Feb, DOI: 10.1002/14651858.CD000230.pub5
32. A.S. Prasad, Clinical, immunological, anti-inflammatory and antioxidant roles of zinc, *Exp Gerontol,* 43(5), pp. 370–7, 2008.
33. A.S. Prasad, Zinc in human health: effect of zinc on immune cells, *Mol Med,*14(5–6), pp. 353–7, 2008.
34. M.J. Tuerk, N. Fazel, Zinc deficiency, *CurrOpin Gastroenterol,* 25(2), pp. 136–43, 2009.
35. A.S. Prasad, Discovery of human zinc deficiency: its impact on human health and disease *Adv Nut,* 4(2), pp. 176–90, 2003.
36. M.Y. Yakoob, E. Theodoratou, A. Jabeen, A. Imdab, T.P. Eisele, J. Ferguson, Preventive zinc supplementation in developing countries: impact on mortality and morbidity due to diarrhea, pneumonia and malaria, *BMC Public Health,*11 (Suppl 3), pp. 3–23, 2011.
37. A.S. Prasad, Impact of the discovery of human zinc deficiency on health, *J Am CollNutr,* 28(3), pp. 257–65, 2009.
38. J. Hui, N.L. Tang, Wilson's disease: a review of treatment options with a focus on zinc therapy, *Orphan Drugs* ,2, pp. 35–45, 2012.
39. B. Szewczyk, Zinc homeostasis and neurodegenerative disorders, *Front Ageing Neurosci,* 5, pp. 33, 2013. doi: 10.3389/fnagi2013.00033.
40. M.L. Plum, L. Rink, H. Haase, The essential toxin: impact of zinc on human health, *Int J Environ Res Public Health,* 7(4), pp. 1342–65, 2010.
41. G.L. Brewer, S.H. Kanzer, E.A. Zimmerman, E.S. Molho, D.F. Celmins, S.M. Heckman, Subclinical zinc deficiency in Alzheimer's disease and Parkinson's disease, *Am J Alzheimers Dis Other Demen,* 25(7), pp. 572–5, 2010.
42. R.J. Wood, A, Ronnenberg. Iron, Shils ME, Shike M, Ross AC, Caballero B, Cousins RJ, editors, *Modern Nutrition in Health and Disease,* 10th edition, Baltimore: Lippincott Williams & Wilkins, pp. 248–70, 2005.
43. A.G. Quintero-Gutiérrez, G. González-Rosendo, J. Sánchez-Muñoz, J. Polo-Pozo, J.J. Rodríguez-Jerez, Bioavailability of heme iron in biscuit filling using piglets as an animal model for humans, *Int J Biol Sc,* 4, pp. 58–62, 2008.
44. P. Aisen, C. Enns, M. Wessling-Resnick, Chemistry and biology of eukaryotic iron metabolism, *Int J Biochem Cell Biol,* 33, pp. 940–59, 2001.
45. P.T. Lieu, M. Heiskala, P.A. Peterson, Y. Yang, The roles of iron in health and disease, *Mol Aspects Med,* 2, pp. 1–87, 2001.

46. L.R. McDowell, Amsterdam: Elsevier Science, *Minerals in Animal And Human Nutrition*, 2nd edition, pp. 660, 2003.
47. R.F. Hurrell, Bioavailability of iron, *Eur J ClinNutr,* 51, pp. S4–8, 1997.
48. Institute of Medicine, Iron, In: *Dietary Reference Intakes for Vitamin A, Vitamin K, Arsenic, Boron, Chromium, Copper, Iodine, iron, Manganese, Molybdenum, Nickel, Silicon, Vanadium, and Zinc, Washington, DC*: National Academy Press, pp. 290–393, 2001.
49. FAO/WHO. Expert Consultation on Human Vitamin and Mineral Requirements, Vitamin and mineral requirements in human nutrition: Report of joint FAO/WHO expert consolation, 2nd ed. Bangkok, pp. 341, 2004.
50. J.D. Cook, B.S. Skikne, S.R. Lynch, M.E. Reusser, Estimates of iron sufficiency in the US population, *Blood,* 68, pp. 726–31, 1986.
51. T.H. Bothwell, R.W. Charlton, A general approach of the problems of iron deficiency and iron overload in the population at large, *SeminHematol,* 19, pp. 54–67, 1982.
52. R.S. Gibson, A.C. MacDonald, P.D. Smit-Vanderkooy, Serum ferritin and dietary iron parameters in a sample of Canadian preschool children, *J Can Dietetic Assoc,* 49, pp. 23–8, 1988.
53. WHO/UNICEF/UNU, *Iron Deficiency Anemia Assessment, Prevention, and Control*, p. 114, 2001.
54. J.L. Beard, J.R. Connor, Iron status and neural functioning, *Annu Rev Nutr,* 23, pp. 41–58, 2003.
55. M.L. Failla, Trace elements and host defense: Recent advances and continuing challenges, *JNutr,* 133, S1443–7, 2003.
56. F.E. Viteri, B. Torun, Anemia and physical work capacity, In: Garby L, editor, *Clinics in Hematology,* Vol. 3. London: WB Saunders, pp. 609–26, 1974.
57. CDC, Breastfeeding Report Card, United States: Outcome Indicators Publication, from Centers for Disease Control and Prevention, National Immunization Survey, 2010.
58. E.S. Cooper, D.A. Bundy, Trichuriasis. *Ballieres, Clin Trop Med Commun Dis,* 2, 629–43, 1987.
59. R. Larocque, M. Casapia, E. Gotuzzo, T.W. Gyorkos, Relationship between intensity of soil-transmitted helminth infections and anemia during pregnancy, *Am J Trop Med Hyg,* 73, pp. 783–9, 2005.
60. M.B. Zimmermann, R.F. Hurrell, Nutritional iron deficiency, *Lancet,* 370, pp. 115–20, 2007.
61. T. Cavalli-Sforza, J. Berger, S. Smitasiri, F. Viteri, Weekly iron-folic acid supplementation of women of reproductive age: Impact overview, lessons learned, expansion plans, and contributions toward achievement of the millennium development goals, *Nutr Rev,* 63, pp. S152–8, 2005.
62. CDC, Iron Deficiency, Centers for Disease Control and Prevention, *MMWR Weekly,* 51, 897–9, 2002. Available from: http://www.cdc.gov/mmwr/preview/mmwrhtml/mm5140a1.htm#tab1.

63. Sports nutrition and hydration guidelines for excellence in sports performance. National Institute of Nutrition, 2007.
64. R.L. Bailey, J.J. Gahche, P.E. Miller, P.R. Thomas, J.T. Dwyer, Why US adults use dietary supplements, *JAMA Intern. Med,* 173, pp. 355–61, 2013.
65. J.J.Y. Woo, Adverse event monitoring and multivitamin-multimineral dietary supplements, *Am. J. Clin. Nutr,* 85, pp. 323S–24S, 2007.
66. C.M Velicer, C.M. Ulrich, Vitamin and mineral supplement use among US adults after cancer diagnosis: a systematic review, *J. Clin. Oncol,* 26, pp. 665–73, 2008.
67. J.J. Knapik, R.A. Steelman, S.S. Hoedebecke, E.K. Farina, K.G. Austin, H.R. Lieberman, A systematic review and meta-analysis on the prevalence of dietary supplement use by military personnel, *BMC Complement. Altern. Med,* 14, pp. 143, 2014.
68. C.A. Mulholland, D.J. Benford, What is known about the safety of multivitamin-multimineral supplements for the generally healthy population? Theoretical basis for harm, *Am. J. Clin. Nutr,* 85, pp. 318S–22S, 2007.
69. E.E. Ziegler, L.J. Filer, *Present Knowledge in Nutrition,* 7th Edition, International Life Sciences Institute-Nutrition Foundation, 1996.
70. G.S. Omenn, G.E. Goodman, M.D. Thornquist, J. Balmes, M.R. Cullen, A. Glass, J.P. Keogh, F.L. Meyskens Jr, B. Valanis, J.H. Williams Jr., S. Barnhart, S. Hammar, Effects of a combination of beta carotene and vitamin A on lung cancer and cardiovascular disease, *N. Engl. J. Med,* 334, pp.1150–55, 1996.
71. The Alpha-Tocopherol Beta Carotene Cancer Prevention Study Group. 1994. The effect of vitamin E and beta carotene on the incidence of lung cancer and other cancers in male smokers. *N. Engl. J. Med* 330:1029–3, 1994.
72. B.W. Gross, M. Gillio, C.D. Rinehart, C.A. Lynch, F.B. Rogers, Omega-3 fatty acid supplementation and warfarin: a lethal combination in traumatic brain injury, *J. Trauma Nurs,* 24, pp.15–18, 2017.
73. M.S. Buckley, A.D. Goff, W.E. Knapp, Fish oil interaction with warfarin, *Ann. Pharmacother,* 38, pp. 50–53, 2004.

2

Prebiotics, Probiotics and Synbiotics

Vikram Kumar[1], Ananya Rana[1], Prajakta Jagtap[1], Tejpal Dhewa[2] and Neetu Kumra Taneja[1,3]*

[1]Department of Basic and Applied Sciences, NIFTEM, Sonepat, Haryana, India
[2]Department of Nutritional Biology, Central University of Haryana, Jant/Pali, Haryana, India
[3]Centre for Advanced Translational Research in Food Nanobiotechnology (CATR-FNB), NIFTEM, Sonepat, Haryana, India

Abstract

Prebiotics and probiotics are tools for managing the microbiota and enhancing human health. They aim for gastrointestinal effects via the gut, although direct administration to other sites, such as the oral cavity, vaginal canal, and skin has been investigated. Here, we discuss human gut-related impacts. Over the past ten years, there has been a fast surge in studies on the gut microbiome, which has corresponded with an increase in interest in probiotics and prebiotics as a strategy to alter gut microbiota. Despite the relevance of these strategies for public health, it is important to reaffirm the details and evidence that support their clinical usage. We review the scientific research on probiotics and prebiotics in this chapter, together with mechanistic insights into potential health benefits. However, *Lactobacillus, Bifidobacterium*, and *Saccharomyces* strains have a long history of safe and effective use as probiotics. *Roseburia, Akkermansia, Propionibacterium*, and *Faecalibacterium* species also show promise for the future. More information is also emerging regarding the prebiotic effects of other compounds, such as oligomers of mannose, glucose, xylose, pectin, starches, human milk, and polyphenols. Glucans and fructans are well-established prebiotics.

***Keywords*:** Probiotics, prebiotics, xylulose, glucose

**Corresponding author*: drneetulab.niftem@gmail.com; ORCID ID: 0000-0002-2390-7647

Tejpal Dhewa, Anil Kumar Puniya and Anil Panghal (eds.) Nutritional Science and Technology: Concept to Application, (21–62)

2.1 Introduction

According to Sumerian tablets, the information on the history of probiotics reverts to 2000 BC when humans used probiotics for the first time by converting milk to fermented dairy products to extend its preservation time. As stated by Roman history, the barbarians converted milk into thickened sour products and used it to treat diarrhea and vaginal discharge in the Palearctic region for many years [1]. During the 13th century BC, nearly all civilizations consumed fermented milk and its products daily for its taste and therapeutic effects. Henry Tissier, a French paediatrician, found that *Bifidobacteria* were largely in the gastrointestinal tract of humans with good health and suggested the ingestion of *Bifidobacteria* to control diarrhea in infants. Elie Metchnikoff, who received the Nobel Prize in 1908 for research on the intake of live microorganisms, wrote a book called "The Prolongation of Life". It gives the earliest scientific information on beneficial gut flora alteration by ingesting substances now commonly known as probiotics [2].

The food habits of people are continuously changing. It may be due to their changing lifestyle, convenience, urbanization, globalization and eating habits. Since the last decade, there has been an inclination towards a diet that confers health benefits. Functional foods benefit health in addition to the nutritional benefits they provide to the consumer [3]. Probiotics belong to the functional foods category that provides health benefits, which can be defined as the live bacteria that, when administered in suitable quantity, confer health benefits to the host [4]. Probiotic bacteria exert positive health effects on the host by improving their intestinal microbial balance, treating chronic inflammation, control of infective diseases, improvement in lactose and cholesterol digestion [5]. Most of these probiotic lactic acid bacteria fit into the genus *Lactobacillus* and *Bifidobacterium,* even though other microorganisms have probiotic characteristics. Examples of probiotics include *Lactobacillus plantarum, L acidophilus, L. johnsonii, L. casei, L. delbrueckii ssp. bulgaricus, L. reuteri, L. brevis, L. cellobiosus, L. curvatus, L. fermentum, and Lactococcus lactis ssp. cremoris.*

Some strains of Lactic acid bacteria also produce functional ingredients, which may be called as nutraceuticals [6]. Nutraceuticals are products obtained from food sources that provide both nutritional and medical benefits, including micro and macro nutrients [7]. Vitamins are the vital micronutrients required during the digestion and metabolism of every living organism and play an important role in regulating the cell's biochemical reactions. Most of the vitamins cannot be synthesized by human beings, so they have to be included as a part of the diet [6]. Riboflavin is an essential vitamin soluble in water that helps in cell growth, and insufficient vitamin B_2 causes ulcers, cheilosis, anaemia and dementia [8, 9]. Deficiency of riboflavin is common in many developed and developing countries [10] even though riboflavin is present in many foods such as dairy products, meat, egg and green vegetables [6]. Hence microorganisms such as LAB that produce riboflavin can be used in fermented foods to obtain riboflavin-rich products [11].

Lactiplantibacillus plantarum is a lactic acid bacterium, which ferments sugars such as pentose and hexose, yielding lactic acid as a main product and CO_2 ethanol/acetate as by-products. Historically, *L. plantarum* has been employed in the fermentation of livestock, vegetables, and dairy products. It can resist acidic and bile conditions in the gut, which helps to pass through the stomach and upper gastrointestinal tract and stick to the digestive epithelial cells by inhibiting the adhesion of pathogens in the lower gastrointestinal tract [12]. *L. plantarum* RYPR1 strain isolated from raabadi by [13] exhibited the highest probiotic potential by showing the maximum cholesterol reduction of 59% and also presented a hypocholesterolemic effect which could help in the prevention of hypercholesterolemia.

To resist unavoidable food handling practices and adverse ecological conditions, probiotics can be protected by encapsulating them inside a protective framework [14]. Microencapsulation is a technique of stuffing solids, liquids or gases inside the coating materials, which releases encapsulated material only under specific stimulus [15]. The viability of the encapsulated probiotics depends on the encapsulation techniques used, the number of cells presents initially, the coating material used, the strain used for encapsulation, the size of the capsule formed, the concentration of coating material and various other factors [15]. There are numerous techniques for encapsulation, such as spray drying, freezing drying, emulsion, extrusion, fluidized bed drying etc. [16]. Compared to all other techniques, encapsulation by spray drying is used mainly because of its higher productivity, minimal structural and functional changes in product, [17] reduced cost, less energy requirement, and storage stability [14]. In spray drying, atomization of the liquid or emulsion along with the encapsulating agent takes place when dried and heated air at a higher temperature is circulated inside the chamber, which results in the evaporation of moisture from the product. As a result, the final product obtained is in powder form, which can be stored for a more extended period [18]. Although spray drying has many advantages, microorganisms' inactivation cannot be ruled out because of higher drying temperatures. Therefore, coating materials are required to protect the bacteria from higher temperatures and increase the chances of survival [19].

Maltodextrin, a carbohydrate, acts as a prebiotic [20] and protective coating against elevated temperature and storage stress [14]. Inulin is prebiotic, and according to the studies carried out by [21], higher encapsulation efficiency was shown when encapsulated with inulin and resistant starch, with an encapsulation efficiency of 93.12% and 94.26%, respectively. Reconstituted skim milk can be used as carrier material as it has improved the survival rate of microorganisms during spray drying [22]. Ming and Anand *et al.* demonstrated that conjugated whey protein, along with maltodextrin, is a suitable carrier for probiotics with viable counts of 8.98 log 10 cfu/g and encapsulation yield of 84.87% [23].

2.2 Prebiotics and Its Types

2.2.1 Prebiotics: Definition

Glenn Gibson and Marcel Roberfroid used the prebiotics approach for the first time in 1995 [24]. The prebiotic is characterized as "an indigestible food constituent that has a beneficiary effect on the host by selectively stimulating/regulating the growth and activity of an individual or group of microorganisms of the gastrointestinal tract and hence improves the host health" [25]. According to this explanation, only a few ingredients can act as prebiotics, such as some oligosaccharides and polysaccharides [Fructooligosaccharides (FOS) and inulin], some proteins and peptides, lactulose, Galacto-oligosaccharide (GOS) etc. In 2008, the 6th convention of the Council of the International Scientific Association of Probiotics and Prebiotics (ISAPP) defined "dietary prebiotics" as "a specific fermented food ingredient that is responsible a sensible changes in the composition and/or activity of the colon microflora, thus deliberating host health benefits" [26]. A compound can classify as a prebiotics based on the following criteria:

1. The compound should not absorb by the gastrointestinal tract, resist low pH of the stomach, resist hydrolyzation, and be non-reactive to mammalian enzymes.
2. It could use as a fermentative product by gastrointestinal microflora.

3. Growth and/or activity of the gastrointestinal microbiota should be selectively altered from a prebiotic compound, leading to improved host health [25, 26].

However, not all types of prebiotics are defined as carbohydrate-based prebiotics. These can be distinguished from the fiber based on the following characteristics:

I. Fibers consist of carbohydrates by polymerization [27].
II. The enzymes located inside the small intestine do not hydrolyze these fibers. It indicated that the fermentability and solubility of fibers are not crucial [28].

Some other revised definitions are also published in scientific reports [29]. The potential of a prebiotic or "selectivity" to stimulate a particular gastrointestinal microbiota is another important part of the definition, although this concept came to be questioned recently [30]. As previously reported, the prebiotic effect was intensified via cross-feeding, stated as a product of one species ingested by a different one [31]. This indicates that the word "selectivity" raises doubt, and debate is still ongoing on the definition of prebiotic [32].

2.2.2 Sources of Prebiotics

The principle sources of prebiotics are classified into many groups. Most are a subset of carbohydrates and are predominantly oligosaccharide carbohydrates (OSCs). This chapter mainly focused on OCSs, but some other research proves that prebiotics is not only carbohydrates.

2.2.2.1 Galacto-Oligosaccharide

Galacto-oligosaccharide is the product of extended forms of lactose, which are categorized into two main subclasses (1) Galacto-oligosaccharide with an abundance of galactose at position C_3, C_4 or C_6 and (2) Galacto-oligosaccharide are synthesized by lactose with enzymatic reaction trans-glycosylation [33]. In this reaction, the end product is generally a mixture of trisaccharide to pentasaccharide in linkages β (1, 6), β (1, 3) and β (1, 4) with galactose. This Galacto-oligosaccharide is also known as trans-galacto-oligosaccharides [34]. Galacto-oligosaccharides can significantly stimulate *Lactobacilli* and *Bifidobacterium*. In infant, *Bifidobacteria* have indicated more incorporation with GOSs [35]. GOSs also stimulate other probiotic bacteria such as *Firmicutes, Enterobacteria* and *Bacteriodetes* but lesser than *Bifidobacteria* [36]. The nutritional properties of GOSs are also important; these are low-calorie sweeteners, contribute to maintaining body weight, mitigate constipation, possess low glycemic index and are not involved in cariogenic activities [37]. GOSs are widely used at the industrial level to formulate dairy products, bakery products, beverages and different types of sweets, all converted into functional food [38]. In addition, GOSs are extensively used in infant prescription to promote the establishment of new born microbiome [39, 40].

2.2.2.2 Fructo-Oligosaccharides

Fructo-oligosaccharides, sometimes known as oligofructose or oligofructan, consist of linear short chains of fructose linked by (2, 1) β-glycosidic bond and possessing only a single unit of D-glucosyl at the non-reducing end [41]. Especially, short chains FOS are making a mixture of the smallest oligosaccharide known as 1-ketose [degree of polymerization (DP) equal 3], nystose (DP4) and 1^F-fructofuranosylnystose (DP5) [42]. The production of FOS can acquire either by enzymatic method or hydrolysis of inulin. Inulin can be obtained from natural sources such as chicory roots, dahlia or agave, artichoke and yacon etc. [43]. In the enzymatic synthesis method of FOS production, the transfructosylation reaction is involved [44] in which fructosyltransferases (β-fructofuranosidase or β-D-fructosyltransferase perform as biocatalysts [45]. In transfructosylation reactions, the β-2,1 glycosidic bond breaks and fructosyl moieties are transferred from carbohydrates as donors onto any other acceptor except water [46]. Almost all fructosyltransferases also possess hydrolytic activity; hence the synthesis of FOS is a complicated process in which various reactions are catalyzed simultaneously in parallel or series [47]. In this manner, the production of FOS is carried out by first steps in which fructosyl acts as donors and acceptors, leading to short-chain formation of FOS by the DP (DP_{n+1}, $DP_{n+2,......,}$ DP_{n+N}) immediately [48].

2.2.2.3 Starch and Glucose-Derived Oligosaccharide

In the upper part of the gastrointestinal tract, the starch is resist to digestion, so this kind of starch is known as resistant starch (RS). RS are capable of producing a high level of butyrate, which influence health benefits. Hence it has been proposed as prebiotics [49]. Different species of *Firmicutes* exhibit maximum incorporation with a high number of RS [50].

As was previously mentioned, *Ruminococcus bromii, Bifidobacterium adolescentis, Eubacterium rectale*, and *Bacteroides thetaiotaomicron* may all use to digest minimally the RS. Although RS cannot breakdown in the absence of *Ruminococcus bromii* in mixed bacterial and faecal incubations [51].

Polydextrose (PDX) is a glucose-derived oligosaccharide, not sweet, neutral in taste. Its commercial applications as a low-calorie bulking agent include dairy products, confectionary, functional beverages, and baked goods. According to studies, PDX has been approved for use in food in 60 countries and is recognised in 20 countries as a dietary fibre and prebiotic [52, 53]. PDX is a randomly bonded glucose-based oligosaccharide with an average of DP 12, extending to 120. PDX is a highly branched molecule with all possible combinations of α and β linkages, such as 1-2, 1-3, 1-4 and 1-6 glycosidic linkages [54]. Due to its complicated structure, mammalian digestive enzymes cannot hydrolyze it in the small intestine, instead sending it into the colon, where gastrointestinal microbes progressively ferment it, excreting the remaining 60 per cent in faeces [55]. Another study conducted by Costabile *et al.* reported some shreds of evidence that PDX can stimulate the growth of *Bifidobacteria,* but it has not been proved yet [56].

2.2.3 Other Oligosaccharides

A polysaccharide formerly known as pectin is the source of several oligosaccharides. Pectin is a diverse mixture of polysaccharides found in higher plants' main cell walls and intercellular spaces [57]. Pectin consists of a group of acidic polymers called rhamnogalacturonan (RG) and homogalacturonan (HG) linked with several neutral sugars/polymers like galactans, arabinans and arabinogalactans to the side chain [58]. These neutral and acidic polymers are obtained as pectic oligosaccharides (POS) from agricultural waste and employed in the production of prebiotics [59]. Pectic oligosaccharides, as has been previously documented, limit the growth of several entero-putrefactive and other pathogenic bacteria, including Enterobacteriaceae and Clostridia spp (*Clostridium difficile* and *Clostridium perfringens* [60, 61]. As previously reported, Pectic oligosaccharides inhibit the growth of some entero-putrefactive and other pathogenic bacteria such as *Enterobacteriaceae* and *Clostridia spp.* (*Clostridium difficile* and *Clostridium perfringens*) respectively [62, 63]. The gastrointestinal fermentation of pectin oligosaccharide results in the synthesis of short-chain fatty acids (SCFA), which contribute a number of health benefits to the host, such as inhibition of pathogens, respite constipation, enhancement in mineral absorption, reduced chances of colon cancer, reduction of glucose level in blood and modulate immune system [64].

2.2.4 Health Benefits of Prebiotics

Prebiotics may be used potentially as chemopreventive agents based on observations and reported data that health-influencing bacteria like *Bifidobacteria* do not synthesis carcinogens or genobiotics compound, but rather synthesise SCFAs that promote health benefits to the host. As reported 10 g/day short-chain *fructo-oligosaccharide* in adenoma and adenoma-free individuals observed more healthy signs in adenoma-free individuals [27]. Clinical severity indices, pro-inflammatory immune responses, and calprotectin, a protein generated by neutrophils in both plasma and stool that increases inflammatory bowel disease in patients, have all been demonstrated to significantly decrease when FOS and inulin are consumed together [65]. Ingestion of GOS considerably enhanced the number of fecal *Bifidobacteria* consequently a remarkable changed is observed in fecal consistency, improved flatulence, bloating problem and subjective global assessment score (SGA) [27]. Hence, it can be concluded, based on reported data, that the prebiotic are non-digestible compounds promoting health benefits to the host in various aspects. Moreover, it has been recommended that prebiotic ingestion may:

- Prebiotics are associated with reducing the prevalence and period of infectious and antibiotic-related diarrhea.
- Reduce the symptoms and inflammatory responses elicited by the host immune system in inflammatory bowel disease persons.
- Exclusively participate in preventive effects of colon cancer.
- Elevate the bioavailability and absorption capacity of minerals such as magnesium, calcium and might be iron.
- Reduce some risk factors associated with cardiovascular disease.
- Encourage satiety weight loss and intercept obesity.

2.3 Probiotics

The term "probiotics" used for live microorganisms that provide health benefits to the host when consumed in sufficient amount [66]. But the definitions of probiotics have evolved over time simultaneously with a continuously developing interest in using live bacterial cells as supplements and understanding their mechanisms [67]. The term was used to depict substances synthesized by a specific microorganism that enhance the growth of others and exert health benefits by contributing balance of intestinal flora [67]. The definition of probiotic used at present was proposed by the Food and Agriculture Organization of the United Nations and World Health Organization (FAO/WHO), based on that probiotics are reconsidered as "live microorganisms which when administered in sufficient amount confer the health benefit on the host" [68]. In concerned to food the definition might be modified by stand on that the favorable effect is applied by the microorganisms "when administered in a sufficient amount as a part of diet" [69]. The strains of lactic acid bacteria (LAB) and *Bifidibacteria* are the most extensively used bacteria that reveal probiotic characteristics and are comprised in various functional foods and nutritional supplements [70]. In 2014 the International Scientific Association for probiotics and prebiotics conference declared that the production of by-products from bacterial metabolic pathways, dead microorganisms and other bacterial-associated nonviable components has potential; but do not categorize as a probiotic constructor [71]. The guidelines for the safety evaluation of probiotics for human use are shown in Figure 2.1. However, many studies have been reported that dead bacteria and their nonviable components show probiotic characteristics [72, 73]. The consequence of probiotics related to host health has been published in numerous research, review and systematic review articles [74, 75]. As previously well documented the, probiotics are involved in the prevention of medical

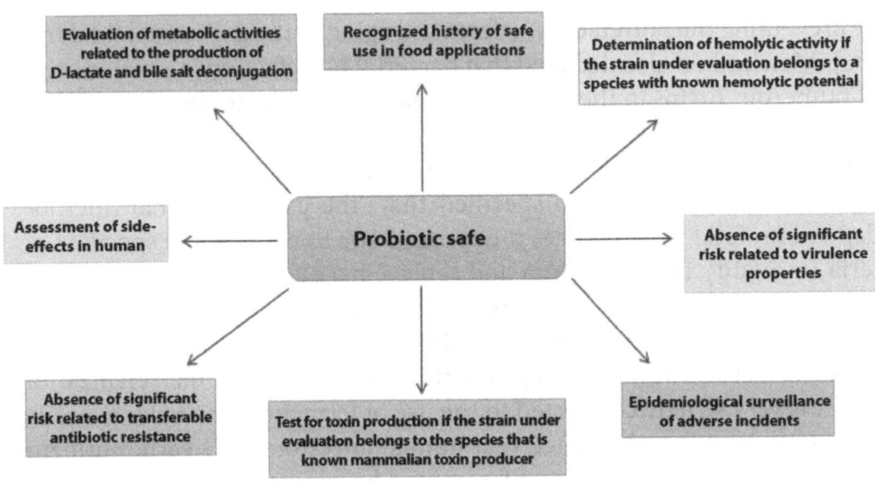

Figure 2.1 Guidelines for the safety evaluation of human probiotics.

complications such as antibiotic-associated diarrhea [70], inflammatory bowel disease (IBD), ulcerative colitis [76], irritable bowel syndrome (IBS) [77], eczema [78] and allergic rhinitis [79].

2.3.1 History of Probiotics

The term "probiotic" derived from Latin *Pro* and Greek *bio,* exactly meaning "for life" was first introduced by German scientist Werner Kollath in 1953 to appoint "active compounds that are mandatory for surviving and development of a healthy life" [80]. In 1965 the term probiotic was used by two other scientists, Lilly and Stillwell, in different references to describe "metabolic compounds produced by one organism which stimulate the growth of other" [81]. More correctly, in 1992, Fuller defined probiotic as "a live microbial feed supplement which beneficially affects the host animal by improving balance" [82].

In the early 1900s the modern era of probiotics was initiated with the pioneering studies by the Nobel laureate Elie Metchnikoff and a Russian scientist working in Paris at Pasteur institute [83]. Louis Pasteur recognized numerous microorganisms responsible for various fermentation processes, while Metchnikoff continuously tried to discover feasible effects of these microorganisms on human health [84]. He was connected with the Bulgarian rural people's regularly consumed fermented dairy products such as yoghurt, and he was identified that they possess enhanced longevity [85]. He attempted to establish a connection between this and the Bulgarian *Bacillus*, which was identified by the Bulgarian physician Stamen Grigorov (27 years old). Later, it was hypothesized that *Lactobacillus* might be preventing the putrefaction effects and aging of the gastrointestinal tract [86]. Metchnikoff also declared that bacterial putrefaction is a major origin of numerous toxins in the gastrointestinal tract. These toxins migrate from there into the circulatory system and cause aging effects [87]. He designated such bacteria as *putrefying bacteria*, at present accepted as *proteolytic clostridia.* Metchnikoff also specified that "the gastrointestinal microbes depend on food which makes it opportunistic to adopt some specific bacteria to modify intestinal flora and replace harmful bacteria by beneficial bacteria [88]. This explanation is very clear in the way of "probiotic concept" [89]. Metchnikoff also favoured those *Lactobacilli* as probiotics.

Nevertheless, the history of probiotics is as ancient as human civilization, and it is very strictly connected to the use of fermented food [1]. However, recent research has elucidated that human ancestors were earlier utilizing yeasts in the manufacturing of fermented beverages a long time ago than

2000 BC [90]. Based upon the remnant of antiquated Egypt, it might be observed that dairy-based fermented products such as 'Leban Rayad' and 'Leban Khed' are still common in the Middle East, present-day were used as earlier as 3500 BC [91]. At the same time as human civilization, much more trials were going on, and all observations were recorded carefully; different cultures were used to produce fermented beverages [92].

2.3.2 Mongolian History

It stated that during Ghengis Khan's conquest of the Mongolian Empire, a prophet approached a town across the desert and requested a bag of water. However, the villagers were angered by the Mongolian invasion; they filled a sack with milk and gave it to him instead of water. When he felt thirsty, the harbinger found some mousse-like liquid in his backpack instead of water, but he had no alternative option, so he drank the liquid to relieve his thirst. When Ghengis Khan heard this tale, he allegedly added sour milk to his army's diet [1].

Two Armenian scientists, Sarkis and Rose Colombosian, who relocated from the United States to Armenia after World War I, began selling their handmade yoghurt called madzoon (yoghurt in Armenian). Due to the unfamiliarity of the product's name, it was initially unpopular, so they opted to offer it under the name yoghurt. In 1929, the company "Colombo and Sons Creamery" was founded in a kitchen in Andover, Massachusetts, United States of America; it was the first yoghurt brand in American history [93]. The yoghurt introduced in 1929 gave the first taste to American people, and 80 years later, demand and popularity of this yoghurt continued to grow [94]. As increased demand and popularity day by day, they introduced their brand, 'Colombo Yoghurt', and then it started to sell out at General Mills in 1993.

Meanwhile, in 1930, Dr Minoru Shirota isolated the first culture of *Lactobacillus casei* strain Shirota from the human gut in the Microbiology laboratory of Kyoto University, Japan. Shirota's strain is resistant to the acidic environment of the intestine, bile acid, and is able to reach the lower intestine following oral delivery, as established by multiple experiments. Dr. Shirota invented Yakult, a type of dairy product utilising probiotic bacteria, in 1935 and successfully introduced it to the market. Its name 'Yakult' was originally obtained from yoghurt in Esperanto, the suggested worldwide language of the 1880s. By regulating the intestinal flora, it was postulated that daily consumption of Yakult (a fermented product) could assist in enhancing intestinal health and contribute to a longer lifespan.

2.3.3 Health Benefits of Probiotics

Numerous evidences are available in support of the beneficial effects of probiotics including enhancement of immune response, reduction of cholesterol level from serum, improved intestinal health and cancer prevention [95]. As reported earlier, these health benefits are strain-specific and influenced by various mechanisms [70]. Although the mechanisms of the several health benefits are well explained, others need additional studies to confirm. There is considerable evidence that probiotics may be used in the prevention of inflammatory gastrointestinal conditions, including antibiotic-associated diarrhea (AAD) [88], ulcerative colitis, enhancement of lactose metabolism, Crohn's disease, infectious diarrhea, non-specific ileitis and infectious diarrhea [67, 81]. Various benefits of probiotics to the host are depicted in Figure 2.2. The etiology of these infectious diseases is not completely accepted, but it is reported that they are appertinent with chronic and current infections or intestinal inflammation [96]. Several clinical studies are exhibited that probiotics influence the remission of ulcerative colitis, but not much effective is observed in the case of Crohn's disease [97].

According to popular belief, probiotics inhibit some pathogen enzymes and reduce the risk of colon cancer in animals. However, the same results are not observed in humans in clinical trials [98]. However, a positive effect on the urogenital system in women [prevention and treatment of bacterial vaginosis and Urinary Tract Infections (UTIs)] is an excellent illustration of the positive effect linked to the usage of probiotics [99]. Probiotics are also

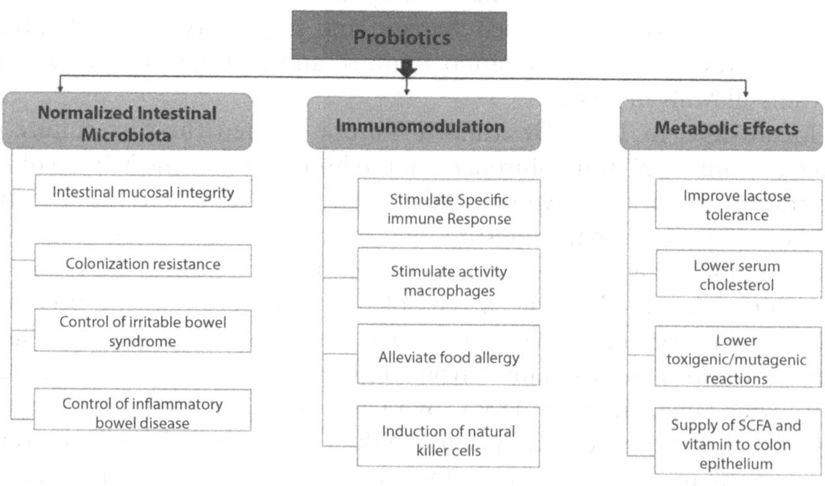

Figure 2.2 Probiotics as functional foods in enhancing gut immunity.

administered in pregnant women and neonates to see the preventive measures for allergic diseases such as atopic dermatitis. However, this type of case study is always controversial [100]. Some evidence is available that the intake of probiotics (dairy products) leads to the reduction of blood cholesterol, which may help overcome diabetes, obesity, cerebral stroke and cardiovascular disease [101]. Some researches document that animals to which probiotic strain *Lactobacillus acidophilus* is orally administered, showed an increased expression of an intestinal receptor for cannabinoid and μ-opioid so as to mediate analgesic functions in the intestine, and observations were comparable with the effects of morphine [102]. Furthermore, some well documented animal studies in which probiotic strain *Lactobacillus plantarum* is orally administered, observed the positive immune system regulation due to high expression of cytokines IL-2, IL-6 [103]. The effectiveness of the probiotic strains is well documented to reduce the risk of healthcare-associated infections (HAI), viral diarrhea and non-nosocomial infections. Probiotics also prevent from intestinal viral infection by increasing the level of immunoglobulin IgA [104].

Antibiotic-associated diarrhoea (AAD) is a common side effect of most medicines, and *Clostridium difficile* disease (CDD), which is caused by antibiotics, contributes to outbreaks of diarrhoea and colitis in healthcare settings. The consumption of probiotics to reduce the risk of these two related diseases is still controversial. However, some specific probiotic strains might be used as an effective treatment for these two diseases. Using meta-analyses, the probiotic strains *Saccharomyces boulardii, Lactobacillus rhamnosus* GG, and a mixture of probiotics remarkably reduced the risk of antibiotic-associated diarrhoea while *Saccharomyces boulardii* potentially reduced the risk of *Clostridium difficile* disease [105]. Studies in Helsinki (Finland) reported that regular consumption of probiotic strain *Lactobacillus rhamnosus* GG is helpful in reducing respiratory tract infections [106]. As previously reported, the regular consumption of yoghurt significantly reduces *Enterobacteriaceae* and improves galactosidase activity in the gastrointestinal tract [107].

2.3.4 Desired Characteristics of Probiotics

Consecutively for a possible probiotic organism allow exerting its beneficial effects, it is looking forward to certain desirable characteristics. The following desirable characteristics are currently determined by *in vitro* studies. The desirable characteristics of the newly isolated probiotic strain are represented in Figure 2.3.

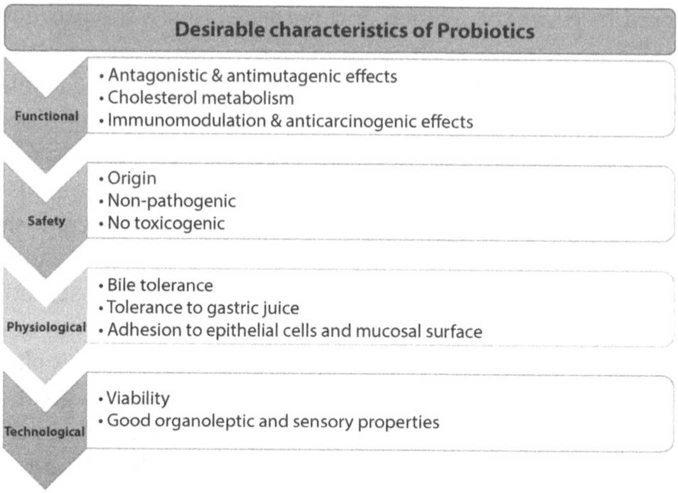

Figure 2.3 Characteristics of probiotics.

I. The probiotic strain should be acid and bile tolerant, which makes it suitable for oral administration.
II. To successfully modulate the immune system, competitive exclusion of pathogens and prevent the pathogen's binding and colonization, the probiotic strain must be able to adhere to mucosal and epithelial surfaces.
III. Possess antagonistic effects against cariogenic and pathogenic bacteria.
IV. Produce antimicrobial substances against pathogenic bacteria.
V. Probiotic strain must be non-pathogenic, non-toxicogenic and should be of human origin.
VI. Persistency and viability is also important characteristic of a probiotic strain.
VII. Possess anticarcinogenic and antimutagenic properties.
VIII. Able to hydrolyze bile salts.
IX. Probiotic strains should be resisting to bacteriophage.
X. Ensure that the probiotic stain does not carry transmissible antibiotic-resistance genes.

2.4 Synbiotics

Synbiotics, a combination of prebiotics and probiotics, exert both prebiotic and probiotic beneficial effects. Synbiotics are not only involved in improving the survival of beneficial microbes but also stimulate and proliferate the growth of specific native microorganisms associated with the gastrointestinal tract [108]. The health benefits of synbiotics are still not well documented. As reported, synbiotics' health benefits are perhaps related to the specific combination of prebiotics and probiotics [109]. Numerous possible combinations of synbiotics can be promising to modulate gastrointestinal microbiota in humans [110].

2.4.1 Mechanism of Action of Synbiotics

As previously reviewed, probiotics' effects are mainly observed in small and large intestines, and the effects of prebiotics are predominantly detected in the large intestine, so the combination of these two may have synergistic effects [111]. Prebiotics are mostly selected based on their efficiency as a medium for a probiotic strain, easily fermentable and intestinal passage. As reported in the literature in the presence of prebiotics, probiotic microbes acquire more stability to survive in intestinal environmental conditions, including oxygenation, temperature, pH etc [112]. However, the mechanism of action to get excess energy and adapt them to survive such harsh conditions is not well explained. The combinations of various components ensure a suitable dietary supplement and provide an appropriate environment for the probiotic strain's growth, leading to beneficial health effects on the host [113]. There are two modes of action of synbiotics that are explained [114].

I. By improving the viability of the probiotic strain.
II. By the provision of a particular positive health effect.

Prebiotic stimulate the growth of probiotics strain, resulting in the modulation of gut metabolism without affecting the basic biological structure of the intestine, the development of good microbiota and the inhibition of pathogenic microbes present in the gastrointestinal tract [115].

2.4.2 Benefits of Synbiotics

Synbiotics possesses the following health-beneficial effects on human [116].

I. Balance the gastrointestinal microflora with the increased number of *Lactobacillus* and *Bifidobacterium* genera.
II. Synbiotics improve hepatic functions in-patient who is suffering from cirrhosis.
III. Stimulate the immunomodulatory system.
IV. Decreased the chances of bacterial translocation and reduced the risk of nosocomial infections during post-surgical procedures and similar obstructions.

2.5 Encapsulation of Probiotics

The encapsulation of probiotics requires the cells with an appropriate wall material to protect and prevent the discharge of encapsulated cells [117]. Therefore, encapsulation of probiotics is a technique that may produce resistant cells to unfavourable circumstances [118]. In encapsulation, generally, physiochemical or mechanical processes are used. The probiotic cells are entrapped in different encapsulating materials that minimize the risk of injuries or cellular losses of encapsulated bacterial cells [119]. Numerous ways are currently employed to encapsulate probiotic cells. When choosing a technique, it is essential to evaluate its suitability for a certain probiotic strain, its simplicity, the viability and stability of the cells, and its compatibility with the intended use [120]. Current probiotic encapsulation techniques include spray drying, spray chilling, extrusion, spray-freeze drying, emulsion, electro spraying, fluidized bed, and coacervation. [121] some of these are briefly discussed in the next section:

2.5.1 Emulsion

In the food and pharmaceutical industries, emulsions are commonly employed to increase desired chemicals' stability, solubility, and physiological activity. Two immiscible liquids are scattered to create emulsions, which are then stabilized by adding a stabilizing agent that has a stronger affinity for the continuous phase than the dispersed phase [122]. However, adding a solidifying agent can separate the dispersed phase droplets from the continuous phase [123]. These emulsions are utilized to encapsulate probiotics to improve the protection and stability of encapsulated bacterial cells. Due to the hydrophilic character of bacterial cell walls, the dispersed aqueous phase is preferred and utilized more frequently [124]. As reported earlier, *Lactobacillus plantarum* 299v was encapsulated by using simple emulsion and metronidazole with aqueous phase (xanthan and gaur gum) and lipidic phase (sunflower oil) [125]. This encapsulation method enhances encapsulated cells' viability and prevents the encapsulated material's release during storage. The use of double emulsion (de Man, Rogosa and Sharp broth, grape oil) also has been reported in the encapsulation strategy of *Lactobacillus rhamnosus* Lc705. The protection provided by the emulsifying agents to the encapsulated cells is well documented under the osmotic stress by the satisfactory viable cells count (≥ log CFU/ml) in sucrose hypertonic solution [126]. Emulsions also improved the possibility of using encapsulated probiotic cells in liquid form. The encapsulation of *Lactobacillus acidophilus* AS 1.2686 was done by double steps emulsification strategy, using water-in-oil-in-water emulsion (w/o/w) that improved the viability of encapsulated cells for approximately 14 days as compared to free cells [127]. First, a water-in-oil emulsion is prepared using a homogenizer. In the second step, the emulsion strongly passed through a silicon plate containing uniform micro-sized pores, including an external water phase. As a result, w/o/w droplets formed, some of which may be cross-linked. As a result, the microsphere could be form. These microspheres were used for the encapsulation of probiotics [128]. The hydrocolloids are also used as emulsifying agents to encapsulate probiotic strains, known as the internal ionic gelation technique. This technique used hydrocolloid material as an encapsulator with a solidifying agent. Thus, the microbial cells blended in hydrocolloid solution and a non-ionized solidifying agent was added to the oil, which acts as a continuous phase of water-in-oil-emulsion. So, as the pH is changed of the medium the solidifying agent is ionized which leads to the formation of particles that can be detached from the continuous phase [129]. The internal ionic gelation technique uses alginate and calcium carbonate as encapsulating and solidifying agents. *Bifidobacterium* BB-12 is effectively encapsulated using this method. The encapsulated particles effectively resisted the probiotic strain from the exposure of gastric and enteric juices and enhanced the survivability and stability of cells during storage for up to 60 days at 25 °C [130].

2.5.2 Extrusion

Extrusion is a widely used technique to encapsulate bacterial cells; it is a simple, easy-to-use, economically feasible and relatively mild approach that assures encapsulated bacterial cells' high viability and stability [131]. Apart from the favourable conditions, this technique has several drawbacks, including it is slow, inefficient in the production of microsphere less than 500 μm, low to moderate viscosity of polymer solution is essential and requires large diameter size of nozzle [132].

Generally, in this technique, the hydrocolloid solutions used with microbial cultures, extrusion of hydrocolloid through a nozzle in the cross-linking solution, leads to the immediate transition of hydrocolloid solution into a gel, finally converted into the microsphere. The gel is highly stable in an acidic medium, although it disintegrates under alkaline conditions [133]. The size of the microsphere is affected by several factors, such as the diameter of the nozzle, solution viscosity, flow rate of polymeric solution, the distance from the drop to the cross-linking solution and the temperature of the polymer [134]. Furthermore, immediate cross-linking and toughening on the outer surface of the microsphere cause hindrance in the movement of ions inside the inner core, influencing the instability of the microsphere [135]. However, the production of the microsphere is carried out at a laboratory scale; the large-scale production is usually difficult due to the slow formation of microsphere [136]. Several methods have been reported to defeat the drawbacks of this simple extrusion technique, including the precision particle fabrication (PPF) method, co-extrusion and coaxial flow method [137], multiple nozzle system [138], droplet generation with electrostatic [139], rotator disc atomization [140], liquid jet cutting method and vibration or acoustic energy technique [141]. If the production of droplets is controlled by a pulse, jet or vibrating nozzle, then the technique is known as laminar jet break-up or prilling. In this technique, the liquid jet is converted into droplets using vibration frequency generated by a vibrator [142]. The size distribution of the microsphere is highly dependent on the polymer solution's viscosity and the solution's flow rate to the encapsulation of probiotic train *Saccharomyces boulardii* using the laminar jet break-up technique [143]. The prepared microsphere of *Saccharomyces boulardii* can be further coated with chitosan. Although, the chitosan coating did not provide any other beneficial characteristic. In fact, the *Saccharomyces boulardii* is more protected inside the free microsphere as compared to coated microsphere [138]. In the coaxial flow technique, the alginate can be used in two different concentrations and introduced separately into the outer surface and inner core chambers of the coaxial nozzles, which lead to the formation of a stable microsphere [144]. In the acoustic excitation technique, the generated polymer droplets are usually favourable to cross-link in the calcium chloride solution, so this technique is appropriate for producing a size-controlled microsphere. The leakage of encapsulated microbial cells from the microsphere can be resisted by improving the concentrations of outer and inner core materials [120]. The other innovative apparatus fabricated by Haunge consisted of a pneumatic micro-vibrator that could be used continuously to produce alginate microspheres in the range of 30 to 70 μm [145]. This apparatus offered many advantages, including flexibility, compactness and minimising the encapsulated cells' damage risk.

2.5.3 Spray Drying

Spray drying technology is commonly used for microencapsulation in the food industry with low cost, easy to use, fast processing and high productivity rate. In general, this technique involves atomising a solution of microbial cells into hot drying air, followed by instant evaporation of water [146]. The encapsulated cells are separated out as a powder from the conveying air. In this technique, different natural polymers can be used, particularly starches and Arabic gum, due to their capability to produce microspheres after the drying process. Furthermore, Fructooligosaccharides, inulin, gums and alginates can also use as encapsulated agents [147]. Several parameters need to be optimised to produce a well-detached microsphere, including inlet and outlet air temperature, feed temperature, airflow, feed rate etc. [148]. Regulation of temperatures is most important because low temperatures can reduce the rate of water evaporation that leads to the formation of clusters of microsphere. However, high temperatures can cause damage to bacterial cell wall, which directly affect the cellular viability of probiotic cells [120]. The schematic representation of the spray-drying process has been represented in Figure 2.4. However, the survivability of probiotic cells encapsulated by the spray drying method is highly affected by the outlet temperature of the system rather than inlet temperature [149]. For instance, as previously reported, in the encapsulation of *Lactobacillus paracasei* NFBC 338 by spray drying technique using skimmed milk as an encapsulated agent, a 97% survival rate was reported at 70 to 75°C outlet temperature and 0% at 120 °C [150]. *Lactobacillus acidophilus* La-5 is currently used for encapsulation by spray drying technique, using inulin as encapsulated material at inlet and outlet temperatures of 120°C and 55°C, respectively. The survivability of bacterial cells measured after completing the encapsulation process reached 86.5% [151].

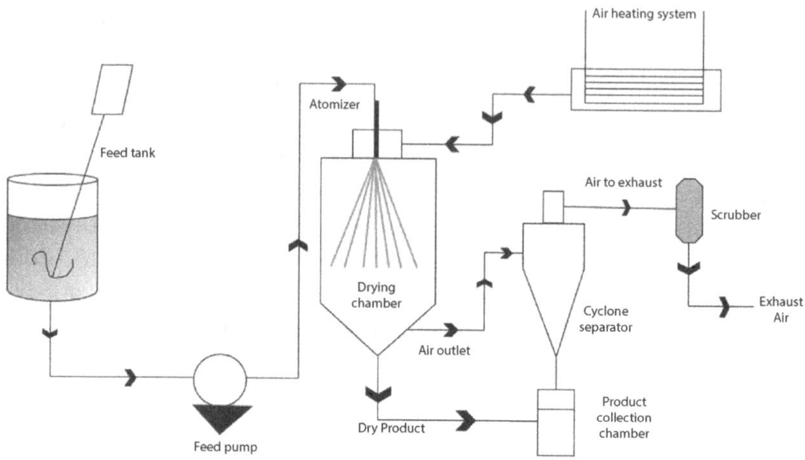

Figure 2.4 Schematic representation of spray-drying mechanism.

Similarly, the cells of *Bifidobacterium* BB-12 encapsulated by spray drying technique with prebiotics using inlet and outlet temperatures 150 °C and 55 °C, respectively, resulted in the survival rate of microbial cells higher than 70% measured [19]. To defeat the adverse effects of high temperatures and improve the stability as well as the viability of encapsulated cells, the addition of soluble fibers, prebiotics, gums and mucilages with encapsulating material act as a thermal protector and provide additional strength to the cells [152]. The cells of the *Bifidobacterium bifidum* BB02 encapsulated by a spray drying process with the combination of whey protein (concentrate), maltodextrin, and mosquita gum, the symbiotic effects of the combination is well documented to improve the resistance and survivability of the encapsulated cells against environmental conditions [153].

2.5.4 Spray Chilling

The spray chilling technique, known as spray cooling or spray congealing, is parallel and similar to spray drying because it produces small droplets. Spray chilling is a suitable technique to encapsulate bacterial cells and other food ingredients because it is economically feasible, continuous and easy to scale-up process [154]. Encapsulation by spray chilling has shown in Figure 2.5. In addition, this technique does not require organic solvents, including alcohols or ethers, and it can be used effectively for temperature-sensitive ingredients such as omega-3 fatty acids and enzymes [155]. However, this technique comprises several technological disadvantages,

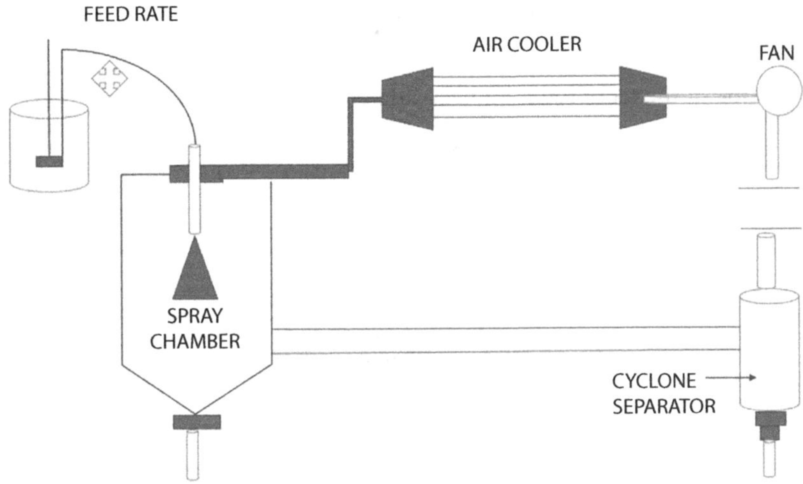

Figure 2.5 Encapsulation by spray chilling.

such as less efficiency in encapsulating the microbial cells and an increased possibility of expulsing microbial cells during storage. In the chilling technique, the encapsulated material is diffused on a semi-liquid matrix. The matrix consists of lipids and is atomized in a cold air chamber, enabling it to solidify the particles [156]. Spray chilling is not new, but it is a less exploiting method than the others, such as spray drying and ion gelling [157]. The lipids, triglycerides, waxes, and fatty acids can be used for matrix preparation, but the hydrophilic compounds, such as polysaccharides, can be employed only in an emulsion with an emulsion appropriate melting temperature [158]. Although in spray chilling, a specific carrier agent is needed that must have hydrophobic nature and lower melting temperature from those have negative effects on encapsulated cells.

In combination, this is challenging for the broad application of this technique in the food sector [156]. Despite the demerits, *Lactobacillus acidophilus* and *Bifidobacterium animalis* subsp *lectis* were functionally encapsulated by this technique using vegetable oil as encapsulated agent. Some principal techniques used for microencapsulating different probiotic strains are depicted in Table 2.1. Moreover, the encapsulated probiotic cells showed high viability during storage at refrigerated conditions for around 90 days [159]. Similarly, the spray chilling method is used to create the single- and double-layered microspheres of the bacteria *Lactobacillus acidophilus, Saccharomyces boulardii*, and *Bifidobacterium bifidum* [160].

Table 2.1 Principal methods for encapsulating probiotic microorganisms [117].

Name of the technique	Encapsulated microorganisms
Emulsion	*Akkermansia muciniphila*
	Lactobacillus rhamnosus GG
	Saccharomyces boulardii
	Enterococcus faecium
	Lactobacillus paracasei spp. *paracasei*
	Lactobacillus acidophilus LA-5
Extrusion	*Lactobacillus casei* 01
	Lactobacillus casei BGP93
	Lactobacillus acidophilus La3
	Lactobacillus lactis cremoris
	Lactobacillus acidophilus KBL409
	Lactobacillus casei ATCC 393
	Faecalibacterium prausnitzii
Spray drying	*Lactobacillus rhamnosus* GG
	Bifidobacterium infantis ATCC15679
	Lactobacillus acidophilus NCDC 016
	Lactobacillus plantarum NCIM 2083
	Kluyveromyces VM004
	Lactobacillus plantarum ATCC 8014
	Lactobacillus fermentum K73
Spray chilling	*Bifidobacterium animalis* spp. *lactis* BI-01
	Lactobacillus acidophilus Lac-04
	Lactobacillus acidophilus
	Bifidobacterium animalis spp. *lactis*
	Saccharomyces boulardii
	Lactobacillus acidophilus LA-5
	Bifidobacterium bifidum BB12
	Lactobacillus acidophilus La3
	Bifidobacterium animalis spp. *lactis* BLC1

2.5.5 Encapsulating Material for Probiotic Cells

To maintain the viability and stability of probiotic cells, selecting an appropriate encapsulating material is essential [161]. The encapsulating material must not be toxic for microbial cells to infect it should provide protection against adverse environmental conditions and be sufficient to prevent leakage of encapsulated cells. During the storage of the probiotic, both the relative humidity (RH) and temperature can affect the stability and viability of cells [162]. Thus, the used material contains water retention capacity to improve the viable count of the encapsulated cells. Furthermore, the material does not completely release microbial cells when exposed to gastric juice otherwise might not be capable of protecting cells [163]. Production of a microsphere for encapsulation using water-soluble polymers provide optimal conditions for the reactivation of the microbial cells. So, both natural as well as synthetic water-soluble polymers are widely used to encapsulate microbial cells. While synthetic polymers provide better mechanical strength and chemical stability than natural polymers [164]. Gelation of polymers can be one by several mechanisms, such as thermal gelation, ionotropic gelation, thermal gelation, polymerization and cross-linking between polymers. Hydrogen bonding, hydrophobic interactions and/or

Table 2.2 Different encapsulating materials used to encapsulate probiotic bacteria [117].

Encapsulating material	Name of technique	Probiotic bacteria
Chitosan-alginate-inulin	Extrusion	*Lactobacillus rhamnosus* GG
Alginate-shellac	Fluidized bed	*Lactobacillus paracasei* BGP-1
Cellulose alginate	Fluidized bed	*Lactobacillus plantarum* IS-10506
Alginate-arabinoxylan	Extrusion	*Lactobacillus plantarum*
Alginate-goats' milk-inulin	Extrusion	*Bifidobacterium animalis* spp. *lactis* BB12
Alginate	Extrusion	*Lactobacillus casei* ATCC 393
Alginate-chitosan	Emulsification; internal gelation	*Bifidobacterium longum* DD98

electrostatic interactions develop the intermolecular and intramolecular interactions during the cross-linking of polymers [165]. A real-time *in situ* method is used to measure the oxygen transportation in an oil/water interface of an emulsion solution. In this method, tris ruthenium (II) bis hexafluorophosphate dye is used for reversible fluorescence quenching reversible fluorescence quenching [166]. The dye encapsulated in emulsion's oil phase by cross-linking with the help of oxygen. This technique can be used to check the efficiency and evaluate the barrier properties of the different polymers used for microencapsulation. Therefore, an appropriate encapsulated material can be selected by using this technique [167]. Some examples of encapsulating material used to encapsulate probiotic strain have been listed in Table 2.2.

2.5.6 Challenges with Current Adopted Methods

Whatever type of encapsulation technique is used, the necessity to maintain sterile conditions is universal for all techniques [168]. Moreover, numerous probiotic strains have a gastrointestinal origin, requiring anaerobic conditions for optimum growth. The presence of oxygen can remarkably harm the survivability and activity of the bacterial cells throughout the encapsulation technique, particularly for *Bifidobacterium* because this group is strictly anaerobes [169]. In addition, the antioxidants are used with different encapsulating materials to prevent the formation of free radicals (oxygenic toxicity), and the anaerobic conditions should be maintained during the entire encapsulation process, including encapsulating instruments [170]. As previously reported, no appropriate deoxygenation technique has been clearly defined except for the addition of antioxidants together with encapsulating materials. An anaerobic chamber is used to maintain anaerobic environmental conditions during encapsulation. However, the utilization of an anaerobic chamber is not appropriate for large-scale processing such as spray drying and other applications of encapsulation used by industries [171]. Although the cost of the encapsulation process is a significant concern, including innovative programming with the device as well as the procedure adopted for encapsulation, which should maximally reduce the risk of damage to the cell's viability due to the oxygenic toxicity throughout the encapsulation process [172]. It has been reported that *in vitro* assessment of encapsulated cells cannot properly replace by *in vivo* assessment because *in vitro* systems cannot perfectly mimic the physiological conditions as in the human gastrointestinal tract [173]. The existence of food constituents *in vivo* may transiently raise the pH of gastric juice.

In contrast, *in vitro* assessment may not be able to consider some biological factors, such as peristaltic movement between the stomach (colon) [174]. The mucin production by epithelial tissue and its function on the surface of the gastrointestinal tract is consider as another factor which is very complicated to mimic *in vitro* assessment [175].

Maintaining the maximum survival rate during the consumption of encapsulated cells through the gastrointestinal tract is censorious for the successful delivery of probiotic cells; enhancing the survival rate and stability during the storage period can be consider as another challenges in the development of probiotic-encapsulated techniques [176]. During the storage, several parameters such as storage temperature, moisture content, power composition, degree of oxygen, encapsulated material, storage material, light exposure and presence of antioxidants significantly affect the survival rate of probiotics in dried powder form [177]. Another previously discussed challenge is the probiotic leakage from the microsphere throughout the storage. The unnecessary leakage of probiotic cells influences the growth of undesired microbes in the food matrix, which may alter the sensory properties and perishability of the food negatively [178]. Triggering the microsphere to release the encapsulated probiotic cells in a specific region of the gastrointestinal tract may consider another challenge for the targeted delivery of encapsulated probiotics. Some basic approaches are mentioned here to release encapsulated probiotic cells in the targeted region [179].

Approach 1: breaking the matrix mechanically.
Approach 2: by developing a pH gradient in the GI tract.
Approach 3: by developing a time-dependent system.
Approach 4: uses of carriers, degradable by microbiota or enzymes produced by microbiota.

While the encapsulation of probiotics in the form of microspheres, used different food matrices for incorporation. Researchers study the relationship between probiotic survival and food matrices, still, the use of safety issues/challenges of encapsulation of probiotics may require more discussion [180].

2.6 Probiotic Foods Developed

The lactic acid-based foods involved a crucial place nowadays in our regular diet. More than 80% populations use fermented probiotic food "yoghurt" in their daily diet as a food supplement [181]. These food supplements have specific characteristics due to the addition of pure starter cultures from a group of Lactic Acid Bacteria (LAB). Using a suitable method, any fermented food product can be obtained with appropriate characteristics such as aroma, texture, taste and other physiological and biological properties [182]. These foods with traditional fermentative LAB have some positive effects on human health, including reducing the risk of putrefaction, inhibiting the growth of pathogens and improving metabolic activities [183]. Lactobacilli are an essential in many types of fermented food and play a crucial role in the production of probiotic meals. Some species of *Lactobacillus* is used as a starter culture in the production of curd, yoghurt, cheese and other dairy-based fermented foods [184]. The classification and types of probiotic foods have been shown in Figure 2.6. Again, it is very important that the properties of a strain should be applicable to large-scale fermentation. Because not all strains can be cultivated at an industrial scale due to the low reproductive rate in providing medium or less survival efficiency throughout the freezing and freeze-drying processes [185], that is why the LAB used in manufacturing probiotics/fermented foods at the industrial level must need some additional requirements. The selection of a probiotic strain to develop probiotics/fermented foods depends on the microbiological criteria for food safety of the final product achieved by using a non-pathogenic strain with proper hygienic conditions [186]. The high number of live cells and best survival rate when passing through the intestine and stomach permitted them to fulfil their health benefits to the host. In the combinations of different probiotic strains, improve the probiotic supplements' biological role [81]. Dairy-based fermented products

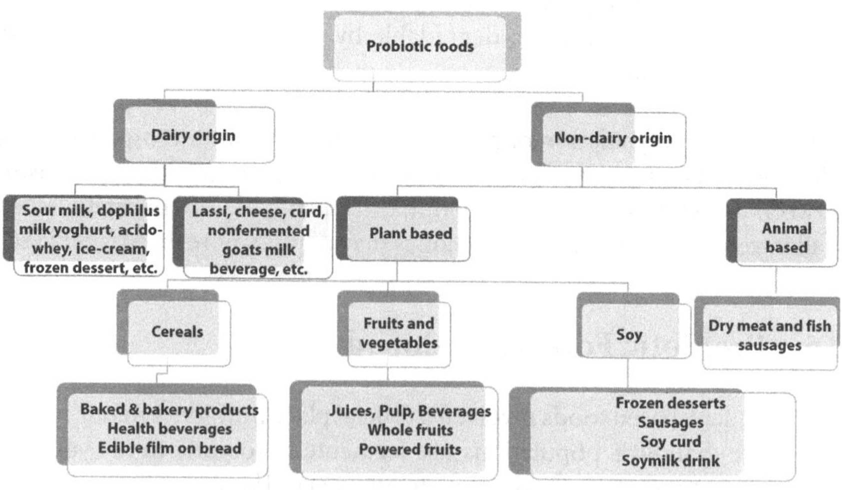

Figure 2.6 Classification and types of probiotic foods.

with probiotic properties need to designed according to the criteria for probiotic development. After bread, yoghurt is the most prominent food used worldwide, developed using a starter culture with the combinations of probiotic strain *Lactobacillus delbrueckii* subsp. *Bulgaricus* NBIMCC 3607 have reproductive capability and fulfils all the requirements as a probiotic culture [187]. The maximum number of viable cells of *Lactobacilli* improves yoghurt's healing and preventive properties.

2.6.1 Viability of Probiotic Cells

Probiotics could fall into every product category regulated by FDA; up to now, FDA does not have any central office or route that deals particularly with probiotics or any regulatory definition of probiotics. Probiotics are also probably losing their viability and degrade under unspecified conditions [188]. The viability of encapsulated probiotic cells refers to the potential of a cell to grow and eventually procreate a colony of cells under specified habitation conditions [189]. Generally, viability is considered the imperative efficacy of probiotics related to the consumers health-promoting benefits. The researcher demonstrated that to stimulate the gut immune system of the host viability, persistency of bacteria was necessary, and in lactose digestion, the effectiveness of viable cells is observed to be high as compared to non-viable probiotic cells [190]. Recently, scientific documentation is focused on various innovative techniques used to enhance the viability of probiotic cells throughout the product shelf-life [191]. Food constituents such as food preservatives (sugars, salts, antimicrobials, bacteriocins or even aroma compounds etc.) can positively or negatively affect probiotic cells' viability [192]. Bacteriocins, antimicrobial compounds and/or sometimes salts are remarkably challenged to maintain the viability of probiotic cells in the food matrix, particularly in the course of storage, although prebiotics known as enhancers, the viability of probiotic cells [193]. Enhancing probiotic cell viability, the isolation, characterization and selection of new species (e.g. *Bifidobacteriaum, Propionibacterium,* and *Lactobacillus*) comprise a major field of research. The criteria for a strain to be characterised and selected as a probiotic could be focused on either or both for food or neutraceutical applications [194]. The major aspects of these criteria can be divided into different categories: 1. Technological 2. Safety 3. Functional 4. Physiological characteristics. Viability of probiotic cells throughout the food processing and/or storage and should be able to survive after transfer into the upper intestinal, along with health benefits to the host, can be the major criteria for the new probiotic selection [195]. As previously reported, losses of 6-8 log cfu/gm of probiotic cells *in vitro* gastric digestion [196, 197] suggested that the remaining counts of probiotic cells are not sufficient to make health benefits for the host [198]. Guidelines and directions released by the Food and Agriculture Organization of the United Nations (FAO)/World Health Organization (WHO) indicate the prerequisite for the probiotic strain to persist in the upper intestinal tract to ensure a health-promoting effect for the host. For instance, to persuade that, the "minimum therapeutic" number of viable probiotic cells should be at least 10^6 cfu/g during the product shelf life [112].

References

1. Ozen M, Dinleyici EC (2015) The history of probiotics: The untold story. Benef Microbes 6:159–165. https://doi.org/10.3920/BM2014.0103
2. Gogineni VK (2013) Probiotics: History and Evolution. J Anc Dis Prev Remedies 01:1–7. https://doi.org/10.4172/2329-8731.1000107
3. Frakolaki G, Giannou V, Kekos D, *et al.* (2021) A review of the microencapsulation techniques for the incorporation of probiotic bacteria in functional foods. Crit Rev Food Sci Nutr 0:1–22. https://doi.org/10.1080/10408398.2020.1761773
4. Hotel A, Cordoba A (2001) Health and nutritional properties of probiotics in food including powder milk with live lactic acid bacteria. Prevention 5:1–34
5. Zhu Z, Luan C, Zhang H, *et al.* (2016) Effects of spray drying on Lactobacillus plantarum BM-1 viability, resistance to simulated gastrointestinal digestion, and storage stability. Dry Technol 34:177–184. https://doi.org/10.1080/07373937.2015.1021009
6. Valle MJ, Vannini V, Sinderen D Van, *et al.* (2011) B-Group vitamin production by lactic acid bacteria – current knowledge and potential applications. 1297–1309. https://doi.org/10.1111/j.1365-2672.2011.05157.x
7. Nasri H, Baradaran A, Shirzad H, Kopaei MR (2014) New Concepts in Nutraceuticals as Alternative for Pharmaceuticals. 5:1487–1499.
8. Narayan KS, Gaurkhede S, Sharma V, *et al.* (2021) Technological and functional assessment of riboflavin enriched probiotic soycurd. Fermentation 7:. https://doi.org/10.3390/fermentation7020047
9. Vijendra V, Ananya M, Jayesh R (2022) Riboswitch - mediated regulation of riboflavin biosynthesis genes in prokaryotes. 3 Biotech 1–14. https://doi.org/10.1007/s13205-022-03348-3
10. Kumar V, Amrutha R, Ahire JJ, Taneja NK (2022) Techno-Functional Assessment of Riboflavin-Enriched Yogurt-Based Fermented Milk Prepared by Supplementing Riboflavin-Producing Probiotic Strains of Lactiplantibacillus plantarum. Probiotics Antimicrob Proteins. https://doi.org/10.1007/s12602-022-10026-6

11. Thakur K, Lule VK, Rajni CS, *et al.* (2016) Riboflavin producing probiotic lactobacilli as a biotechnological strategy to obtain riboflavin-enriched fermented foods. J Pure Appl Microbiol 10:161–166.
12. Liu YW, Liong MT, Tsai YC (2018) New perspectives of Lactobacillus plantarum as a probiotic: The gut-heart-brain axis. J Microbiol 56:601–613. https://doi.org/10.1007/s12275-018-8079-2
13. Yadav R, Puniya AK, Shukla P (2016) Probiotic properties of Lactobacillus plantarum RYPR1 from an indigenous fermented beverage Raabadi. Front Microbiol 7:1–9. https://doi.org/10.3389/fmicb.2016.01683
14. Arepally D, Reddy RS, Goswami TK (2020) Encapsulation of: Lactobacillus acidophilus NCDC 016 cells by spray drying: Characterization, survival after *in vitro* digestion, and storage stability. Food Funct 11:8694–8706. https://doi.org/10.1039/d0fo01394c
15. Rajam R, Subramanian P (2022) Encapsulation of probiotics: past, present and future. Beni-Suef Univ J Basic Appl Sci. https://doi.org/10.1186/s43088-022-00228-w
16. Krasaekoopt W, Bhandari B, Deeth H (2003) Evaluation of encapsulation techniques of probiotics for yoghurt Evaluation of encapsulation techniques of probiotics for yoghurt. 6946:2–13. https://doi.org/10.1016/S0958-6946(02)00155-3
17. Seth D, Mishra HN, Deka SC (2017) Functional and reconstitution properties of spray-dried sweetened yogurt powder as influenced by processing conditions. Int J Food Prop 20:1603–1611. https://doi.org/10.1080/10942912.2016.1214965
18. Huang S, Vignolles M, Chen XD, *et al.* (2017) Spray drying of probiotics and other food-grade bacteria: A review. Trends Food Sci Technol. https://doi.org/10.1016/j.tifs.2017.02.007
19. Fritzen-Freire CB, Prudêncio ES, Pinto SS, *et al.* (2013) Effect of microencapsulation on survival of Bifidobacterium BB-12 exposed to simulated gastrointestinal conditions and heat treatments. LWT - Food Sci Technol 50:39–44. https://doi.org/10.1016/j.lwt.2012.07.037
20. Barro NPR, da Silva LM, Hassemer G de S, *et al.* (2021) Microencapsulation of probiotic Lactobacillus helveticus with different wall materials by spray drying. Biointerface Res Appl Chem 11:11221–11232. https://doi.org/10.33263/BRIAC114.1122111232
21. Nunes GL, Etchepare M de A, Cichoski AJ, *et al.* (2018) Inulin, hi-maize, and trehalose as thermal protectants for increasing viability of Lactobacillus acidophilus encapsulated by spray drying. LWT - Food Sci Technol 89:128–133. https://doi.org/10.1016/j.lwt.2017.10.032
22. Fritzen-Freire CB, Prudêncio ES, Amboni RDMC, *et al.* (2012) Microencapsulation of bifidobacteria by spray drying in the presence of prebiotics. Food Res Int 45:306–312. https://doi.org/10.1016/j.foodres.2011.09.020

23. Minj S, Anand S (2022) Development of a spray-dried conjugated whey protein hydrolysate powder with entrapped probiotics. J Dairy Sci 105:2038–2048. https://doi.org/10.3168/jds.2021-20978
24. Gibson GR, Roberfroid MB (1995) Dietary modulation of the human colonic microbiota: Introducing the concept of prebiotics. J Nutr 125:1401–1412. https://doi.org/10.1093/jn/125.6.1401
25. Davani-Davari D, Negahdaripour M, Karimzadeh I, *et al.* (2019) Prebiotics: Definition, types, sources, mechanisms, and clinical applications. Foods 8:1–27. https://doi.org/10.3390/foods8030092
26. Gibson GR, Scott KP, Rastall RA, *et al.* (2010) Dietary prebiotics: current status and new definition. Food Sci Technol Bull Funct Foods 7:1–19. https://doi.org/10.1616/1476-2137.15880
27. Slavin J (2013) Fiber and prebiotics: Mechanisms and health benefits. Nutrients 5:1417–1435. https://doi.org/10.3390/nu5041417
28. Howlett JF, Betteridge VA, Champ M, *et al.* (2010) The definition of dietary fiber - Discussions at the Ninth Vahouny Fiber Symposium: Building scientific agreement. Food Nutr Res 54:1–5. https://doi.org/10.3402/fnr.v54i0.5750
29. Rastall B (2006) Prebiotics
30. Bindels LB, Delzenne NM, Cani PD, Walter J (2015) Opinion: Towards a more comprehensive concept for prebiotics. Nat Rev Gastroenterol Hepatol 12:303–310. https://doi.org/10.1038/nrgastro.2015.47
31. Scott KP, Gratz SW, Sheridan PO, *et al.* (2013) The influence of diet on the gut microbiota. Pharmacol Res 69:52–60. https://doi.org/10.1016/j.phrs.2012.10.020
32. Hutkins RW, Krumbeck JA, Bindels LB, *et al.* (2016) Prebiotics: Why definitions matter. Curr Opin Biotechnol 37:1–7. https://doi.org/10.1016/j.copbio.2015.09.001
33. Martins GN, Ureta MM, Tymczyszyn EE, *et al.* (2019) Technological Aspects of the Production of Fructo and Enzymatic Synthesis and Hydrolysis. 6:. https://doi.org/10.3389/fnut.2019.00078
34. Macfarlane GT, Steed H, Macfarlane S (2008) Bacterial metabolism and health-related effects of galacto-oligosaccharides and other prebiotics. J Appl Microbiol 104:305–344. https://doi.org/10.1111/j.1365-2672.2007.03520.x
35. O'Callaghan A, van Sinderen D (2016) Bifidobacteria and their role as members of the human gut microbiota. Front Microbiol 7:. https://doi.org/10.3389/fmicb.2016.00925
36. Zhang D, Li S, Wang N, *et al.* (2020) The Cross-Talk Between Gut Microbiota and Lungs in Common Lung Diseases. Front Microbiol 11:1–14. https://doi.org/10.3389/fmicb.2020.00301
37. Nichol AD, Holle MJ, An R (2018) Glycemic impact of non-nutritive sweeteners: A systematic review and meta-Analysis of randomized controlled trials. Eur J Clin Nutr 72:796–804. https://doi.org/10.1038/s41430-018-0170-6

38. Mudgil D, Barak S (2019) Dairy-based functional beverages. Milk-Based Beverages Vol 9 Sci Beverages 67–93. https://doi.org/10.1016/B978-0-12-815504-2.00003-7
39. Romano N, Santos M, Mobili P, *et al.* (2016) Effect of sucrose concentration on the composition of enzymatically synthesized short-chain fructo-oligosaccharides as determined by FTIR and multivariate analysis. Food Chem 202:467–475. https://doi.org/10.1016/j.foodchem.2016.02.002
40. Romano N, Schebor C, Mobili P, Gómez-Zavaglia A (2016) Role of mono- and oligosaccharides from FOS as stabilizing agents during freeze-drying and storage of Lactobacillus delbrueckii subsp. bulgaricus. Food Res Int 90:251–258. https://doi.org/10.1016/j.foodres.2016.11.003
41. Kelly G (2009) Inulin-type prebiotics: A review (Part 2). Altern Med Rev 14:36–55
42. Bali V, Panesar PS, Bera MB, Panesar R (2015) Fructo-oligosaccharides: Production, Purification and Potential Applications. Crit Rev Food Sci Nutr 55:1475–1490. https://doi.org/10.1080/10408398.2012.694084
43. Singh RS, Singh T, Larroche C (2019) Biotechnological applications of inulin-rich feedstocks. Bioresour Technol 273:641–653. https://doi.org/10.1016/j.biortech.2018.11.031
44. Kashyap R, Palai T, Bhattacharya PK (2015) Kinetics and model development for enzymatic synthesis of fructo-oligosaccharides using fructosyltransferase. Bioprocess Biosyst Eng 38:. https://doi.org/10.1007/s00449-015-1478-4
45. Vega-Paulino RJ, Zúniga-Hansen ME (2012) Potential application of commercial enzyme preparations for industrial production of short-chain fructooligosaccharides. J Mol Catal B Enzym 76:44–51. https://doi.org/10.1016/j.molcatb.2011.12.007
46. Plou FJ, Segura AG De, Ballesteros A (2007) Application of glycosidases and transglycosidases in the synthesis of oligosaccharides. Ind Enzym Struct Funct Appl 141–157. https://doi.org/10.1007/1-4020-5377-0_9
47. Meyer TSM, Miguel ASM, Fernández DER, Ortiz GMD (2015) Biotechnological Production of Oligosaccharides — Applications in the Food Industry. Food Prod Ind. https://doi.org/10.5772/60934
48. Fuller R (2001) Handbook of Probiotics
49. Fuentes-Zaragoza E, Sánchez-Zapata E, Sendra E, *et al.* (2011) Resistant starch as prebiotic: A review. Starch/Staerke 63:406–415. https://doi.org/10.1002/star.201000099
50. Walker AW, Ince J, Duncan SH, *et al.* (2011) Dominant and diet-responsive groups of bacteria within the human colonic microbiota. ISME J 5:220–230. https://doi.org/10.1038/ismej.2010.118
51. Ze X, Duncan SH, Louis P, Flint HJ (2012) Ruminococcus bromii is a keystone species for the degradation of resistant starch in the human colon. ISME J 1535–1543. https://doi.org/10.1038/ismej.2012.4
52. Probert HM, Apajalahti JHA, Rautonen N, *et al.* (2004) Polydextrose, Lactitol, and Fructo-Oligosaccharide Fermentation by Colonic Bacteria

in a Three-Stage Continuous Culture System. 70:4505–4511. https://doi.org/10.1128/AEM.70.8.4505
53. do Carmo MMR, Walker JCL, Novello D, *et al.* (2016) Polydextrose: Physiological function, and effects on health. Nutrients 8:1–13. https://doi.org/10.3390/nu8090553
54. Lahtinen SJ, Knoblock K, Drakoularakou A, *et al.* (2010) Effect of molecule branching and glycosidic linkage on the degradation of polydextrose by gut microbiota. Biosci Biotechnol Biochem 74:2016–2021. https://doi.org/10.1271/bbb.100251
55. Holscher HD, Gregory Caporaso J, Hooda S, *et al.* (2015) Fiber supplementation influences phylogenetic structure and functional capacity of the human intestinal microbiome: Follow-up of a randomized controlled trial. Am J Clin Nutr 101:55–64. https://doi.org/10.3945/ajcn.114.092064
56. Costabile A, Fava F, Röytiö H, *et al.* (2012) Impact of polydextrose on the faecal microbiota: A double-blind, crossover, placebo-controlled feeding study in healthy human subjects. Br J Nutr 108:471–481. https://doi.org/10.1017/S0007114511005782
57. Chen J, Liang RH, Liu W, *et al.* (2013) Pectic-oligosaccharides prepared by dynamic high-pressure microfluidization and their *in vitro* fermentation properties. Carbohydr Polym 91:175–182. https://doi.org/10.1016/j.carbpol.2012.08.021
58. Øbro J, Harholt J, Scheller HV, Orfila C (2004) Rhamnogalacturonan I in Solanum tuberosum tubers contains complex arabinogalactan structures. Phytochemistry 65:1429–1438. https://doi.org/10.1016/j.phytochem.2004.05.002
59. Lama-Muñoz A, Rodríguez-Gutiérrez G, Rubio-Senent F, Fernández-Bolaños J (2012) Production, characterization and isolation of neutral and pectic oligosaccharides with low molecular weights from olive by-products thermally treated. Food Hydrocoll 28:92–104. https://doi.org/10.1016/j.foodhyd.2011.11.008
60. Garthoff JA, Heemskerk S, Hempenius RA, *et al.* (2010) Safety evaluation of pectin-derived acidic oligosaccharides (pAOS): Genotoxicity and sub-chronic studies. Regul Toxicol Pharmacol 57:31–42. https://doi.org/10.1016/j.yrtph.2009.12.004
61. Mussatto SI, Mancilha IM (2007) Non-digestible oligosaccharides: A review. Carbohydr Polym 68:587–597. https://doi.org/10.1016/j.carbpol.2006.12.011
62. Baldan B, Bertoldo A, Navazio L, Mariani P (2003) Oligogalacturonide-induced changes in the developmental pattern of Daucus carota L. somatic embryos. Plant Sci 165:337–348. https://doi.org/10.1016/S0168-9452(03)00193-6
63. Manderson K, Pinart M, Tuohy KM, *et al.* (2005) *In vitro* determination of prebiotic properties of oligosaccharides derived from an orange juice manufacturing by-product stream. Appl Environ Microbiol 71:8383–8389. https://doi.org/10.1128/AEM.71.12.8383-8389.2005

64. Babbar N, Dejonghe W, Gatti M, *et al.* (2016) Pectic oligosaccharides from agricultural by-products: production, characterization and health benefits. Crit Rev Biotechnol 36:594–606. https://doi.org/10.3109/07388551.2014.996732
65. Konikoff MR, Denson LA (2006) Role of fecal calprotectin as a biomarker of intestinal inflammation in inflammatory bowel disease. Inflamm Bowel Dis 12:524–534. https://doi.org/10.1097/00054725-200606000-00013
66. Hill C, Guarner F, Reid G, *et al.* (2014) Expert consensus document: The international scientific association for probiotics and prebiotics consensus statement on the scope and appropriate use of the term probiotic. Nat Rev Gastroenterol Hepatol 11:506–514. https://doi.org/10.1038/nrgastro.2014.66
67. Kechagia M, Basoulis D, Konstantopoulou S, *et al.* (2013) Health Benefits of Probiotics: A Review. ISRN Nutr 2013:1–7. https://doi.org/10.5402/2013/481651
68. Bielecka M (2006) Probiotics in food. Chem Funct Prop Food Components, Third Ed 413–426. https://doi.org/10.1201/9781420009613.ch16
69. FAO/WHO (2002) Guidelines for the Evaluation of Probiotics in Food. 1–11
70. Plaza-Diaz J, Ruiz-Ojeda FJ, Gil-Campos M, Gil A (2019) Mechanisms of Action of Probiotics. Adv Nutr 10:S49–S66. https://doi.org/10.1093/advances/nmy063
71. Salminen S, Collado MC, Endo A, *et al.* (2021) The International Scientific Association of Probiotics and Prebiotics (ISAPP) consensus statement on the definition and scope of postbiotics. Nat Rev Gastroenterol Hepatol 18:649–667. https://doi.org/10.1038/s41575-021-00440-6
72. Plaza-Díaz J, Ruiz-Ojeda FJ, Gil-Campos M, Gil A (2018) Immune-mediated mechanisms of action of probiotics and synbiotics in treating pediatric intestinal diseases. Nutrients 10:1–20. https://doi.org/10.3390/nu10010042
73. Plaza-Díaz J, Robles-Sánchez C, Abadía-Molina F, *et al.* (2017) Gene expression profiling in the intestinal mucosa of obese rats administered probiotic bacteria. Sci Data 4:1–10. https://doi.org/10.1038/sdata.2017.186
74. Hempel S.J.; Maher, A.R.; Wang, Z.; Miles, J.N.V.; Shanman, R.; Johnsen, B.; Shekelle, P.G. H. N (2012) Probiotics for the Prevention and Treatment of Antibiotic-Associated Diarrhea. Jama 307:1959–1969
75. Didari T, Solki S, Mozaffari S, *et al.* (2014) A systematic review of the safety of probiotics. Expert Opin Drug Saf 13:227–239. https://doi.org/10.1517/14740338.2014.872627
76. Saez-Lara MJ, Gomez-Llorente C, Plaza-Diaz J, Gil A (2015) The role of probiotic lactic acid bacteria and bifidobacteria in the prevention and treatment of inflammatory bowel disease and other related diseases: A systematic review of randomized human clinical trials. Biomed Res Int 2015:. https://doi.org/10.1155/2015/505878
77. Moayyedi P, Ford AC, Talley NJ, *et al.* (2010) The efficacy of probiotics in the treatment of irritable bowel syndrome: A systematic review. Gut 59:325–332. https://doi.org/10.1136/gut.2008.167270

78. Rather IA, Bajpai VK, Kumar S, *et al.* (2016) Probiotics and atopic dermatitis: An overview. Front Microbiol 7:1–7. https://doi.org/10.3389/fmicb.2016.00507
79. Berings M, Karaaslan C, Altunbulakli C, *et al.* (2017) Advances and highlights in allergen immunotherapy: On the way to sustained clinical and immunologic tolerance. J Allergy Clin Immunol 140:1250–1267. https://doi.org/10.1016/j.jaci.2017.08.025
80. Johnson D, Thurairajasingam S, Letchumanan V, *et al.* (2021) Exploring the role and potential of probiotics in the field of mental health: Major depressive disorder. Nutrients 13:1–18. https://doi.org/10.3390/nu13051728
81. Markowiak P, Ślizewska K (2017) Effects of probiotics, prebiotics, and synbiotics on human health. Nutrients 9:. https://doi.org/10.3390/nu9091021
82. McFarland L V. (2015) From yaks to yogurt: The history, development, and current use of probiotics. Clin Infect Dis 60:S85–S90. https://doi.org/10.1093/cid/civ054
83. Mackowiak PA (2013) Recycling Metchnikoff: Probiotics, the intestinal microbiome and the quest for long life. Front Public Heal 1:1–3. https://doi.org/10.3389/fpubh.2013.00052
84. Weill R ancient food in the 21 century
85. Veiga P, Miret S, Jimenez L (2020) Danone : The gut microbiome and probiotics – 100 years of shared history. Danon Nutr Res 577:S26–S30
86. Selle KM, Klaenhammer TR, Russell WM (2014) Lactobacillus: Lactobacillus acidophilus, Second Edi. Elsevier
87. Tannock GW (2004) A special fondness for lactobacilli. Appl Environ Microbiol 70:3189–3194. https://doi.org/10.1128/AEM.70.6.3189-3194.2004
88. Milner E, Stevens B, An M, *et al.* (2021) Utilizing Probiotics for the Prevention and Treatment of Gastrointestinal Diseases. Front Microbiol 12:1–13. https://doi.org/10.3389/fmicb.2021.689958
89. Gasbarrini G, Bonvicini F, Gramenzi A (2016) Probiotics History. J Clin Gastroenterol 50:S116–S119. https://doi.org/10.1097/MCG.0000000000000697
90. Sicard D, Legras JL (2011) Bread, beer and wine: Yeast domestication in the Saccharomyces sensu stricto complex. Comptes Rendus - Biol 334:229–236. https://doi.org/10.1016/j.crvi.2010.12.016
91. Science D, Agric F (2008) Origin, History and Manufacturing Process of Egyptian Dairy Products: An Overview. Alexandria J Food Sci Technol 5:51–62. https://doi.org/10.21608/ajfs.2008.20141
92. Luisa Alba-Lois & CS-K (2017) Yeast Fermentation and the Making of Beer and Yeast Fermentation and the Making of Beer and Wine. Nat Educ 3:17
93. Shurtleff W, Aoyagi A (2012) History of Soy Yogurt, Soy Acidophilus Milk And Other Cultured Soymilks (1918-2012): Extensively Annotated Includes: Fermented and Nonfermented Soy Yogurts , Fermented Spoonable and by
94. Hansen RG (1974) Milk in Human Nutrition
95. Nazir Y, Hussain SA, Abdul Hamid A, Song Y (2018) Probiotics and Their Potential Preventive and Therapeutic Role for Cancer, High Serum

Cholesterol, and Allergic and HIV Diseases. Biomed Res Int 2018:. https://doi.org/10.1155/2018/3428437

96. Hendrickson BA, Gokhale R, Cho JH (2002) Clinical aspects and pathophysiology of inflammatory bowel disease. Clin Microbiol Rev 15:79–94. https://doi.org/10.1128/CMR.15.1.79-94.2002
97. Geier MS, Butler RN, Howarth GS (2007) Inflammatory bowel disease: Current insights into pathogenesis and new therapeutic options; probiotics, prebiotics and synbiotics. Int J Food Microbiol 115:1–11. https://doi.org/10.1016/j.ijfoodmicro.2006.10.006
98. Team R, Guarner F, Khan AG, *et al.* (2012) World Gastroenterology Organisation Global Guidelines Probiotics and Prebiotics October 2011. 46:119–129
99. Falagas ME, Betsi GI, Tokas T, Athanasiou S (2006) Probiotics for Prevention of Recurrent Urinary Tract Infections in Women. Drugs 66:1253–1261. https://doi.org/10.2165/00003495-200666090-00007
100. Wang H, Anvari S, Anagnostou K (2019) The Role of Probiotics in Preventing Allergic Disease. Children 6:24. https://doi.org/10.3390/children6020024
101. Simons LA, Amansec SG, Conway P (2006) Effect of Lactobacillus fermentum on serum lipids in subjects with elevated serum cholesterol. Nutr Metab Cardiovasc Dis 16:531–535. https://doi.org/10.1016/j.numecd.2005.10.009
102. Lieske JC, Goldfarb DS, De Simone C, Regnier C (2005) Use of a probioitic to decrease enteric hyperoxaluria. Kidney Int 68:1244–1249. https://doi.org/10.1111/j.1523-1755.2005.00520.x
103. Rousseaux C, Thuru X, Gelot A, *et al.* (2007) Lactobacillus acidophilus modulates intestinal pain and induces opioid and cannabinoid receptors. Nat Med 13:35–37. https://doi.org/10.1038/nm1521
104. Parvez S, Malik KA, Ah Kang S, Kim HY (2006) Probiotics and their fermented food products are beneficial for health. J Appl Microbiol 100:1171–1185. https://doi.org/10.1111/j.1365-2672.2006.02963.x
105. McFarland L V. (2006) Meta-analysis of probiotics for the prevention of antibiotic associated diarrhea and the treatment of Clostridium difficile disease. Am J Gastroenterol 101:812–822. https://doi.org/10.1111/j.1572-0241.2006.00465.x
106. Hatakka K, Savilahti E, Pönkä A, *et al.* (2001) Effect of long term consumption of probiotic milk on infections in children attending day care centres: Double blind, randomised trial. Br Med J 322:1327–1329. https://doi.org/10.1136/bmj.322.7298.1327
107. Alvaro E, Andrieux C, Rochet V, *et al.* (2007) Composition and metabolism of the intestinal microbiota in consumers and non-consumers of yogurt. 126–133. https://doi.org/10.1017/S0007114507243065
108. Gourbeyre P, Denery S, Bodinier M (2011) Probiotics, prebiotics, and synbiotics: impact on the gut immune system and allergic reactions. J Leukoc Biol 89:685–695. https://doi.org/10.1189/jlb.1109753

109. Guiné R de PF, Silva ACF (2016) Probiotics, prebiotics and synbiotics. Funct Foods Sources, Heal Eff Futur Perspect 143–207. https://doi.org/10.1201/b15561-2
110. BM S, Fc H, Lhs M, *et al.* (2014) Impact of Prebiotics, Probiotics and Synbiotics on Components of the Metabolic Syndrome. Ann Nutr Disord Ther 1:1–13
111. Hamasalim HJ (2016) Synbiotic as Feed Additives Relating to Animal Health and Performance. Adv Microbiol 06:288–302. https://doi.org/10.4236/aim.2016.64028
112. Terpou A, Papadaki A, Lappa IK, *et al.* (2019) Probiotics in food systems: significance and emerging strategies towards improved viability and delivery of enhanced beneficial value. Nutrients 11:. https://doi.org/10.3390/nu11071591
113. Nagpal R, Kumar A, Kumar M, *et al.* (2012) Probiotics, their health benefits and applications for developing healthier foods: A review. FEMS Microbiol Lett 334:1–15. https://doi.org/10.1111/j.1574-6968.2012.02593.x
114. Wang X, Yang J, Qiu X, *et al.* (2021) Probiotics, Pre-biotics and Synbiotics in the Treatment of Pre-diabetes: A Systematic Review of Randomized Controlled Trials. Front Public Heal 9:. https://doi.org/10.3389/fpubh.2021.645035
115. Cunningham M, Azcarate-Peril MA, Barnard A, *et al.* (2021) Shaping the Future of Probiotics and Prebiotics. Trends Microbiol 29:667–685. https://doi.org/10.1016/j.tim.2021.01.003
116. Zhang MM, Cheng JQ, Lu YR, *et al.* (2010) Use of pre-, pro-and synbiotics in patients with acute pancreatitis: A meta-analysis. World J Gastroenterol 16:3970–3978. https://doi.org/10.3748/wjg.v16.i31.3970
117. Rodrigues FJ, Cedran MF, Bicas JL, Sato HH (2020) Encapsulated probiotic cells: Relevant techniques, natural sources as encapsulating materials and food applications – A narrative review. Food Res Int 137:. https://doi.org/10.1016/j.foodres.2020.109682
118. Kim JU, Kim B, Shahbaz HM, *et al.* (2017) Encapsulation of probiotic Lactobacillus acidophilus by ionic gelation with electrostatic extrusion for enhancement of survival under simulated gastric conditions and during refrigerated storage. Int J Food Sci Technol 52:519–530. https://doi.org/10.1111/ijfs.13308
119. Wu Y, Zhang G (2018) Synbiotic encapsulation of probiotic Latobacillus plantarum by alginate -arabinoxylan composite microspheres. Lwt 93:135–141. https://doi.org/10.1016/j.lwt.2018.03.034
120. Rathore S, Desai PM, Liew CV, *et al.* (2013) Microencapsulation of microbial cells. J Food Eng 116:369–381. https://doi.org/10.1016/j.jfoodeng.2012.12.022
121. Frakolaki G, Giannou V, Kekos D, Tzia C (2021) A review of the microencapsulation techniques for the incorporation of probiotic bacteria in functional foods. Crit Rev Food Sci Nutr 61:1515–1536. https://doi.org/10.1080/10408398.2020.1761773

122. Alemzadeh I, Hajiabbas M, Pakzad H, *et al.* (2020) Encapsulation of food components and bioactive ingredients and targeted release. Int J Eng Trans A Basics 33:1–11. https://doi.org/10.5829/ije.2020.33.01a.01
123. Zhang Y, Lin J, Zhong Q (2016) S/O/W emulsions prepared with sugar beet pectin to enhance the viability of probiotic Lactobacillus salivarius NRRL B-30514. Food Hydrocoll 52:804–810. https://doi.org/10.1016/j.foodhyd.2015.08.020
124. Wang L, Song M, Zhao Z, *et al.* (2020) Lactobacillus acidophilus loaded pickering double emulsion with enhanced viability and colon-adhesion efficiency. Lwt 121:108928. https://doi.org/10.1016/j.lwt.2019.108928
125. Pandey S, Senthilguru K, Uvanesh K, *et al.* (2016) Natural gum modified emulsion gel as single carrier for the oral delivery of probiotic-drug combination. Int J Biol Macromol 92:504–514. https://doi.org/10.1016/j.ijbiomac.2016.07.053
126. Huerta-Vera K, Flores-Andrade E, Pérez-Sato JA, *et al.* (2017) Enrichment of Banana with Lactobacillus rhamnosus Using Double Emulsion and Osmotic Dehydration. Food Bioprocess Technol 10:1053–1062. https://doi.org/10.1007/s11947-017-1879-2
127. Hadidi M, Majidiyan N, Jelyani AZ, *et al.* (2021) Alginate/fish gelatin-encapsulated lactobacillus acidophilus: A study on viability and technological quality of bread during baking and storage. Foods 10:. https://doi.org/10.3390/foods10092215
128. Sugiura S, Nakajima M, Yamamoto K, *et al.* (2004) Preparation characteristics of water-in-oil-in-water multiple emulsions using microchannel emulsification. J Colloid Interface Sci 270:221–228. https://doi.org/10.1016/j.jcis.2003.08.021
129. Zhang R, Zhou L, Li J, *et al.* (2020) Microencapsulation of anthocyanins extracted from grape skin by emulsification/internal gelation followed by spray/freeze-drying techniques: Characterization, stability and bioaccessibility. Lwt 123:109097. https://doi.org/10.1016/j.lwt.2020.109097
130. Holkem AT, Raddatz GC, Nunes GL, *et al.* (2016) Development and characterization of alginate microcapsules containing Bifidobacterium BB-12 produced by emulsification/internal gelation followed by freeze drying. LWT - Food Sci Technol 71:302–308. https://doi.org/10.1016/j.lwt.2016.04.012
131. Krasaekoopt W, Bhandari B, Deeth H (2003) Evaluation of encapsulation techniques of probiotics for yoghurt. Int Dairy J 13:3–13. https://doi.org/10.1016/S0958-6946(02)00155-3
132. Reis CP, Neufeld RJ, Vilela S, *et al.* (2006) Review and current status of emulsion/dispersion technology using an internal gelation process for the design of alginate particles. J Microencapsul 23:245–257. https://doi.org/10.1080/02652040500286086
133. Rodrigues FJ, Omura MH, Cedran MF, *et al.* (2017) Effect of natural polymers on the survival of Lactobacillus casei encapsulated in alginate microspheres.

J Microencapsul 34:431–439. https://doi.org/10.1080/02652048.2017.1343872
134. Brun-Graeppi AKAS, Richard C, Bessodes M, *et al.* (2011) Cell microcarriers and microcapsules of stimuli-responsive polymers. J Control Release 149:209–224. https://doi.org/10.1016/j.jconrel.2010.09.023
135. Liu XD, Yu WY, Zhang Y, *et al.* (2002) Characterization of structure and diffusion behaviour of Ca-alginate beads prepared with external or internal calcium sources. J Microencapsul 19:775–782. https://doi.org/10.1080/0265204021000022743
136. Burgain J, Gaiani C, Linder M, Scher J (2011) Encapsulation of probiotic living cells: From laboratory scale to industrial applications. J Food Eng 104:467–483. https://doi.org/10.1016/j.jfoodeng.2010.12.031
137. Piazza L, Roversi T (2011) Preliminary study on microbeads production by co-extrusion technology. Procedia Food Sci 1:1374–1380. https://doi.org/10.1016/j.profoo.2011.09.204
138. Kim IY, Pusey PL, Zhao Y, *et al.* (2012) Controlled release of Pantoea agglomerans E325 for biocontrol of fire blight disease of apple. J Control Release 161:109–115. https://doi.org/10.1016/j.jconrel.2012.03.028
139. Zhang Z, Zhang R, Zou L, McClements DJ (2016) Protein encapsulation in alginate hydrogel beads: Effect of pH on microgel stability, protein retention and protein release. Food Hydrocoll 58:308–315. https://doi.org/10.1016/j.foodhyd.2016.03.015
140. Herrero EP, Martín Del Valle EM, Galán MA (2006) Development of a new technology for the production of microcapsules based in atomization processes. Chem Eng J 117:137–142. https://doi.org/10.1016/j.cej.2005.12.022
141. Whelehan M, Marison IW (2011) Microencapsulation using vibrating technology. J Microencapsul 28:669–688. https://doi.org/10.3109/02652048.2011.586068
142. Del Gaudio P, Colombo P, Colombo G, *et al.* (2005) Mechanisms of formation and disintegration of alginate beads obtained by prilling. Int J Pharm 302:1–9. https://doi.org/10.1016/j.ijpharm.2005.05.041
143. Graff S, Hussain S, Chaumeil JC, Charrueau C (2008) Increased intestinal delivery of viable Saccharomyces boulardii by encapsulation in microspheres. Pharm Res 25:1290–1296. https://doi.org/10.1007/s11095-007-9528-5
144. Loftus R (2016) Calcium Alginate Microbead Production via an Air Assisted Shearing Process
145. Huang S Bin, Wu MH, Lee G Bin (2010) Microfluidic device utilizing pneumatic micro-vibrators to generate alginate microbeads for microencapsulation of cells. Sensors Actuators, B Chem 147:755–764. https://doi.org/10.1016/j.snb.2010.04.021
146. Ray S, Raychaudhuri U, Chakraborty R (2016) An overview of encapsulation of active compounds used in food products by drying technology. Food Biosci 13:76–83. https://doi.org/10.1016/j.fbio.2015.12.009

147. Hadzieva J, Mladenovska K, Crcarevska MS, *et al.* (2017) Lactobacillus casei encapsulated in soy protein isolate and alginate microparticles prepared by spray drying. Food Technol Biotechnol 55:173–186. https://doi.org/10.17113/ftb.55.02.17.4991
148. Vega C, Roos YH (2006) Invited review: Spray-dried dairy and dairy-like emulsions - Compositional considerations. J Dairy Sci 89:383–401. https://doi.org/10.3168/jds.S0022-0302(06)72103-8
149. Martín MJ, Lara-Villoslada F, Ruiz MA, Morales ME (2015) Microencapsulation of bacteria: A review of different technologies and their impact on the probiotic effects. Innov Food Sci Emerg Technol 27:15–25. https://doi.org/10.1016/j.ifset.2014.09.010
150. Corcoran BM, Ross RP, Fitzgerald GF, *et al.* (2006) Enhanced survival of GroESL-overproducing Lactobacillus paracasei NFBC 338 under stressful conditions induced by drying. Appl Environ Microbiol 72:5104–5107. https://doi.org/10.1128/AEM.02626-05
151. Łopusiewicz Ł, Bogusławska-Wąs E, Drozłowska E, *et al.* (2021) The application of spray-dried and reconstituted flaxseed oil cake extract as encapsulating material and carrier for probiotic lacticaseibacillus rhamnosus gg. Materials (Basel) 14:. https://doi.org/10.3390/ma14185324
152. Rajam R, Anandharamakrishnan C (2015) Microencapsulation of Lactobacillus plantarum (MTCC 5422) with fructooligosaccharide as wall material by spray drying. Lwt 60:773–780. https://doi.org/10.1016/j.lwt.2014.09.062
153. Rodríguez-Huezo ME, Durán-Lugo R, Prado-Barragán LA, *et al.* (2007) Pre-selection of protective colloids for enhanced viability of Bifidobacterium bifidum following spray-drying and storage, and evaluation of aguamiel as thermoprotective prebiotic. Food Res Int 40:1299–1306. https://doi.org/10.1016/j.foodres.2007.09.001
154. Favaro-Trindade CS, de Matos Junior FE, Okuro PK, *et al.* (2021) Encapsulation of active pharmaceutical ingredients in lipid micro/nanoparticles for oral administration by spray-cooling. Pharmaceutics 13:1–14. https://doi.org/10.3390/pharmaceutics13081186
155. Favaro-Trindade C, Okuro P, de Matos F (2015) Encapsulation via Spray Chilling/Cooling/Congealing. Handb Encapsulation Control Release 71–87. https://doi.org/10.1201/b19038-8
156. Pedroso DL, Dogenski M, Thomazini M, *et al.* (2013) Microencapsulation of Bifidobacterium animalis subsp. lactis and Lactobacillus acidophilus in cocoa butter using spray chilling technology. Brazilian J Microbiol 44:777–783. https://doi.org/10.1590/S1517-83822013000300017
157. Bertoni S, Albertini B, Passerini N (2019) Spray congealing: An emerging technology to prepare solid dispersions with enhanced oral bioavailability of poorly water soluble drugs. Molecules 24:1–21. https://doi.org/10.3390/molecules24193471

158. Queirós M de S, Viriato RLS, Ribeiro APB, Gigante ML (2020) Dairy-based solid lipid microparticles: A novel approach. Food Res Int 131:109009. https://doi.org/10.1016/j.foodres.2020.109009
159. Bampi GB, Backes GT, Cansian RL, *et al.* (2016) Spray Chilling Microencapsulation of Lactobacillus acidophilus and Bifidobacterium animalis subsp. lactis and Its Use in the Preparation of Savory Probiotic Cereal Bars. Food Bioprocess Technol 9:1422–1428. https://doi.org/10.1007/s11947-016-1724-z
160. Arslan-Tontul S, Erbas M, Gorgulu A (2019) The Use of Probiotic-Loaded Single- and Double-Layered Microcapsules in Cake Production. Probiotics Antimicrob Proteins 11:840–849. https://doi.org/10.1007/s12602-018-9467-y
161. Liliana SC, Vladimir VC (2013) Probiotic encapsulation. African J Microbiol Res 7:4743–4753. https://doi.org/10.5897/ajmr2013.5718
162. Librán CM, Castro S, Lagaron JM (2017) Encapsulation by electrospray coating atomization of probiotic strains. Innov Food Sci Emerg Technol 39:216–222. https://doi.org/10.1016/j.ifset.2016.12.013
163. Yao M, Xie J, Du H, *et al.* (2020) Progress in microencapsulation of probiotics: A review. Compr Rev Food Sci Food Saf 19:857–874. https://doi.org/10.1111/1541-4337.12532
164. ter Horst B, Moiemen NS, Grover LM (2019) Natural polymers. Biomater Ski Repair Regen 151–192. https://doi.org/10.1016/B978-0-08-102546-8.00006-6
165. Hazrati Z, Madadlou A (2021) Gelation by bioactives: Characteristics of the cold-set whey protein gels made using gallic acid. Int Dairy J 117:104952. https://doi.org/10.1016/j.idairyj.2020.104952
166. Tikekar R V., Johnson A, Nitin N (2011) Real-time measurement of oxygen transport across an oil-water emulsion interface. J Food Eng 103:14–20. https://doi.org/10.1016/j.jfoodeng.2010.08.030
167. Bakry AM, Abbas S, Ali B, *et al.* (2016) Microencapsulation of Oils: A Comprehensive Review of Benefits, Techniques, and Applications. Compr Rev Food Sci Food Saf 15:143–182. https://doi.org/10.1111/1541-4337.12179
168. Chen J, Wang Q, Liu CM, Gong J (2017) Issues deserve attention in encapsulating probiotics: Critical review of existing literature. Crit Rev Food Sci Nutr 57:1228–1238. https://doi.org/10.1080/10408398.2014.977991
169. Ruiz L, Ruas-Madiedo P, Gueimonde M, *et al.* (2011) How do bifidobacteria counteract environmental challenges? Mechanisms involved and physiological consequences. Genes Nutr 6:307–318. https://doi.org/10.1007/s12263-010-0207-5
170. Lobo V, Patil A, Phatak A, Chandra N (2010) Free radicals, antioxidants and functional foods: Impact on human health. Pharmacogn Rev 4:118–126. https://doi.org/10.4103/0973-7847.70902
171. Chavarri M, Maranon I, Carmen M (2012) Encapsulation Technology to Protect Probiotic Bacteria. Probiotics. pp. 501–540, https://doi.org/10.5772/50046. 501–540

172. Singh MN, Hemant KSY, Ram M, Shivakumar HG (2010) Microencapsulation: A promising technique for controlled drug delivery. Res Pharm Sci 5:65–77
173. Costa J, Ahluwalia A (2019) Advances and Current Challenges in Intestinal *in vitro* Model Engineering: A Digest. Front Bioeng Biotechnol 7:1–14. https://doi.org/10.3389/fbioe.2019.00144
174. Wojtunik-Kulesza K, Oniszczuk A, Oniszczuk T, *et al.* (2020) Influence of *in vitro* digestion on composition, bioaccessibility and antioxidant activity of food polyphenols—a non-systematic review. Nutrients 12:. https://doi.org/10.3390/nu12051401
175. Linden SK, Sutton P, Karlsson NG, *et al.* (2008) Mucins in the mucosal barrier to infection. Mucosal Immunol 1:183–197. https://doi.org/10.1038/mi.2008.5
176. Calinoiu LF, Ştefanescu BE, Pop ID, *et al.* (2019) Chitosan coating applications in probiotic microencapsulation. Coatings 9:1–21. https://doi.org/10.3390/COATINGS9030194
177. Anekella K (2012) Microencapsulation of probiotics (*Lactobacillus acidophilus* and *Lactobacillus rhamnosus*) in raspberry powder by spray drying: optimization and storage stability studies. ProQuest Diss Theses 129
178. Rad AH, Aghebati-Maleki L, Kafil HS, *et al.* (2021) Postbiotics, as dynamic biomolecules, and their promising role in promoting food safety. Biointerface Res Appl Chem 11:14529–14544. https://doi.org/10.33263/BRIAC116.1452914544
179. Solanki HK, Pawar DD, Shah DA, *et al.* (2013) Development of microencapsulation delivery system for long-term preservation of probiotics as biotherapeutics agent. Biomed Res Int 2013:. https://doi.org/10.1155/2013/620719
180. Silva DR, Sardi J de CO, Pitangui N de S, *et al.* (2020) Probiotics as an alternative antimicrobial therapy: Current reality and future directions. J Funct Foods 73:104080. https://doi.org/10.1016/j.jff.2020.104080
181. Gómez-Gallego C, Gueimonde M, Salminen S (2018) The role of yogurt in food-based dietary guidelines. Nutr Rev 76:29–39. https://doi.org/10.1093/nutrit/nuy059
182. Bintsis T (2018) Lactic acid bacteria as starter cultures: An update in their metabolism and genetics. AIMS Microbiol 4:665–684. https://doi.org/10.3934/microbiol.2018.4.665
183. Sharma R, Garg P, Kumar P, *et al.* (2020) Microbial fermentation and its role in quality improvement of fermented foods. Fermentation 6:1–20. https://doi.org/10.3390/fermentation6040106
184. Dimidi E, Cox S, Rossi M, Whelan K (2019) Fermented Foods : Definitions and Characteristics , Gastrointestinal Health and Disease. Nutrients 11:1–26
185. Kim S Il, Kim JW, Kim KT, Kang CH (2021) Survivability of collagen-peptide microencapsulated lactic acid bacteria during storage and simulated gastrointestinal conditions. Fermentation 7:. https://doi.org/10.3390/fermentation7030177

186. Marco ML, Sanders ME, Gänzle M, *et al.* (2021) The International Scientific Association for Probiotics and Prebiotics (ISAPP) consensus statement on fermented foods. Nat Rev Gastroenterol Hepatol 18:196–208. https://doi.org/10.1038/s41575-020-00390-5
187. Widyastuti Y, Febrisiantosa A, Tidona F (2021) Health-Promoting Properties of Lactobacilli in Fermented Dairy Products. Front Microbiol 12:1–8. https://doi.org/10.3389/fmicb.2021.673890
188. Hoffman FA, Heimbach JT, Sanders ME, Hibberd PL (2008) Executive summary: Scientific and regulatory challenges of development of probiotics as foods and drugs. Clin Infect Dis 46:53–57. https://doi.org/10.1086/523342
189. Wilkinson MG (2018) Flow cytometry as a potential method of measuring bacterial viability in probiotic products: A review. Trends Food Sci Technol 78:1–10. https://doi.org/10.1016/j.tifs.2018.05.006
190. Pelletier X, Laure-Boussuge S, Donazzolo Y (2001) Hydrogen excretion upon ingestion of dairy products in lactose-intolerant male subjects: Importance of the live flora. Eur J Clin Nutr 55:509–512. https://doi.org/10.1038/sj.ejcn.1601169
191. Champagne CP, Gomes da Cruz A, Daga M (2018) Strategies to improve the functionality of probiotics in supplements and foods. Curr Opin Food Sci 22:160–166. https://doi.org/10.1016/j.cofs.2018.04.008
192. Narvhus JA, Axelsson L, Gene N (1993) Lactic acid bacterium starter. Process Biochem 28:126. https://doi.org/10.1016/0032-9592(93)80020-h
193. Kumar H, Salminen S, Verhagen H, *et al.* (2015) Novel probiotics and prebiotics: Road to the market. Curr Opin Biotechnol 32:99–103. https://doi.org/10.1016/j.copbio.2014.11.021
194. Ku S, Park MS, Ji GE, You HJ (2016) Review on bifidobacterium bifidum bgn4: Functionality and nutraceutical applications as a probiotic microorganism. Int J Mol Sci 17:. https://doi.org/10.3390/ijms17091544
195. Hansen LT, Allan-Wojtas PM, Jin YL, Paulson AT (2002) Survival of Ca-alginate microencapsulated Bifidobacterium spp. in milk and simulated gastrointestinal conditions. Food Microbiol 19:35–45. https://doi.org/10.1006/fmic.2001.0452
196. Brinques GB, Ayub MAZ (2011) Effect of microencapsulation on survival of Lactobacillus plantarum in simulated gastrointestinal conditions, refrigeration, and yogurt. J Food Eng 103:123–128. https://doi.org/10.1016/j.jfoodeng.2010.10.006
197. Sabikhi L, Babu R, Thompkinson DK, Kapila S (2010) Resistance of microencapsulated Lactobacillus acidophilus LA1 to processing treatments and simulated gut conditions. Food Bioprocess Technol 3:586–593. https://doi.org/10.1007/s11947-008-0135-1
198. Vijayakumar M, Ilavenil S, Kim DH, *et al.* (2015) *In-vitro* assessment of the probiotic potential of Lactobacillus plantarum KCC-24 isolated from Italian rye-grass (Lolium multiflorum) forage. Anaerobe 32:90–97. https://doi.org/10.1016/j.anaerobe.2015.01.003

3

Mechanistic Insights of Dietary Modulation on Gut Microflora and Associated Physiological Changes

Aarti Yadav[1*,†], S. Hamsa[2†], Ruby Tiwari[2†], Asha Bharti[3†], Tejpal Dhewa[4], Rekha Mehrotra[1] and Preeti Verma[1]

[1]*Department of Microbiology, Shaheed Rajguru College of Applied University of Delhi, New Delhi, India*
[2]*Department of Genetics, University of Delhi, South Campus, New Delhi, India*
[3]*Department of Microbiology, IARI, Delhi, India*
[4]*Department of Nutrition Biology, School of Interdisciplinary and Applied Sciences, Central University of Haryana, India*

Abstract

The gut microbiome is a collection of bacteria that live throughout the gastrointestinal system. This microbial community has a host-specific composition that changes over time and is sensitive to both exogenous and endogenous alterations. From dietary patterns to behaviour and stress response, the microbiota is intricately implicated in many facets of normal host physiology. Furthermore, they might be a primary or secondary cause of a variety of disorders, affecting both nearby and distant organ systems. The total equilibrium of the gut microbiome, as well as the presence or exclusion of essential species capable of eliciting specific reactions, are crucial in maintaining homeostasis at the intestinal mucosa and distinct pathways. The processes by which the microbiota exerts its good or harmful effects are not yet completely understood, although they include the development of signalling molecules and the identification of bacterial epitopes by both intestinal epithelial and mucosal immune cells. Advances in gut microbiota will expand our understanding of their function in health and disease, enabling more personalised treatment.

***Keywords*:** IBS, gut microbiota, short-chain fatty acids, inflammatory bowel disease, microbiome, polysaccharide

**Corresponding author*: aartiyadav003@gmail.com
†All the authors have worked equally.

Tejpal Dhewa, Anil Kumar Puniya and Anil Panghal (eds.) Nutritional Science and Technology: Concept to Application, (63–82) © 2023 Scrivener Publishing LLC

3.1 Introduction to Flora in the GIT Tract of Humans-Genus Metabolites

From the days of their discovery, microorganisms have been known to play a role in modulating various cellular and extracellular processes. The gastrointestinal tract (GIT) is known to begin from our buccal cavity and ends at the rectum which is part of our large intestine. The whole passage of the GIT is colonized by a variety of microbes (Figure 3.1).

To begin with, the buccal cavity of an adult has the *Streptococcus* species, *Prevotella* sp., *Veillonella* sp. majorly; however, a few species of actinomycetes and Lactobacilli are also found [1]; although the stomach has an acidic pH, it still has some viable microbes including bacteria like *Streptococcus, Enterococcus, Diptheroids, Helicobacter, Staphylococcus, Peptostreptococcus, Lactobacillus,* and yeasts like *Candida*; and the last part of the passage constituting the large intestine is colonized by *Escherichia coli, Clostridium perfingens, Bifidobacterium, Bacteroides, Streptococcus fecalis, Candida albicans, Entamoeba hartmanni, Endoliniax nana, Trichomonas hominis.* The composition of the gut is highly dynamic and it is known to vary with age, gender, usage of medicinal drugs, changes in the physiology of the gut, intoxication habits, genetics, environment and the diet consumed by the host [2] (Figure 3.2).

The gut possesses a trillion microbes, thus creating a complex microbial population consisting of approximately 1,000–1,100 different bacterial species containing 10^{14}–10^{15} microbes altogether. The newborn infant has very few colonizers, acquiring them during the passage through the birth canal whereas during adolescence and middle age it remains more or less consistent, forming

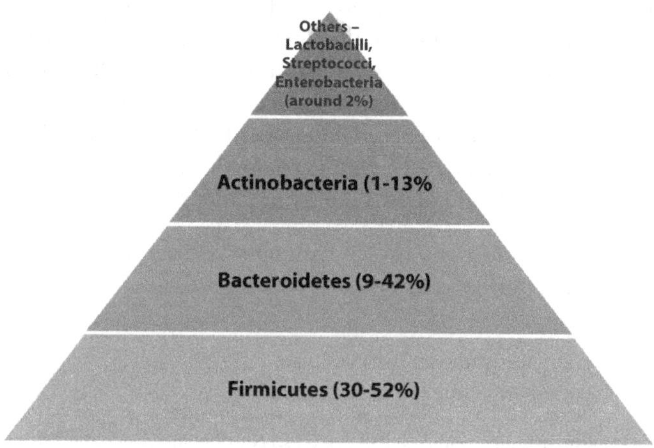

Figure 3.1 Composition of microflora in healthy gut.

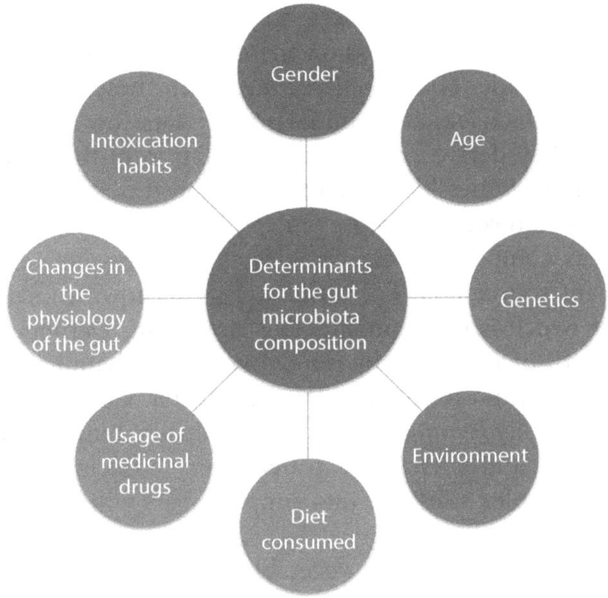

Figure 3.2 Determinants for the gut microbiota.

a consortium predominated by a few microbes [3]. Subsequently the change in the gut flora happens during old age when dietary habits change and body metabolism is not able to maintain its dynamics. Thus there is always a succession happening in the gut flora by the normal resident flora and the transient microflora. Consequently, for a fit body, a perfect and responsive balanced microbe interaction with the host is required.

Diet has the most significant role to play in defining the composition of gut microbiota over other factors. It is the key point which connects directly the macroelements present in the environment to the internal system and thus regulating the homeostasis of the system. In brief, anaerobic microbes in the intestine typically derive their energy from the fermentation of indigestible oligosaccharides and other carbohydrates that are consumed in diet by the host [4–6]. The consumption of a western diet which is generally rich in macromolecules like sugar (sucrose, xylose, glucose), protein, fat and low in dietary fibre has been associated with enrichment of microbes that can ferment these sugars to lactate, glucose, valerate and caproate [7]. However, people consuming the diet rich in polysaccharides harbour the microbes like *Prevotella* sp., *Bacteroides* that can degrade the complex carbohydrates like cellulose, xyloglucan, starch, β-(1,4)-mannan using the carbohydrate degrading enzymes such as glycoside hydrolases, thus enabling them to digest the complex plant fibres and the associated molecules [8, 9]. A study done by Arumugam *et al.* (2011) [10] divided the gut flora into three clusters based on the diet being consumed. Cluster 1 was rich in Bacteroides (Bacteroides: *Prevotella*); these people consumed the western diet which is rich in animal protein, fats, low fibre and nutrient chlorine [11]. They derive energy by fermentation of these saccharolytic rich diets using enzymes like galactosidases, proteases, hexosaminidases or by glycolysis followed by pentose phosphate pathway. Cluster 2 was rich in *Prevotella* as compared to Bacteroides as these people consumed the high-fibre plant source diet. *Prevotella* is a well-known mucin glycoprotein degrader which is found in cluster 2 along with *Desulfovibrio* which helps the growth of *Prevotella* by removing the sulphate being produced during the degradation in the gut. The last cluster, Cluster 3, was rich in Firmicutes, especially *Ruminococcus* species. Cluster 3 microbes are potent mucin degraders and membrane transporters especially sugars. They degrade these complex substrate into simple sugars which then readily be utilized by these microbes.

A diet rich in fruits, legumes, vegetables, minimally processed cereals and nuts, comparatively high intake of fish, low consumption of saturated fat, dairy products and meat with a routine, but moderate, alcohol consumption constitutes the mediterranean diet. This diet has been found to have a positive effect on disease like diabetes, obesity, cardiovascular and inflammation by increasing the short chain fatty acids (SCFAs) by gut flora like *Prevotella* and fibre-degrading Firmicutes [7, 12].

People with a vegan-rich diet have more bacteria falling in the class Bacteroides, Lactobacilli, and Firmicutes over the people who prefer non-vegan diet. Diets rich in fats, either saturated or polyunsaturated, were found to show higher levels of *Akkermansia, Bilophila* and *Lactobacillus* in the case of mice. The consortium of microbiota and expression profiles vary as per the diet consumed. A study done by Kong *et al.* (2014) [13] showed that people on a low-fat healthy diet had the lowest C-reactive protein and the highest microbial richness and diversity than the high-fat diet. The healthy low-fat diet also reduces the risk to metabolic defects and resistance to insulin.

The modern paleolithic diet, which is rich in lean meat, fruits, vegetables, seeds, nuts, fish and eggs and lacks the dairy products, sugar, grains has been associated with numerous health benefits due to the growth of saccharolytic bacterial species like *Odoribacter*, *Sutterella*, saturated fat-consuming species like *Biophila*, unsaturated fat-consuming species like *Akkermansia* and bile tolerant species like *Collinsella, Dorea* [9].

As a trend now the consumption of emulsifiers like Polysorbate-80, carboxymethylcellulose and artificial sweeteners has increased but all this has been associated with reduction in microbial richness and dis-balance of the microbial homeostasis in the GIT, therefore paving the way to develop metabolic syndromes, to begin with rise in fasting sugar, glucose intolerance and further complicated diseases [5, 14]. However the usage of polyphenols (tea, coffee, olives, berries) boost the SCFAs producing bacteria and therefore reduces the risk of metabolic syndromes [15–17].

A ketogenic diet, which is low on carbohydrate and dietary fibre, is gaining importance among the people who want to shed their extra weight and enhances the blood profiles thereby lowering the risk of cardiovascular disease. However, a ketogenic diet reduces the microbial diversity and is associated with constipation [5, 18]. Both the high and low carbohydrate diet possess mortality risk but in the case of low carbohydrates it depends on the type of protein and fat source being consumed; an animal product poses a higher risk than the plant-based products [19] (Table 3.1).

As an age-old practice in major parts of the world, people have consumed prebiotics and probiotics in their diet at some stage of their life span. These prebiotics are the supplements rich in short chain carbohydrates which are processed by the gut bacteria. The very foremost in this category is the mother's breast milk which favours the growth of *Bifidobacteria* and has the antiadhesive, antibacterial and antiviral activities to prevent the infection in infants. The prebiotics have been associated to favour the growth *of Bifidobacterium, Lactobacillus, Ruminococcus bromii, Roseburia intestinalis, Faecalibacterium prausnitzi* and *Eubacterium rectale*. Probiotics refers to consumption of bacterial and yeast strains belonging to genus *Bacillus, Lactobacillus, Enterococcus, Bifidobacterium, Saccharomyces,* and *Streptococcus* and are found to have a positive impact on human health. They mediate these affects by increasing the release of hormones like glucagon-like peptide (GLP)-1, GLP-2, and peptide tyrosine; increasing the production of SCFA especially butyrate [20, 21].

The ageing process affects various processes and creates numerous changes in the body, including functioning of the immune system, and alterations in the intestinal microbiota, dietary pattern, metabolism and gastrointestinal physiology [22, 23]. In general, there is a decline in the variety of gut microbiota and the commensals such as bifidobacteria, bacteroids, and lactobacilli, whereas rise in opportunistic pathogens level such as enterobacteria, *C. perfringens*, *C. difficile* in older have been documented [24]. These changes are attributed to disturbances in functionality and composition of microbiota thus paving the way for vulnerability and disease. Starch degradation enzymes such as glucan phosphorylases and glycosidases; and secA preprotein translocase, for example, escalate with age which might be in response to reduced host dietary carbohydrate

Table 3.1 Different types of diets.

Diet	Composition of diet	Effect on gut flora	Effect on host	Reference
Western diet	sugar rich, protein, fat and low in dietary fibre	higher levels of Firmicutes, *Akkermansia, Bilophila* and *Lactobacillus;* Decrease in Bacteroidetes	the risk to metabolic defects, obesity, and resistance to insulin	[6, 7, 10]
High-fibre plant source diet	rich in polysaccharides	Increase in *Prevotella* spp., Bacteroides and decrease in Firmicutes	digest the complex plant fibres and the associated molecules	[6, 8, 9]
Mediterranean diet	rich in fruits, legumes, vegetables, fish, moderate alcohol minimally processed cereals and nuts, saturated fat, dairy products and meat	increasing the SCFAs by *Prevotella* and Firmicutes	positive effect on disease like diabetes, obesity, cardiovascular and inflammation	[7, 12]
Paleolithic diet	rich in lean meat, fruits, vegetables, seeds, nuts, fish and eggs	abundance of bile-tolerant and fat-loving microbes like *Odoribacter, Sutterella, Biophila, Akkermansia, Collinsella, Dorea*	improvement in sensitivity to insulin, glycemic index control and leptin levels, and decrease in triglyceride levels and total fat mass	[9]
Ketogenic diet	low on carbohydrate and dietary fibre	reduced microbial diversity	lowering the risk of cardiovascular disease, associated with constipation	[19]

breakdown efficiency with age [10]. Alternatively, the expression of RNA polymerase facultative sigma-24 subunit, decreases with age that regulates expression during different stress responses and is correlated with intestinal gut flora survival. This could be attributed to the fact that with age the host undergoes immunosenescence which reduces the burden of the need of stress response [10]. Furthermore, decreased intestinal motility contributes to constipation and increases the duration of intestinal transit. As a consequence of this, altered availability and absorption of nutrients occurs, leading to decreased stability and diversity in microbial communities, bleed through across the mucosal lining and intestinal barrier dysfunction, and inflammation creating the conditions for the opportunistic infecting microorganisms. The intestinal microbiota of the elderly is marked by a decrease in diversity, and beneficial microorganisms, with shifts towards facultative anaerobic bacteria, a decrease in total SCFA levels and its ratio as compared to those found in healthy adults [22].

3.2 Association of Gut Flora and its Impact on Human Health and Associated Diseases

The microbiome present in the gut gets translated into metagenome affecting the gut flora as well the health of the host. This alters the metabolome physiology of the microenvironment. The gut microbes especially obligate anaerobes like firmicutes, bacteroids also help in breaking down of complex dietary molecules like galacto, fructo oligosaccharides, and inulin to the simpler versions like mannans, glycans, etc., by using their enzymes like mannases, pectinases to produce energy for themselves and at the same time converting the complex dietary molecule in a more beneficial metabolite like SCFA (acetate (C2), propionate (C3), butyrate (C4)) [25], neurotransmitters (γ-aminobutyric acid (GABA)), bile acids that can be utilized by host.

Acetate, butyrate and propionate constitute 90-95% of the SCFAs [26]. SCFAs display a broad array of functions ranging from regulation of immune response by modulating the differentiation of regulatory T cells; production of inflammatory molecules like tumor necrosis factor (TNF)-alpha. Thus, having a direct and indirect impact on our bodies. During lipogenesis and glucogenesis, acetate is converted to acetyl-CoA which is then shunted to tricarboxylic acid cycle (TCA) cycle for generation of energy, whereas propionate and butyrate act as immunomodulators and regulate the gut physiology [27]. The absence of butyrate in the gut silences metabolic signalling. Butyrate activates the nuclear receptor peroxisome proliferator-activated receptor gamma (PPAR-γ) which is responsible for maintaining the microbial homeostasis in the gut. This receptor is found in the colonocytes and the adipocytes [28]. In the absence of butyrate the beta-oxidation in the mitochondria within colonocytes becomes ineffective and results in transfer or diffusion of oxygen into the cavity of the gut, resulting in the growth of pathogenic facultative anaerobes like *E. coli*. It also results in the increase in the nitrate production (NOS2 expression) by inhibiting the signalling *via* PPAR-γ [28]. These nitrate molecules act as the electron acceptor and results in breakdown of carbohydrates into carbon dioxide instead of fermentation, *i.e.*, nitrate respiration [29].

In epithelial cells, adipocytes and immune cells numerous receptors for the SCFAs like G-protein coupled receptor 43/free fatty acid receptor2 (GPR43/FFAR2), GPR109A, GPR41/FFAR3, and olfactory receptor 78 have been identified through which they mediate their effects [27, 30]. GPR41 has been found to induce expression of leptin pathway, decrease food intake, and expend more energy, heightened regulatory T cells. GPR43 confers protection against IBD, has increased insulin sensitivity, anti-lipolysis, preadipocyte and T regulatory cells differentiation. GPR109A is also anti-lipolysis, and plays a role in reduction of triglycerides and protection against colitis. Olfr78 also has regulatory function, regulating the hormone secretion (GLP-1, PYY), and blood pressure (renin-angiotensinaldosterone pathway) [27, 31].

Apart from degradation of complex polysaccharides and carbohydrates into readily absorbable forms, the gut flora also generates the essential vitamin B12 and K, thiamine, biotin, ascorbate, pantothenate riboflavin, pyridoxine, and various signal transmitting molecules (serotonin, tryptophan and catecholamines) in the intestine which are then absorbed by the host cells, showing a close symbiotic association of the host to the microbes [10, 25]. The cluster 1 and 2 microbes were found to be enriched in biosynthetic pathways of vitamins. Cluster 1 enterotypes were enriched for the biosynthetic pathways for biotin, riboflavin, ascorbate and pantothenate whereas the latter was enriched in thiamine and folate [10].

A stable microbiota in the intestine regulates regional homeostatic immune responses through exposure to bacterial lipopolysaccharide (especially in the case of gram negative bacteria) or peptidoglycan that became evident in mice models, which completely lacked the intestinal microbes and therefore suffered from defects associated with lymph nodes and lower secretion of immunoglobulin (Ig) A. The butyrate produced by the gut microbiota also aids in the natural defence mechanism by inhibiting the signal transduction via NF-κB thus blocking the mucosal entry of the unwanted pathogens. Many studies have illustrated the need of fine balance of the gut microbiota and host's health to maintain a healthy lifestyle.

However, due to certain conditions like antibiotics, unhealthy food habits, and intoxication there might be disruption of this equilibrium, which is termed dysbiosis, often linked to a range of diseases including metabolic disorders (obesity, diabetes, cardiovascular diseases), inflammatory bowel disease (IBD), chronic obstructive pulmonary disease, skin problems (acne, eczema), periodontitis (dental caries), cancer (colorectal, pancreatic, hepatobiliary, gastric), cystic fibrosis, asthma, allergies, auto-immune disorders non-alcohol fatty liver disease and neurological disorders (neurodegenerative disease-alzheimer's, parkinson's, other disorders of the brain).

The high-fat diet (HFD) and high sugar has been found to cause insulin sensitivity, and thus can cause diabetes and increase in body weight leading to obesity [32]. HFD favours the growth of the bacteria belonging to family Clostridiaceae, Firmicutes and inhibits the growth of Bacteroidacece members, *Akkermansia muciniphila* and *F. prausnitzii*. People with obesity have microbial consortium composed of *Bacteroids, Ruminococcus, Parabacteroids, Dialister, Campylobacter, Staphylococcus, Porphyromonas*, and *Anaerostipes* however people with lean phenotype have *Faecalibacterium, Bifidobacterium, Lactobacillus, Butyrivibrio, Alistipes, Akkermansia, Coprococcus* and *Methanobrevibacteristics*. There has been a positive correlation between weight gain and the prevalence of *Lactobacillus reuteri*; on the contrary, opposite effects have been observed for *Ruminicoccus. intestinalis, Bifidobacterium animalis, Lactobacillus paracasei, Lactobacillus plantarum*, and *Christensenella minuta*. Dysregulation of genes associated with lipid metabolism acetyl-CoA carboxylase 1 and fatty acid synthase and inhibition of lipoprotein lipase, is one of the mechanisms by which gut flora found in the obese promotes the obesity. A high amount of circulating lipopolysaccharides (LPS) due to their translocation from cell wall of gram negative bacteria, has been found in people consuming a HFD; and obesity. This rise in the level of LPS triggers the host immune responses mediated through toll-like receptors (TLRs) especially TLR4 and causes the state of inflammation and development of other diseases [33–35]. *A. muciniphila* has also been shown to reverse these obesity defects and insulin resistance in type II diabetic patients. Metformin, the most common antidiabetic drug, improves the SCFA production by gut flora by enhancing their growth. *Prevotella copri* and *Bacteroides* negatively impacts the host response to insulin and thus insulin resistance; however, after anti-diabetic treatment *Akkermansia spp.* were found in great abundance.

With respect to metabolic disorders associated with the gut, the most common associated disease is IBD which is manifested in two forms that differ based on the part of the intestine being infected – ulcerative colitis or crohn's disease. IBD is the mounted immune response in reaction to the alterations in metabolite production by the gut microbes (Figures 3.3 and 3.4).

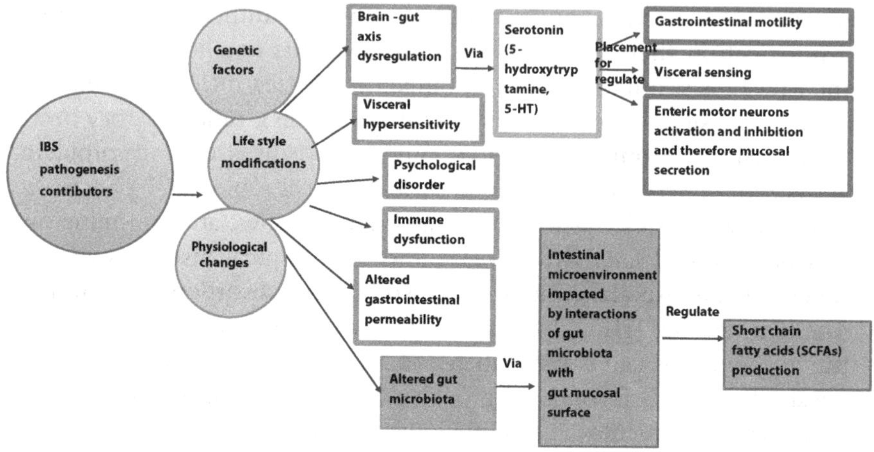

Figure 3.3 Potential contributors to the pathogenesis of IBS/IBD [48]. IBS: Irritable bowel syndrome; IBD: Inflammatory bowel disease; SCFA: Short chain fatty acids; 5-HT: 5-hydroxytryptamine. Luo M, Zhuang X, Tian Z, Xiong L. *BMC Gastroenterol. 2021;21(1):14.*

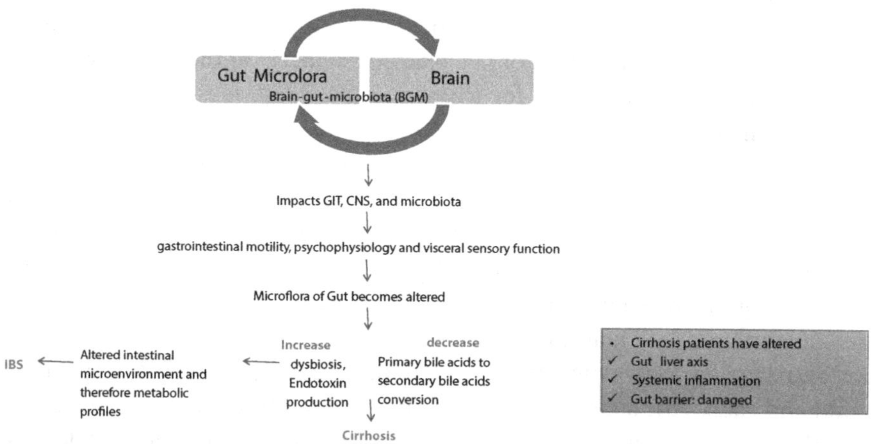

Figure 3.4 Alterations in metabolome and dysregulation of the brain-gut axis [49, 50]. IBS: Irritable bowel syndrome; BGM: brain-gut-microbiota; CNS: central nervous system; GIT: gastrointestinal tract. Yoshimatsu Y, Mikami Y, Kanai T. Inflamm Regen. 2021;41(1):3. Lee NY, Suk KT. Int J Mol Sci. 2020;22(1):199.

The condition of stress is generated due to the production of respiratory electron acceptors as a by-product of the inflammatory response of the host which becomes detrimental to the microbes. These stressed conditions lead to gut dysbiosis and digression in the microbial richness and their growth and favouring the growth of facultative anaerobes like invasive strains of *E. coli, Citrobacter* and inflammation inducing anaerobes, for instance

Ruminococcus gnavus [5, 16, 29, 36]. *R. gnavus* produces inflammatory oligosaccharide Rhamanose which induces the inflammatory cytokine TNFα by dendritic cells and promoting the IBD [36]. The notable producer of SCFA in the gut *Faecalibacterium prausnitzii* is shown to reduce in numbers in individuals suffering from IBD. It has also been seen that in the case of ulcerative colitis the PPAR-gamma receptor synthesis is inhibited due to extensive repair in the epithelial cells, therefore inhibiting the beta-oxidation and favouring the growth of facultative anaerobes [5]. The phospholipids (phosphatidylcholine) and amino acids such as L-carnitine present in non-vegan diets are converted to trimethylamine (TMA gas) by gut bacteria and further processed by liver to trimethylamine-N-oxide (TMAO) which is reported to be a potential agent to cause cardiovascular diseases and fatty liver. Non-alcoholic fatty liver disease (NAFLD) is now being caused due to change in our dietary practices and found to be associated with increase in Bacteroidetes *(Porphyromonas, Prevotella), Bradyrhizobium, Propionibacterium acnes, Anaerococcus, Dorea, Peptoniphilus,* and Proteobacteria *(Escherichia, Alcaligenaceae)* numbers, and reduction in Firmicutes *(Blautia, Coprococcus, Eubacterium, Roseburia, Faecalibacterium, Oscillospira, Ruminococcus),* and Actinobacteria (*Bifidobacterium*) (Figure 3.5).

Potential reasons for the linkage between microbiome and NAFLD could be attributed to differences found in SCFA metabolism, production of endogenous ethanol, or decrease in choline and trimethylamine and

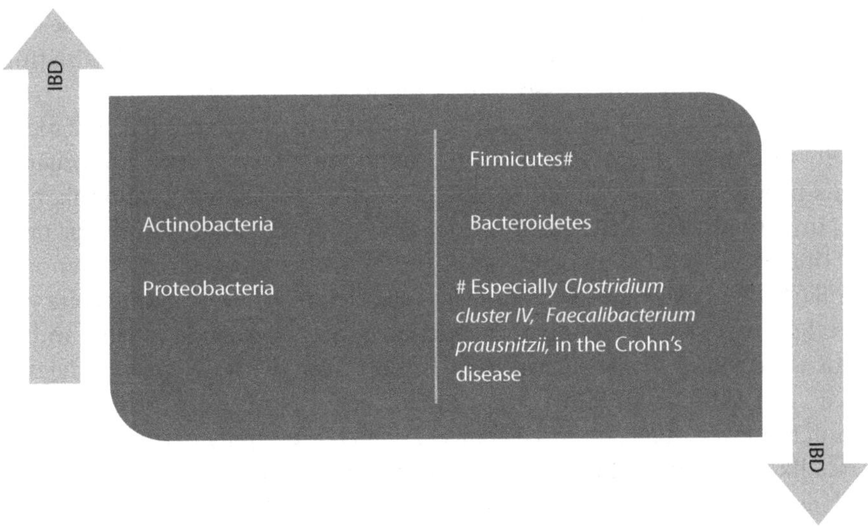

Figure 3.5 Alterations in gut microbiota phyla: IBD and NAFLD [49, 51]. IBD: Inflammatory bowel disease; NAFLD: Non-alcoholic fatty liver disease. Yoshimatsu Y, Mikami Y, Kanai T. *Inflamm Regen.* 2021;41(1):3.

discharge of LPS from intestinal microbiota. Dysbiosis may also trigger de novo lipogenesis, increase bowel permeability and translocation of both bacteria and endotoxins from the lumen to the extra intestinal area. Even fructose has been shown to cause inflammatory response and aiding the process of NAFLD by altering the permeability of junctions in the gut.

The metabolic changes associated with the displacement of gut microflora can be used as biomarkers for speculating the health status and therapeutic intervention like antimicrobial peptides, inflammatory proteins, and SCFA levels. The bacterial metabolic products such as branched SCFA-isobutyric acid, isovaleric acid and ratio of acetic acid to propionic acid have been associated with hypercholesterolemia [26, 37, 38]. The higher the Acetic acid: Propionic acid ratio, the higher is the cholesterol synthesis and the lipogenesis. Lower abundance of *Anaeroplasma*, *Phascolarctobacterium faecium*, *Phascholarctobacterium succinatutens* and *Haemophilus* and abundance of bacteria like *Odoribacter* (Bacteroidetes) and *Ruminococcus* (Firmicutes) in faeces have also been associated with unusual lipid profiles and hypercholesterolemia.

3.3 Dietary Modulation of Gut Flora for Therapeutic Usage

Soon after the human genome project 2008 the impact of host microbes on human health was highlighted and resulted in the development of new therapeutic approaches that basically focus on the alteration of the gut microbiota, perhaps by retaining the impaired functions or by eliminating the harmful or undesired associated functions (Table 3.2).

Alterations in level and types of metabolites can be exploited therapeutically to indicate the health status of an individual, such as SCFA, organic acids-propanoic, oleic, butanoic, amino acids–phenylalanine, valine alanine etc. in patients with NAFLD dysfunctional pathways were shown for production of metabolites as 1-butanol, 4-methyl-2-pentanone and 2 butanone. Similarly for obesity altered levels of acetate, lactate, urine hippurate, phenylacetylglycine, 4-hydroxylphenylacetic acid and branched chain amino acids were found. *Fusobacterium nucleatum* along with *Campylobacter spp* have been associated with colorectal cancer. Inflammatory proteins such as interleukin 6, TNF, cyclooxygenase 2 which are highly expressed in diseases like IBD and colorectal cancer can also be used as biomarkers [39]. Inflammatory markers as C-reactive protein and faecal calprotectin can also be tested to check the inflammatory host response [40].

Levels of choline, betaine and TMAO generally found in animal protein can be monitored to predict the development of cardiovascular diseases [5].

Table 3.2 Effects of modulating gut microbiome: Probiotics [48, 52, 53].

Effects
Regulates metabolic syndrome
Obesity-related comorbidities including dyslipidemia and insulin resistance
Reduction in obesity and improvement in insulin sensitivity
Lowers liver lipid accumulation
Lowers cholesterol especially serum and improving liver lipid metabolism
Increases the excretion of bile acids from feces
Improves hepatic steatosis
Restores epithelial barrier function
Reduces endotoxin leakage from the blood
Beneficial effects on mood and cognition
Release of Short chain fatty acid

Total cholesterol, Triglycerides, Low Density Lipoprotein-C, non-High Density Lipoprotein-C, apolipoprotein B, and apoB/apoA-1 can also be monitored to check the health status and development of disease [41].

3.4 Data from Human and Animal Studies

The recent study performed by McGinley *et al.* (2020) [42] on C57BL/6 mice showed a 3-fold increment in the total bacterial count and a 2-5 fold increase in the bacteria belonging to class firmicutes when mice were fed with pulses, rich in dietary fibres. *Akkermansia municiphila* bacterium was found to be 49 times higher when mice were fed on pulses. Significant reduction in lipid accumulation was also seen, indicating overall healthy well-being [42].

Wu *et al.* (2019) [41] performed the clinical trials with 2,400 Chinese patients and found elevated levels of HDL-C and apoA-1 while the age progresses, and women featuring high concentration of total cholesterol, HDL-C, LDL-C, and apoB/apoA-1 as compared to men [41].

The study in rats as reported by Bauer *et al.* (2018) [43] highlighted the fact that a high-fat diet leads to reduction in sensing of glucose by Sodium Glucose cotransporter1 (SGLT1). They also pointed out the intake of sucralose and Fructose might impact the glucose sensing and its tolerance. The SGLT1 levels were restored upon the faecal transplantation of the gut microbiota [43].

Studies published by De Filippis *et al.* (2016) [7] and Barone *et al.* (2019) [9] on Italians consuming the mediterranean diet and the modern paleothic diet, respectively, outlined the utmost importance of the diet on gut flora and association with the diseases. People who did not restrict themselves to the mediterranean diet had high levels of TAMO, which is associated with cardiovascular diseases. In their study there was significant rise in SCFAs, with increase in microbial diversity; *Prevotella* (Bacteroidetes) and *Lachnospira* (Firmicutes) were the most prominent carbohydrate fermenters.

The impact of *L. Rhamnosus* GG, was studied by Luoto *et al.* (2010) [44] on obesity in children up to 10 years and it was seen that when this probiotic was given to children from early childhood, it altered the gut microflora and restrained the weight gain in them. Another trial was executed by by Larsen *et al.* (2013) [45] with obese adolescence in an attempt to examine the effect of *L. salivarius* Ls-33 on faecal microflora resulted in significant upsurge in the numbers of *Bacteroides*, *Prevotella*, *Porphyromonas.*

Further study by Lee *et al.* (2014) [46] showed a positive correlation in reduction of body weight with consumption of combined probiotics comprising *L. acidophilus*, *L. plantarum*, *L. rhamnosus*, *B. lactis*, *B. breve*, *B. longum*, and *Streptococcus thermophiles* and herbal medicines. The conclusion confirmed that probiotics play an important role in the deterrence of obesity by reducing the production of LPS through altered gut microflora.

A later study reported by Rajkumar *et al.* (2015) [47] the effect of prebiotic (fructo-oligosaccharide) and probiotic *L. salivarius* UBLS22 consumption either individually or in combination in healthy individuals as biomarkers of obesity and the gut flora. They found a rapid increase in the number of Lactobacilli; however, significant reduction in the number of coliforms was noted.

Probiotics species (Figure 3.6) have shown to have benefits of modulation of gut-flora. Alteration in the resident colonic flora with administration of probiotics may lead to the following effects [54]:

- Modify the colonic metabolism of nutrient
- Alter colonic transit and fluid fluxes
- Reduce flatulence and abdominal pain
- Possess anti-inflammatory effects: *Lactobacillus casei*, decreased TNF-α release in ileal tissues, Increase in anti-inflammatory cytokines such as IL-8 or decrease in TNF-α or other pro-inflammatory cytokines.
- *Bifidobacterium* improvement in symptoms with bifidobacteria was associated with changes in the relative production of anti-inflammatory to proinflammatory cytokines.

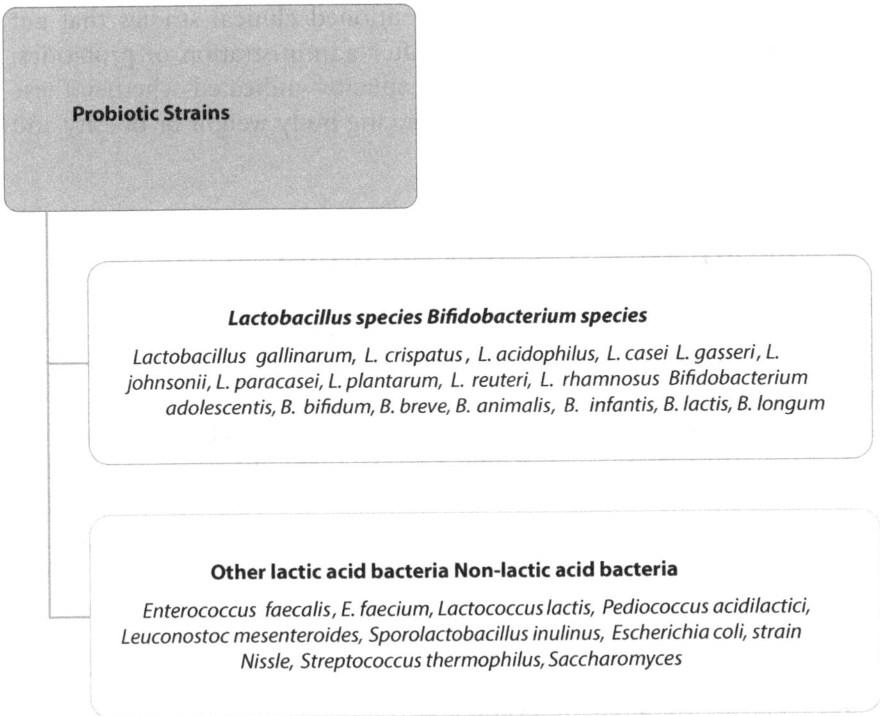

Figure 3.6 Different probiotics species to treat IBS and liver diseases [55]. IBS: Irritable bowel syndrome, *L.: Lactobacillus; B.: Bifidobacterium; E.: Enterococcus*. Thayumanavan L, Tarakeswari M, Muthukumaran K. *Gastroenterology, Hepatology and Endoscopy Practice* 2021 Jan 1;1(1):11.

- *Lactobacillus* and *Bifidobacterium ssp.* can deconjugate and absorb bile acids
- Alter the colonic milieu or motility
- Reduce Colon's reflex motor responses to balloon distension

The targets for probiotic therapy are identified as clinical conditions involving impaired mucosal barrier function, particularly gastrointestinal infectious and inflammatory diseases [56]. Lactobacilli, bifidobacteria, and nonpathogenic yeasts, such as *Saccharomyces boulardii*, are the most common probiotics and may influence gastrointestinal disorders including acute infectious diarrhoea, IBD, allergic disorders, and IBS [57]. Probiotics appear to have potential in relieving specific symptoms of IBS, such as: abdominal pain (e.g., Visbiome or *Bacillus coagulans* GBI-30, 6086), bloating (e.g., *Lactobacillus acidophilus* NCFM + *Bifidobacterium lactis* Bi-0737 or *B. bifidum* MIMBb75) and flatulence (e.g., *S. boulardii*) [58].

It can be inferred from the aforementioned clinical studies that gut microbiota displays dramatic change after administration of probiotics; however, none of these studies have explicitly indicated whether these changes are entirely responsible for reducing body weight or obesity and modulation of host metabolome.

3.5 Future Prospects and Conclusion

The above documented studies have underlined the importance and role of gut flora in the human physiology and health; however; some of the facets need to be further examined in detail to complete the underlying principle and involved mechanism. The lifestyle and physiological changes contribute to gastrointestinal disorders and liver diseases due to imbalance of the intestinal flora. Alterations in gut microflora may be associated with the pathogenesis of IBS, liver diseases and affect the symptoms of IBS and liver disease patients. There are potential benefits of the specific probiotic combination in IBS and liver disease. Specific probiotic properties might be exploited in prophylactic and therapeutic intervention for IBS and liver diseases. The application of probiotics currently lies in reducing the risk of diseases associated with gut barrier dysfunction. A comparative and quantitative study would be an integral approach while dealing with different probiotic strains and their role in altering the permeability of the gut and hormones. Relating all clues from various research reports, we can conclude that prebiotics and probiotics has the ability to address obesity and associated metabolic disorders.

References

1. Aas JA, Paster BJ, Stokes LN, Olsen I, Dewhirst FE. Defining the normal bacterial flora of the oral cavity. *Journal of Clinical Microbiology*. 43(11):5721-32, 2005.
2. Sanz Y, Romaní-Perez M, Benítez-Páez A, Portune KJ, Brigidi P, Rampelli S, *et al.* Towards microbiome-informed dietary recommendations for promoting metabolic and mental health: Opinion papers of the MyNewGut project. *Clinical nutrition (Edinburgh, Scotland)*. 37(6 Pt A):2191-7, 2018.
3. Conlon MA, Bird AR. The impact of diet and lifestyle on gut microbiota and human health. *Nutrients*. 7(1):17-44, 2014.
4. Jandhyala SM, Talukdar R, Subramanyam C, Vuyyuru H, Sasikala M, Nageshwar Reddy D. Role of the normal gut microbiota. *World Journal of Gastroenterology*. 21(29):8787-803, 2015.

5. Hills RD, Jr., Pontefract BA, Mishcon HR, Black CA, Sutton SC, Theberge CR. Gut Microbiome: Profound Implications for Diet and Disease. *Nutrients.* 11(7), 2019.
6. Riaz Rajoka MS, Shi J, Mehwish HM, Zhu J, Li Q, Shao D, *et al.* Interaction between diet composition and gut microbiota and its impact on gastrointestinal tract health. *Food Science and Human Wellness.* 6(3):121-30, 2017.
7. De Filippis F, Pellegrini N, Vannini L, Jeffery IB, La Storia A, Laghi L, *et al.* High-level adherence to a Mediterranean diet beneficially impacts the gut microbiota and associated metabolome. *Gut.* 65(11):1812-21, 2016.
8. Vangay P, Johnson AJ, Ward TL, Al-Ghalith GA, Shields-Cutler RR, Hillmann BM, *et al.* US Immigration Westernizes the Human Gut Microbiome. *Cell.* 175(4):962-72.e10, 2018.
9. Barone M, Turroni S, Rampelli S, Soverini M, D'Amico F, Biagi E, *et al.* Gut microbiome response to a modern Paleolithic diet in a Western lifestyle context. *PloS one.* 14(8):e0220619, 2019.
10. Arumugam M, Raes J, Pelletier E, Le Paslier D, Yamada T, Mende DR, *et al.* Enterotypes of the human gut microbiome. *Nature.* 473(7346):174-80, 2011.
11. Wu GD, Chen J, Hoffmann C, Bittinger K, Chen YY, Keilbaugh SA, *et al.* Linking long-term dietary patterns with gut microbial enterotypes. *Science* (New York, NY). 334(6052):105-8, 2011.
12. Garcia-Mantrana I, Selma-Royo M, Alcantara C, Collado MC. Shifts on Gut Microbiota Associated to Mediterranean Diet Adherence and Specific Dietary Intakes on General Adult Population. *Frontiers in Microbiology.* 9:890, 2018.
13. Kong LC, Holmes BA, Cotillard A, Habi-Rachedi F, Brazeilles R, Gougis S, *et al.* Dietary patterns differently associate with inflammation and gut microbiota in overweight and obese subjects. *PloS one.* 9(10):e109434, 2014.
14. Skodje GI, Sarna VK, Minelle IH, Rolfsen KL, Muir JG, Gibson PR, *et al.* Fructan, Rather Than Gluten, Induces Symptoms in Patients with Self-Reported Non-Celiac Gluten Sensitivity. *Gastroenterology.* 154(3):529-39.e2, 2018.
15. Valdes AM, Walter J, Segal E, Spector TD. Role of the gut microbiota in nutrition and health. *BMJ (Clinical research ed.).*361:k2179, 2018.
16. Martinez-Medina M, Garcia-Gil LJ. Escherichia coli in chronic inflammatory bowel diseases: An update on adherent invasive Escherichia coli pathogenicity. *World Journal of Gastrointestinal Pathophysiology.* 5(3):213-27, 2014.
17. Kim Y, Je Y, Giovannucci E. Coffee consumption and all-cause and cause-specific mortality: a meta-analysis by potential modifiers. *European Journal of Epidemiology.* 34(8):731-52, 2019.
18. Zhang Y, Zhou S, Zhou Y, Yu L, Zhang L, Wang Y. Altered gut microbiome composition in children with refractory epilepsy after ketogenic diet. *Epilepsy research.* 145:163-8, 2018.

19. Seidelmann SB, Claggett B, Cheng S, Henglin M, Shah A, Steffen LM, *et al.* Dietary carbohydrate intake and mortality: a prospective cohort study and meta-analysis. *The Lancet Public Health.* 3(9):e419-e28, 2018.
20. Wang Y, Wang B, Wu J, Jiang X, Tang H, Nielsen OH. Modulation of Gut Microbiota in Pathological States. *Engineering.* 3(1):83-9, 2017.
21. McBurney MI, Davis C, Fraser CM, Schneeman BO, Huttenhower C, Verbeke K, *et al.* Establishing What Constitutes a Healthy Human Gut Microbiome: State of the Science, Regulatory Considerations, and Future Directions. *Journal of Nutrition.* 149(11):1882-95, 2019.
22. Salazar N, González S, Nogacka AM, Rios-Covián D, Arboleya S, Gueimonde M, *et al.* Microbiome: Effects of Ageing and Diet. *Current Issues in Molecular Biology.* 36:33-62, 2020
23. de Diego I, Peleg S, Fuchs B. The role of lipids in aging-related metabolic changes. *Chemistry and Physics of Lipids.* 222:59-69, 2019.
24. Nagpal R, Mainali R, Ahmadi S, Wang S, Singh R, Kavanagh K, *et al.* Gut microbiome and aging: Physiological and mechanistic insights. *Nutrition and Healthy Aging.* 4(4):267-85, 2018.
25. Andoh A. Physiological Role of Gut Microbiota for Maintaining Human Health. *Digestion.* 93(3):176-81, 2016.
26. Granado-Serrano AB, Martín-Garí M, Sánchez V, Riart Solans M, Berdún R, Ludwig IA, *et al.* Faecal bacterial and short-chain fatty acids signature in hypercholesterolemia. *Scientific Reports.* 9(1):1772, 2019.
27. Ohira H, Tsutsui W, Fujioka Y. Are Short Chain Fatty Acids in Gut Microbiota Defensive Players for Inflammation and Atherosclerosis? *Journal of Atherosclerosis and Thrombosis.* 24(7):660-72, 2017.
28. Byndloss MX, Olsan EE, Rivera-Chávez F, Tiffany CR, Cevallos SA, Lokken KL, *et al.* Microbiota-activated PPAR-γ signaling inhibits dysbiotic Enterobacteriaceae expansion. *Science* (New York, NY). 357(6351):570-5, 2017.
29. Winter SE, Bäumler AJ. Dysbiosis in the inflamed intestine: chance favors the prepared microbe. *Gut Microbes.* 5(1):71-3, 2014.
30. Kasubuchi M, Hasegawa S, Hiramatsu T, Ichimura A, Kimura I. Dietary gut microbial metabolites, short-chain fatty acids, and host metabolic regulation. *Nutrients.* 7(4):2839-49, 2015.
31. Cani PD. Human gut microbiome: hopes, threats and promises. *Gut.* 67(9):1716-25, 2018.
32. Sen T, Cawthon CR, Ihde BT, Hajnal A, DiLorenzo PM, de La Serre CB, *et al.* Diet-driven microbiota dysbiosis is associated with vagal remodeling and obesity. *Physiology & Behavior.* 173:305-17, 2017.
33. Gentile CL, Weir TL. The gut microbiota at the intersection of diet and human health. *Science* (New York, NY). 362(6416):776-80, 2018.
34. Miclotte L, Van de Wiele T. Food processing, gut microbiota and the globesity problem. *Critical Reviews in Food Science and Nutrition.* 60(11):1769-82, 2020.

35. Sekirov I, Russell SL, Antunes LCM, Finlay BB. Gut Microbiota in Health and Disease. *Physiol Rev*. 90(3):859-904, 2010.
36. Henke MT, Kenny DJ, Cassilly CD, Vlamakis H, Xavier RJ, Clardy J. Ruminococcus gnavus, a member of the human gut microbiome associated with Crohn's disease, produces an inflammatory polysaccharide. *Proc Natl Acad Sci USA*. 116(26):12672-12677, 2019.
37. Weitkunat K, Schumann S, Nickel D, Kappo KA, Petzke KJ, Kipp AP, *et al.* Importance of propionate for the repression of hepatic lipogenesis and improvement of insulin sensitivity in high-fat diet-induced obesity. *Molecular Nutrition & Food Research*. 60(12):2611-21, 2016.
38. Wong JM, de Souza R, Kendall CW, Emam A, Jenkins DJ. Colonic health: fermentation and short chain fatty acids. *Journal of Clinical Gastroenterology*. 40(3):235-43, 2006.
39. Brennan CA, Garrett WS. Gut Microbiota, Inflammation, and Colorectal Cancer. *Annual Review of Microbiology*. 70:395-411, 2016.
40. Sommer F, Rühlemann MC, Bang C, Höppner M, Rehman A, Kaleta C, *et al.* Microbiomarkers in inflammatory bowel diseases: caveats come with caviar. *Gut*. 66(10):1734-8, 2017.
41. Wu TT, Zheng YY, Yang YN, Li XM, Ma YT, Xie X. Age, Sex, and Cardiovascular Risk Attributable to Lipoprotein Cholesterol Among Chinese Individuals with Coronary Artery Disease: A Case-Control Study. *Metabolic Syndrome and Related Disorders*. 17(4):223-31, 2019.
42. McGinley JN, Fitzgerald VK, Neil ES, Omerigic HM, Heuberger AL, Weir TL, *et al.* Pulse Crop Effects on Gut Microbial Populations, Intestinal Function, and Adiposity in a Mouse Model of Diet-Induced Obesity. *Nutrients*. 12(3), 2020.
43. Bauer PV, Duca FA, Waise TMZ, Rasmussen BA, Abraham MA, Dranse HJ, *et al.* Metformin Alters Upper Small Intestinal Microbiota that Impact a Glucose-SGLT1-Sensing Glucoregulatory Pathway. *Cell Metabolism*. 27(1):101-17.e5, 2018.
44. Luoto R, Laitinen K, Nermes M, Isolauri E. Impact of maternal probiotic-supplemented dietary counselling on pregnancy outcome and prenatal and postnatal growth: a double-blind, placebo-controlled study. *British Journal of Nutrition*. 103(12):1792-9, 2010.
45. Larsen N, Vogensen F, Gøbel R, Michaelsen K, Forssten S, Lahtinen S, *et al.* Effect of Lactobacillus salivarius Ls-33 on fecal microbiota in obese adolescents. *Clinical Nutrition* (Edinburgh, Scotland). 32, 2013.
46. Lee DK, Park JE, Kim MJ, Seo JG, Lee JH, Ha NJ. Probiotic bacteria, B. longum and L. acidophilus inhibit infection by rotavirus *in vitro* and decrease the duration of diarrhea in pediatric patients. *Clinics and Research in Hepatology and Gastroenterology*. 39(2):237-44, 2015.
47. Rajkumar H, Kumar M, Das N, Kumar SN, Challa HR, Nagpal R. Effect of Probiotic Lactobacillus salivarius UBL S22 and Prebiotic Fructo-oligosaccharide on Serum Lipids, Inflammatory Markers, Insulin Sensitivity,

and Gut Bacteria in Healthy Young Volunteers: A Randomized Controlled Single-Blind Pilot Study. *Journal of Cardiovascular Pharmacology and Therapeutics.* 20(3):289-98, 2015.
48. Luo M, Zhuang X, Tian Z, Xiong L. Alterations in short-chain fatty acids and serotonin in irritable bowel syndrome: a systematic review and meta-analysis. *BMC Gastroenterol.* 21(1):14, 2021.
49. Yoshimatsu Y, Mikami Y, Kanai T. Bacteriotherapy for inflammatory bowel disease. *Inflamm Regen.* 41(1):3, 2021.
50. Lee NY, Suk KT. The Role of the Gut Microbiome in Liver Cirrhosis Treatment. *Int J Mol Sci.*22(1), 2020.
51. Kwong EK, Puri P. Gut microbiome changes in Nonalcoholic fatty liver disease & alcoholic liver disease. *Transl Gastroenterol Hepatol.* 6:3, 2021.
52. Lee NY, Shin MJ, Youn GS, *et al.* Lactobacillus attenuates progression of nonalcoholic fatty liver disease by lowering cholesterol and steatosis. *Clin Mol Hepatol.* 27(1):110-124, 2021.
53. D'Mello C, Ronaghan N, Zaheer R, *et al.* Probiotics Improve Inflammation-Associated Sickness Behavior by Altering Communication between the Peripheral Immune System and the Brain. *J Neurosci.* 35(30):10821-10830, 2015.
54. Kim HJ, Vazquez Roque MI, Camilleri M, *et al.* A randomized controlled trial of a probiotic combination VSL# 3 and placebo in irritable bowel syndrome with bloating. *Neurogastroenterol Motil.* 17(5):687-696, 2005.
55. Thayumanavan L, Tarakeswari M, Muthukumaran K. Probiotics in gastroenterology. *Gastroenterology, Hepatology and Endoscopy Practice* 1(1):11, 2021.
56. Isolauri E, Kirjavainen PV, Salminen S. Probiotics: a role in the treatment of intestinal infection and inflammation? *Gut.* 50 Suppl 3(Suppl 3):III54-III59, 2002.
57. Choi CH, Jo SY, Park HJ, *et al.* A randomized, double-blind, placebo-controlled multicenter trial of saccharomyces boulardii in irritable bowel syndrome: effect on quality of life. *J Clin Gastroenterol,* 45(8), 679-683, 2011.
58. Ouwehand AC. The role of probiotics in digestive health. *Nutrition and Dietary Supplements.* 7:103-109, 2015.

4

Nutritional Profile, Functional Characteristics, Health Benefits, and Potential Application of Edible Gum (*Gond*)

Anju Kundalia[1], Angel Mishra[2], Chanchal Rani[3], Deepti Gupta[4], Kalpana Gautam[5], K.M. Sushma[6] and Murlidhar Meghwal[7]*

[1]*MSc (DFSM), Dietitian, Vibes Healthcare, New Delhi, India*
[2]*MSc (DFSM), Holistic Coach-Holistic Nutrition Works, Hyderabad, Telangana, India*
[3]*MBBS, MSc (DFSM), Medical Officer, Shri Jagdamba Charitable Eye Hospital, Sri Ganganagar, Rajasthan, India*
[4]*MSc (DFSM), IGNOU Study Center NIFTEM Kundli, IGNOU Regional Center Karnal, Haryana, Indira Gandhi National Open University, New Delhi, India*
[5]*BHMS, MD, MSc (DFSM), Medical Officer, In-Charge, State Homeopathic Dispensary, Mundakhera, Bulandshahr, Uttar Pradesh Sarkar, Uttar Pradesh, India*
[6]*MSc (DFSM), Consultant Dietician, Urolife Clinic, New Delhi, India*
[7]*PhD, Department of Food Science and Technology, National Institute of Food Technology Entrepreneurship & Management Kundli, Sonepat, Haryana, India*

Abstract

Gond are important food items which can easily dissolve in water. Gond have gels and mucilages formation properties. Gond have xylose, glucose, rhamnose, arabinose, and galacturonic acid. Edible gum has stabilization qualities, thickening, water control, emulsification and filling properties. Polysaccharides are broadly used in preparation of food products, due to their wide creativity. The different uses and applications of edible gum gond are possible because of its chemical structures. Tragacanth gum or edible gum (Gond) is widely used in various food products. Gond is a valuable food comulative. Gond is obtained by slash from the stems of some of the specific trees. Various species of Astragalus which submit gum are

**Corresponding author*: murli.murthi@gmail.com

Tejpal Dhewa, Anil Kumar Puniya and Anil Panghal (eds.) Nutritional Science and Technology: Concept to Application, (83–102) © 2023 Scrivener Publishing LLC

healthily found. Gond contains food fiber are non-digestible carbohydrates and lignin. Food fibers of edible gum are good for health and they are of different types. Some common examples are carbohydrate polymers (edible), synthetic carbohydrate and carbohydrate polymers. Nutritional value of Gond Per 100 g is 70 kcal, 35 g carbohydrates, 30 g fiber, 5 g hydrates, 0 g fat and sodium 9 g. Capacity to form film or gel, Chemical reactivity, Sweetening and taste enhancement. Hygroscopicity, rheological properties, osmotic pressure power to attach flavor compounds. Resorption and water-binding capacity are unique properties of gond. Other health benefits of Gond include Cooling agent, beneficial for digestion, helpful for pregnant and lactating mothers, also several applications in the pharmaceutical industry.

Keywords: Gond, dietary, edible gum, nutritional value, health, food

4.1 Introduction

The ancient physicians had insight into constipation control through the use of food items with fiber-rich food consumption such as edible gum leading to free bowel movement and bulked excretions. Fibre is comprised of plant components that are resistant to enzymes in the human gut. Gond is very good food fibre along with other important nutrients. Gond (edible gum) have oligosaccharides, polysaccharides, lignin, and other nutrients. Dietary fibers of gond have health advantages (blood cholesterol reduction, laxation, and blood glucose reduction [1, 2]. Edible gum/gond fiber are carbohydrate polymers with 10 or higher monomeric units. It acts prebiotics in the food system. Gond edible gum fibres are of different types in its properties [3, 4].

Sources of Gond

Gond is obtained by incision from the stems of some specific trees such as *Astragalus gummifer Labill* and some other species of Astragalus of the family Papilionaceae. Figure 4.1 showing deep fried gond.

Types of *Gond*

Different species of the genus Astragalus including *A. adscendens, A. gummifer,* and *A. tragacanthus.* They have different ratios of the two fractions and different chemical compositions [5].

4.2 Nutritional Facts of Gond

Tragacanth gum/*gond* is widely used in various food products and is considered a valuable additive. Table 4.1 lists the value of nutrients available per 100 g of tragacanth gum [6].

Figure 4.1 In butter, deep-fried edible gum (*Gond*) for sweet preparations.

Table 4.1 Nutritional value of *Gond*.

Nutrients	Value (per 100 g)
Energy	70 kcals
Carbohydrates	35 g
Fibre	30 g
Hydrates	5 g
Fat	0 g
Sodium	9 g

4.3 Properties of *Gond*

Gond have various properties. Some of the major normal properties has been summarised here. Table 4.2 shows the various properties of gond.

Table 4.2 Properties of *Gond*.

Colour	White, pale yellowish, white flakes
Odor	Odorless
Taste	Mucilaginous
Shape, Size	Ribbon or flake
Appearance	Similar to goat thorn

Figure 4.2 Astragalus gummifer.

4.4 Biological Sources of Gond

It is the air-dried gummy exudates that flow naturally or are procured by incision from the stems and branches of Astragalus gummifer Labill and several other Astragalus species in the Leguminosae family [5, 6]. Figure 4.2 showing astragalus gummifer.

4.5 Geographical Sources of Gond

Several trees and plants are good sources of gond or edible gum. Several species of Astragalus that produce gum are abundant in mountainous regions at elevations of 1,000–3,000 m. Persian tragacanth and Smyrana or Anatolian tragacanth are two essential tragacanth varieties. In India, these can be found in the wild in the Kumaon and Garhwal regions and other parts of the country.

4.6 Benefits of Gond

Gond is extensively consumed in the preparation of nutrition and energy dense dishes; Gond laddus are most popular applications in India in the preparation of sweets, which are typically given to pregnant and lactating women, as well as those who are ill. Gond is typically obtained from plant, primarily from Middle Eastern legume or acacia plants. This natural gum, also known as Tragacanth or acacia gum, is procured by drying the sap of a variety of gum-producing plants found in the Middle East as well as parts of Maharashtra, Gujarat, Haryana, Rajasthan, Karnataka, and Himalaya region. Edible gum is water-soluble, making it simple to combine with a variety of ingredients. The gum has traditionally been used in remedies to treat ailments such as diarrhoea, coughing, and congestion. The food and pharmaceutical industries have long used gond, or edible gum, as a binding and thickening agent. It is also used as an emulsifier, stabiliser, and texturing agent. The ingredient is commonly found in bakery goods, cosmetics, fizzy energy drinks, and ice cream. It is primarily used in the winter because of its warming and calorific properties. Consumption ladoos made of gond are primarily consumed during the winter months to provide extra calories to the body [7, 8].

Physiochemical properties of Tragacanth gum and nutritional information

Tragacanth is a natural gum derived from the stems and branches of Astragalus gummifer Labillardiere and other Asiatic Astragalus species (family Leguminosae).

Chemical and Physical Properties

Unground Tragacanth gum can be found in flattened, lamellated, straight or curved fragments or spirally twisted pieces 0.5-2.5 mm thick and up to 3 cm long. It is white to pale yellow in colour, with a red tinge in some pieces. The pieces have a horny texture and a short fracture. It has no odour and has an insipid mucilaginous taste in solutions. Powdered tragacanth ranges in colour from white to pale yellow or pinkish brown (pale tan). The gum is thorny and translucent, with longitudinal and transverse ridges. It is partially soluble in water and completely insoluble in alcohol. Gum tragacanth is an anionic and branched carbohydrate with a heterogeneous structure composed of polysaccharides (galactoarabans and acidic polysaccharides) that, when hydrolyzed, yield galacturonic acid, galactose, arabinose, xylose, and fucose. There may also be trace amounts of rhamnose and glucose (derived from traces of starch and/or cellulose) [9].

It is made up of two major components: tragacanth, which accounts for 8-10% (non-swellable/water-soluble) residues of d-galacturonic acid, d-xylose, l-fucose, and d-galactose, and bassorin, which accounts for 60–70% (swellable/non-water soluble) residues of l-arabinose, d-galactose The ratio of tragacanthin to bassorin varies between gum tragacanth species. Gum tragacanth has different rheological, physicochemical, and pharmaceutical properties as a result of this variation. It contains 15% methoxy group, which accounts for the high viscosity. The viscosity of a 1% tragacanth solution is 250 centipoises [10].

Potential Applications of Gond

Gond is the oldest and best known among all-natural gums to date. The applications of Gond date back to 5,000 years ago. It has numerous applications in food, beverages, cosmetics, pet food, and pharmaceutical industries as an additive, shelf-life enhancer, emulsifier, bulking agent, and a bioactive component and has various health benefits; it is now in nanotechnology as well Table 4.3.

Table 4.3 Applications of different gums [1, 3, 4, 6, 8–17].

Food items	Name	Phytochemical	Properties and utilization
Gums (Fruit)			
Mucilage of date palm	*Phoenix dactylifera*	mannose, fructose, maltose glucose, sucrose	Anticancer
Mucilage from Erva Baleeira	*Cordia ohliqua*	galactose, arabinose, pyrralinose	binding, emulsifier agent, tablet, expectorant
Jackfruit	*A. heterophyllus*	Starch, galactomannan	Stabilizing, binding, mucoadhesive, emulsifier
Gums (Grain/Seed)			
Tamarind gum	*Tamarind us indica*	Galactose, gl ucose, xylose	Bioadhesion, tablet, controlled drug release biodegradable compound
Mucilage of fenugreek	*Trigonella foenum-graceum*	Galactomannan	Enhancer, anthocyanin, encapsulating agent
Tara gum	*Caesalpinia spinosa*	Galactose, Mannose	Fat replacer, used for food packing material

(*Continued*)

Table 4.3 Applications of different gums [1, 3, 4, 6, 8–17]. (*Continued*)

Food items	**Name**	**Phytochemical**	**Properties and utilization**
Honey locust gum	*Gleditsia triacanthos*	Galactomannan	Hydrocolloid in food industry, Matrix formulating agent in tablets
Sickle pod gum	*Cassia tora*	Glucose and arabinose	Suspension stabilizing agent, binding agent
Flamboyant gum	*Mimosa scahrella*	Mannose, Galactose	Microencapsulation, probiotic
Guar gum	*Ocimum americanum*	arabinose, xylose, galacturonicacids, rhamnose	Paper manufacturing, nanocomposite films
Trunk exudate gums			
Silk plant	*Alhizia stipulata*	Arabinose, galactose,	Health uses- anticancer, mental conditions
Almond gum	*Prunus amygdalus*	L-arabinose, A1dobionic acid, D-mannose, L-galactose	Emulsifying, stabilizing, thickening and binding agent
Cashew gum/ cashew nut gum	*Anacardium occidentale*	arabinose, Galactose, Glucuronic acid, glucose, rhamnose	Industrial applications- binding, stabilizer, thicken in chocolates
Cherry gum	*P. avium*	Arabinogalactan	Antioxidant
Raphia palm	*R. hookeri*	Galactose, mannose	Ecofriendly anticorrosive for steel

(*Continued*)

Table 4.3 Applications of different gums [1, 3, 4, 6, 8–17]. (*Continued*)

Food items	Name	Phytochemical	Properties and utilization
Tragacanth gum	gummifer	D-galactose, D-galacturonic acid, L-arabinose, L-rhamnose, D-xylose	Non-fat food formulations
Indian tragacanth gum	Cochlospermum gossypium	Rhamnogalacturonan	Emulsifier in food, scaffold for bandages, matrix for tablets
Gum (Leaves)			
Mucilage of broom creeper	*C. hirsutus*	gelatinous materials	Treatment of skin diseases, fever, sedative
China rose mucilage	*H. rosasinensis*	D-glucuronic acid, L-rhamnose, Dgalactouronic acid, D-galactose	Cosmetic, medicinal uses
Microbes gum			
Curdlan gum	*Agrobacterium* spp.	Glucose	Meat free products, bakery, gelling action
Gellan gum	*Sphingomonas* spp.	Glucuronate, rhamnose, glucose	Food additive, hydrogel
Xanthan gum	*Xanthomonas* spp.	glucuronic acid, D-mannose, D-glucose	Encapsulation, texture modifier
Gums (Tuber)			
Konjac glucomannan	*A. konjac*	D -mannose, D-Glucose	Controlled drug release, gelling effect
Taro	*C. Esculenta*	Arabinose, galactose	Gelling action

Industrial Application

Food Additive

Gond has a low calorific value; therefore, it is considered a suitable option for food fortification, e.g., Gond helps in preserving creamy texture and replaces fat in the dairy alternatives industry, providing us with healthier alternatives Table 4.3 [11].

Emulsifier

Gond can be used as an exceptional emulsifying agent and hence can be used as an additive in sauces, dressings, etc. (Table 4.3).

Binding Agent

Gond can be used as a good binding agent. With this property of keeping all ingredients together without making it messy, Gond extends its applications in various snacks and cereals industries and also improves the shelf life of snacks naturally [11, 12].

Pet Food

The texturing and water retention properties of Gond have immense applications in various formulations of pet food.

Dietary Fiber

The intestinal enzymes do not digest Gond completely, so it gets fermented in the colon and hence is used as dietary fibre. Research studies have shown its use in the production of jack fruit puree. Further, Gondh can be used as a surface response agent as it provides stability to water in oil emulsion. Various compositions of Gond are used in water emulsions for stability, which is very useful for the beverage industry, for example, to provide flavor, color, and turbidity to soft drinks [11–13].

Vegetable Shelf-Life Enhancer

Fruits and vegetables are perishable, which is the major concern of the food industry. To provide freshness, quality and prolong the shelf life and prevent ripening, an edible coating of Gond can be used instead of wax coating. In this way, under controlled temperature storage life can be extended for commercial use [13, 14].

Microencapsulation

Various researchers in their studies have stated that Gond provides stability in the form of microcapsule sprays and is superior to other raisins. To retain the flavor profile and protection against sensitivity to heat, light, and oxygen in dried strawberry powder. The spray drying of Gond, maltodextrin, and modified starch acts as an encapsulator and stabilizer [14].

Print Media and Textile

The binding and stabilizing property of gond is used in the textile industry to control ink in the dyeing industry.

Cosmetic Industry

The emulsification and stabilizing property of gond is used in creams to provide a more natural and organic replacement for harsh chemicals.

Paint Industry

The worldwide use of gond as a layer forming agent is used in paints to replace expensive and toxic agents.

Applications in the Health and Pharmacology Industry

Gut Health

Gond is a non-digestible polymer. This fiber property of Gond benefits in maintaining intestinal health overall. It is fermented in the gut, which helps in the production of good bacteria and formation of fatty acids (short-chain) in the gut. Along with this Gond significantly increases intestinal excretion of calcium and magnesium.

Obesity Drugs

In a few studies, when 40gm of Gond dissolved in water was given to subjects, in a matter of 3 hours a reduced intake of 200 Kcals was observed. This study shows that the principle can be used in the preparation of anti-obesity drugs, as the fiber and bulking agent property of Gond can also be applied in weight-loss diets.

Antimicrobial Drug Trials

Nishi *et al.* investigated the role of Gond as an antimalarial agent and the study showed that it significantly reduced the vigor of malaria in labs. Along with that, for centuries, Gond has been known as an antimicrobial agent. In recent studies, the oxidized Gond was tested in amphotericin B therapy against fungal infections, and the drug conjugate was found to be stable and nontoxic. Gond promises an immense potential for future application in drug development and the pharma industry [14, 15].

Application in Dental Erosion

Gond showcases a tendency to form a protective layer over enamel of teeth which protects from erosion by acids. Not just that, Gond has a high concentration of calcium assisting in the remineralization of teeth.

Anticoagulant

Hadi *et al.* conducted a study that shows the anticoagulant effect of Gond. In a lab, rats were treated with 10gm/100ml and 6gm/100ml, respectively. The rats with 10gm Gond solution demonstrated significant improvement in prothrombin time, which is a crucial blood clotting factor. Hence, it can be tried as an anticoagulant agent in mild cases.

Anti-Inflammatory Agent

The soothing effect of gondh is tested against time in various studies. The gond has been proven as an anti-inflammatory agent and reduces DNA damage and oxidative stress. From the time of ayush practices, Gond has been considered soothing to the body and a balancing factor for vatta, pitta and kapha.

Smart Hydrogel

Hydrogels have hydrophilic properties and gond fulfills the criteria to become a modified hydrogel which has an immense role in tissue engineering. These novel hydrogels can be applied as a biomaterial as well.

Drug Delivery Agent

Various researches are being conducted and it is found that non-toxic drug excipients are required. Gond is an ideal option. It has been tried as a pellet to deliver ferrous sulphate in anemia treatments and found to be efficient. It can be used as a coating agent as well as hydrogel to deliver drugs to the body.

Tumor Marking

Tumor imaging is a tedious and money-draining process. Agents are required which are stable and can be more sensitive to tissue for cell selectivity. Gondh has been tested for imaging as a copolymer in studies conducted in China and shows good uptake by specific cells which can help in tumor imaging. It could be a future application of gond.

Nanotechnology

Along with its application in molecular imaging, gond acts as a redox agent, which is the key application in molecular diagnostics and the functioning of nanoparticles. This property of gondh opens up immense potential in the technology area where gond has applications.

Human Health and Gond

According to Ayurveda, Gond boosts stamina and health, helping in keeping away sickness, cough, and cold. Gond contains many phytoconstituents such as Alkaloid, phenolic compounds, flavonoids, proteins, and amino acids, gum, and mucilage. These phytoconstituents are beneficial for human health and provide nutritive and healing properties to Gond.

Gond as a cooling agent

In summer, gond katira is mixed with water and consumed to keep the body temperature low. Hence, the consumption of gond katira in summer can prevent heatstroke. It is widely recommended by Ayurvedic doctors as a means to maintain the internal body temperature for optimum functioning of the vital organs in our body. Studies have found that it has the potential of acting as a hypotensive (blood pressure lowering) agent, which is likely to contribute to its cooling properties. Incidences of nose bleeding due to excessive heat can be prevented by using gond katira. Crystals of Gond are dried for making churans and kadhas for increasing stamina. Gond ladoos help in gaining weight in weak people and in people who are recovering from illness or some injury [16, 17].

Gond as a warming agent

Edible gum (*Gum acacia*) which is water-soluble is traditionally consumed in winter due to its warming property and calorific character.

Gond is good for digestion and gut health

Gond is highly beneficial to health owing to its digestive property. The usage of gond katira has been found to provide a purgative (cleansing the gut by regularizing bowel movements) effect. Certain enzymes like xylogalacturonan hydrolase present in tragacanth are responsible for promoting digestion. Gond, if included in a regular diet, can prevent constipation. Due to its rich fiber content and soluble fiber content, it acts as a prebiotic and is good for gut health. Acacia fiber is beneficial in improving heart health, diabetes, and IBS [16, 17].

Nutritive properties of Gond

Gond is considered to be very beneficial for pregnant women due to its nutritive properties. Gond fulfills the requirement of calcium and proteins in the body and increases vitamin D levels. Gond is beneficial for people who have problems with their lungs and eliminates weakness and fatigue. The practice of giving sweets containing gond is very common in India. It is believed to be a healthy snack and a good source of nourishment for the mother as well as the child. In India, it is a common practice to give gond laddus to pregnant and lactating women. Patients who are recovering from a disease or injury can also consume these preparations. Gond contains nutrients like protein, amino acids, and calcium; these nutrients are required by pregnant women. Post-pregnancy, gond helps mothers to regain their strength. Studies have shown that gond increases the production of milk in lactating (breastfeeding) mothers. Proteins are the main components of the cells, so they are very much important for human health [11, 13, 14].

Gond improves urinary functions and reproductive functions

The effect of gond against involuntary urination and urinary incontinence (lack of urinary control) is well known. It helps in soothing the muscles of the urinary tract in case of a blockage of the tract. Diabetes-induced urinary complications can also be reduced by the consumption of gond. It treats weakness and male fertility issues.

Gond is good for skin health

Gond katira is believed to have anti-aging properties. It is often mixed with other plant products like aloe vera and used as a face mask. The combination is also found to help promote wound healing. These face masks are very effective in reducing wrinkles and fine lines. It is used as an emollient in cosmetics. It has several beneficial nutrients and thus has an important role in skincare regimes. Gond works magically for skin health. Gond pastes are used for skin nourishment. The clinical trials for use of gond in skincare products are ongoing.

Gond has pain-relieving properties

Gond reduces back pain and other joint pains. It lubricates the joints. Phenols present in gond are known as anti-inflammatory agents. Various studies have been done to analyze the result of gond katira in reducing pain. It is indicated that tragacanth gum could be used as an effective analgesic (painkiller) when given in appropriate amounts. The gum acts upon certain components of the nervous system to reduce the sensation of pain. Further studies *in vivo* are, however, required to study the exact mechanism of how gond acts as an analgesic.

Gond has anticancer properties

Gum acacia contains natural antioxidants, It is effective against cardiac pathologies and is a potent anticancer agent. Gond has anti-aging properties. It decreases angiogenin and beta-catenin expression, which prevents the development of colonic cancer and inflammatory bowel diseases. It stimulates fetal hemoglobin production in sickle cell disease. Gum Arabic presents hepatic and renal protective effects among rheumatoid arthritis patients. Tragacanth gum is suggested to be anti-cancerous. The gum has been found to contain certain biochemicals that slow down the rate of cell division, a property effective against the development of tumors. Studies have established the potential of gond as an anti-cancer agent. Further clinical trials are, however, required to determine the use of gond katira in day-to-day life as a preventive measure against cancer.

Cardiovascular and renal health

Gondh is significantly rich in calcium which in turn helps in decreasing urinary excretion of phosphate and thus increases plasma vitamin D levels. A study showed that Gondh increases creatinine clearance, which in turn decreases daily urine output and reduces the excretion of sodium which favors renal insufficiency.

Gond is used in Ayurvedic preparations

Gond is used in various Ayurvedic preparations. It is found that it can enhance stamina and immunity. It is also found to be useful in treating male infertility. Edible gum is obtained mainly from Babool (keekar) in India. Babool has been shown to increase energy and stamina. Gond, a byproduct of Babool, is helpful for people having psychiatric problems, anxiety, depression, piles, and vitamin D deficiency. Gond boosts their immunity, increases stamina, and promotes health.

Gond is used in the nutraceutical and pharmaceutical preparations

Gond has been utilized as a thickening, jelling, suspending and binding agent in the food and pharmaceutical industry. Gond acts as an emulsifier, stabilizer, and texturing additive. Since edible gum is soluble in water, it is possible to mix it with a lot of ingredients. The gum has long been used for treating diarrhea, cough, and congestion. Edible gum gond is also used for its soothing agent in medicine preparations. Mucilage, which is present in gond, acts as a cohesive in the medicines and as a bulking agent in the tablets. Minerals present in Gond help in regularizing the physiological functions of the human body.

Gond used as powder

Tragacanth gum is extracted from the roots of the Astragalus plant species. The roots of the plant are tapped and the gum oozes out of them in flakes (thin sheets) or twisted ribbons. The gum is then allowed to dry and then crushed to make tragacanth powder. Gond katira is mostly available in the market in powdered form. Crushed Gond is useful as a cohesive. It is used in externally used creams and jellies. Gond is used as an emulsifying and stabilizing agent for preparing ice-cream in 0.2–0.3% concentration and it is also used in making various food products as sauces. Phenols present in gond act as anti-inflammatory agents and natural antioxidants. This is effective against cardiac problems and is a potent anticancer agent. These show anti-aging properties. Minerals in gond powder help in regularizing the physiological functions of the human body.

4.7 Other Uses of Gond

Apart from its health benefits, gond has a wide variety of uses in the food and leather industry. Its gum-like consistency has been explored in numerous ways in the science and research fields. some uses of gond that one should know are described below. Figure 4.3 and Figure 4.4 various ingredients preparation for gond based sweet preparations and various gond based sweets prepared respectively.

Gond as a fat substitute

In the food industry, continuous efforts are being made to replace fats with healthier substitutes. Gond is an effective substitute in several packaged or processed meat products such as sausage. Its gum-like consistency also

Figure 4.3 Procedure for making some of the traditional gond sweets.

Figure 4.4 Various types of sweets prepared from gond (a) Gond ke ladoo, (b) Gond panjiri, (c) Gond churma, (d) Gond peanut chikki.

makes it a good binding agent with processed meats and a more nutritive alternative to foods with higher fat content.

Gond against microbes

Gond has been suggested to be effective against various microbes. This gum has been studied by numerous researchers to evaluate its potential as an anti-microbial agent. In a lab study, the action of gond katira was found to be more effective against microbes than the well-known antibiotic ampicillin. *In vivo*, clinical trials are, however, required to test this property of gond.

Gond as an oral delivery system in diabetics

The Gond extract can be used for encapsulating (containing) insulin. Diabetic patients who are on insulin are required to inject it directly into their blood system. The glue-like property of tragacanth can be used to capture the required insulin dosage. Then, this preparation may be consumed in the form of a tablet. Gond holds a promising potential in this aspect and research studies are still ongoing to develop the required effective product.

Gond as a thickening agent

Gond has thickening properties and is relatively stable to heat and acidity. This makes it a suitable thickening agent in salad dressing, pickles, ketchup, and mayonnaise. The stabilizing property of gond also makes it a suitable preservative. Additionally, its anti-acidic property helps to enhance shelf-life (period when the product is safe for use) of these products. Gond is used as a gelling agent, thickening, binding agent, suspending matter, and as an emulsifying agent.

Care, Concern, and Precaution in Gond Consumptions

Not enough is known about the utilization of gond during pregnancy and for the lactating mother. Stay on the safe side and keep away from use.

Allergy

Gond (Tragacanth) can cause breathing issues in people who are delicate to quillaia allergy.

Conclusion

Gond can be used in both the food industry and pharmaceutical medicines. The natural gum – Gond (Tragacanth) has innumerable benefits, not only health-wise but in various aspects. This natural gum is safe to use, and its wonderful properties definitely can be used in so many ways besides the mentioned ones, such as Paper marbling in Turkey, as a binder in Incense making, pastels for artists, etc. It is already being utilized in countless ways by mankind but still we look forward to discovering more uses for it. Gond can be used in milk-based food items. Gond in ice-cream, water ices, pops, puddings, cheese, chocolate and milk drinks acts as a stabilizer. Gond or edible gum have various applications in confectionery food products, dressings, sauces, beverages and functional food items.

References

1. Meghwal, M., Kadeppagari, R. Dietary Fibers and their Role as Functional Food for Human Health: Food Fibers and Human Health. In S. Benjamin (Ed.),

Examining the Development, Regulation, and Consumption of Functional Foods, IGI Global, 29-44, 2017.
2. Dhingra D., Michael M., Rajput H., Patil R.T. Dietary fibre in foods: a review. *J Food Sci Technol.*, 49(3), 255-66, 2012.
3. Zandleven, J., Beldman, G., Bosveld, M., Benen, J., Voragen, A., Mode of action of xylogalacturonan hydrolase towards xylogalacturonan and xylogalacturonan oligosaccharides. *J. Biochem*, 387 (3), 719-25, 2005.
4. Lupton, J.R., Betteridge, V.A., Pijls, L.T.J. Codex final definition of dietary fibre: issues of implementation. *Quality Assurance and Safety of Crops & Foods*, 1(4), 206–212, 2009.
5. Mortensen A., Aguilar F., Crebelli R., Di Domenico A., Frutos M., Galtier P. Reevaluation of tragacanth (E 413) as a food additive. *EFSA J.*, 2017.
6. Phillips, G.O., Williams, P.A. Essential Chemistry for Formulators of Semisolid and Liquid Dosages. In *Handbook of hydrocolloids*, Woodhead Publishing, 43-69, 2021.
7. Nur, M., Vasiljevic, T. Insulin Inclusion into a Tragacanth Hydrogel: *An Oral Delivery System for Insulin. Materials (Basel).* 1, 79, 2018.
8. Mayhew, E. Roe, E.M.F. Changes in the Permeability of Landschütz Ascites Tumour Cells to Vital Stains after Treatment with Tumour-Inhibitory or modified Samples of Gum Tragacanth or with Gum Karaya. *Br J Cancer*, 18(3), 537–542, 1964.
9. Atashkar, M., Hojjatoleslamy, M., Boroujeni, L.S., The influence of fat substitution with κ-carrageenan, konjac, and tragacanth on the textural properties of low-fat sausage. *Food Sci Nutr.*, 6(4), 1015–1022, 2018.
10. Sharma, H., Sharma, B.D, Talukder, S. Giriprasad Ramasamy. Utilization of gum tragacanth as bind enhancing agent in extended restructured mutton chops. *J Food Sci Technol.*; 52(3), 1626–1633., 2015.
11. Anderson, N.E., Clydesdale, F.M. Effects of processing on the dietary fiber content of wheat bran, pureed green beans and carrots. *Jr of Food Science*, 45(6), 1533–1537, 1980.
12. Arrigoni, E., Caprez, A., Amado, R., Neukom, H. Chemical composition and physical properties of modified dietary fibre sources. *Food Hydrocolloids*, 1(1), 57–64, 1986.
13. Southgate D. The Dietary Fibre Hypothesis: A Historical Perspective. *Dietary Fibre-A Component of Food*, 3-20, 1992.
14. Bjorck, I., Nyman, M., Asp, N.G. Extrusion-cooking and dietary fiber. I. Effects on dietary fiber content and on degradation in the rat intestinal tract. *Cereal Chemistry*, 61, 74-179, 1984.
15. Ralet M.C., Thibault J.F., Della, V.G., Solubilisation of sugar beet cell wall polysaccharides by extrusion-cooking. Òmtreatment on the dietary fibre contents of potato and tomato. *J Food Technol*, 19:485-492, 1991.

16. Ranganna, S., Handbook of Analysis and Quality control for fruit and vegetable project., Tata McGraw-Hill, 1995.
17. Prajapati V, Jani G, Moradiya N, Randeria N. Pharmaceutical applications of various natural gums, mucilages and their modified forms. *Carbohydrate Polymers.* 92(2), 1685-1699, 2013.

5

Omega-3 Fatty Acids: Nutritional Aspects and Their Role in Health and Diseases

Shriya Bhatt[1,2], Rashim Kumari[1,2], Deepika[3], Rajni Chopra[4], Tejpal Dhewa[3] and Anita Kumari[3]*

[1]Dietetics and Nutrition Division, CSIR-Institute of Himalayan Bioresource Technology, Palampur, Himachal Pradesh, India
[2]Academy of Scientific and Innovative Research (AcSIR), Ghaziabad, Uttar Pradesh, India
[3]Department of Nutrition Biology, School of Interdisciplinary and Applied Sciences, Central University of Haryana, Mahendergarh, Haryana, India
[4]Department of Food Science and Technology, National Institute of Food Technology Entrepreneurship and Management, Sonipat, Haryana, India

Abstract

Omega-3 fatty acids are polyunsaturated fatty acids (PUFA) and are designated by the presence of a number of carbon atoms and double bonds. Omega-3 fatty acids are also referred to as "n-3". Most PUFAs have more than 18 carbon atoms, no. of double bonds, and are consequently defined as long-chain PUFAs. The most common omega-3 fatty acids are α-linolenic acid (ALA), docosahexaenoic acid (DHA), and eicosapentaenoic acid (EPA). Omega-3 is found naturally in various foods and is also added to some foods which are then known as fortified foods. The main sources of omega-3 fatty acids are marine (fish and other seafood like bacteria, fungi, algae, etc.), plant sources (flaxseed oil, soybean oil, and canola oil), nuts and seeds (flaxseed, chia seeds, and walnuts), fortified foods (eggs, yogurt, juices, milk, soy beverages, and infant formulas) as well as dietary supplements (fish oil). ALA is mostly found in plant sources while DHA and EPA are found in fish and other seafood. Several studies have reported that consumption of omega-3 fatty acids – eicospaentaenoic acid (EPA), docosapentaenoic acid (DPA), and docosahexaenoic acid (DHA) is helpful in providing protection against cardiovascular diseases (heart disease, cancer, hypertension, and arthritis, etc.). Several clinical trials also revealed that higher consumption of n-3 fatty acids may be toxic.

**Corresponding author*: writetodranita@gmail.com

Tejpal Dhewa, Anil Kumar Puniya and Anil Panghal (eds.) Nutritional Science and Technology: Concept to Application, (103–116) © 2023 Scrivener Publishing LLC

This chapter will give insights related to nutritional aspects, intake, and safety of omega-3 fatty acids along with their role in the management of diseases.

***Keywords*:** Omega-3 fatty acids, nutrition, food sources, safety levels, health benefits

5.1 Introduction

In the past few decades, a substantial amount of research has been carried out in determining the heath-associated benefits of polyunsaturated long-chain omega-3 fatty acids (n-3 PUFA). Omega-3 fatty acids are the group of heterogeneous fatty acids encompassed of double bonds from the methyl end between third and the fourth carbon [1]. The associated health benefits include a reduction in risk of various cardiovascular, neuropsychiatric, neurodegenerative, and other anti-inflammatory diseases including different types of cancer [2]. Various researchers have also revealed the effective role of n-3 PUFA as a major player in modifying the blood lipid profile and its composition, affecting the biosynthesis of eicosanoid, gene expression, and cell signaling mechanism thus influencing the overall health. In addition, various diseases such as cardiovascular disease, cancer, diabetes and age-related mental weakening have also been investigated for the role of n-3 PUFA in the management of these diseases. The n-3 PUFA is regarded as the essential fatty acid (FA) due to the inability of humans to synthesize it naturally; it consequently needs supplementation from external sources for appropriate human functioning. The human body is deficient in desaturase enzyme which is indispensable for the synthesis of α-linolenic acid (ALA) and linolenic acid (LA). The polyunsaturated fatty acids n-3 PUFA encompasses α-linolenic acid (ALA), along with stearidonic acid (SDA), docosahexaenoic acid (DHA), eicosapentaenoic acid (EPA), and docosapentaenoic acid (DPA). Amongst all, the DHA and EPA comprise the critically essential FA obligatory for normal human development. They must be either acquired from diet or synthesized in the body through 18-carbon atom long n-3 PUFA such as ALA obtained from various sources such as walnuts, flaxseed, and canola oil [3]. However, the synthesis rate is limited in the body due to the low activity of Δ6-desaturase enzyme involved in various metabolic pathways. Thereby, it becomes essential to identify the sources for maximum utilization in day-to-day life.

There are numerous plant and animal sources of n-3 PUFA, where oil encompassing these essential fatty acids originates mainly from the plant, animal, and other single-cell sources. Long-chain n-3 PUFA as DHA and EPA are particularly found in fish and sold as n-3 PUFA supplements. However, fungi, algae, and other single-cell entities have recently become quite popular for their novelty and most importantly renewable source for n-3 PUFA. Additionally, soy and various other genetically modified plants are known for the high content of n-3 PUFA. Various plant species, namely *Brassica* and flax, have been genetically modified by incorporating EPA and DHA. Though aquatic organisms are known as the reservoir of n-3 PUFA, some seeds from plant sources also contain n-3 PUFA ineffective amounts, namely chia, canola, and flax seeds as a competent source of ALA [4]. Since production is limited in the body, it becomes essential to supplement it from the outside for proper maintenance of physiological functions in the body when supplemented in proper amount.

Though the particular intake for all n-3 PUFA essential for proper maintenance of physiological conditions is not yet known, in various studies it has been reported for higher intake than the optimal recommendation [5]. The European Food Safety Agency (EFSA) had anticipated the dietary intake value of 0.25g of DHA and EPA, which is in accordance with the recommendation by the Food and Agriculture Organization (FAO) being negatively interrelated to development of cardiovascular disease [6]. In addition, the n-3 PUFA, particularly DHA, has been explored for its role in the development of the fetus and the functioning of brain and eyes in infants [4]. Conversely, various reports have also highlighted the negative impact of increased uptake of n-3 PUFA, in induction or exacerbation of various neoplastic pathology [2]. According to the US Food and Drug Administration, consumption of EPA and DHA combined should not be more than 3g/day inclusive of 2g/day of supplements. Thus, it is essential to understand the role of n-3 PUFA in maintaining the normal physiological condition of the body along with recommended intake either from natural sources or as supplements.

Thus n-3 PUFA exhibits specific functional or dietary properties, making it a nutraceutical.

5.2 Nomenclature and Types of Omega-3 Fatty Acids

The FA typically is comprised of 16-25 long carbon atom chain. The n-3 term is the structural representation for the PUFA family, where n-3 denotes the position of closest double bond from the methyl end amid long FA chain. Among all n-3, the double bond is allocated on the third carbon, when methyl carbon is the first number [7]. The n-3 PUFA possesses two or more double bonds termed according to entire chain length and position of bonds such as DHA (22:6) is n-3 PUFA comprising carbon atoms (22) with double bonds (6). The simplest n-3 FA is α-linolenic acid, i.e., 18:3. It is produced by D15-desaturase from precursor linoleic acid (18:2). Thereby, α-linolenic acid is characteristically an essential FA in addition to linoleic acid, synthesized by plants. However, animals are unable to synthesize α-linolenic acid, metabolizing to various long-chain unsaturated n-3 FA. It includes an arrangement of associated reactions (elongation and desaturation) processing in the liver. The ALA through D6-desaturase can be converted to SDA besides elongation of SDA to eicosatetraenoic acid. Subsequently, the enzyme D5-desaturase yields EPA via desaturation. The ALA afterwards gets converted into EPA along with transformation of ALA into arachidonic acid due to the utilization of the same type of enzymes. EPA via DPA3 (22:5) can be converted further to DHA (22:6). This conversion is due to the D6-Desaturaseenzyme involving limited beta-oxidation in peroxisomes. The transformation of ALA into DPA, DHA and EPA has earlier been studied through diverse approaches which explicate the restricted conversion, particularly for DHA. The conversion of stearidonic acid into EPA is better when compared with ALA, as their conversion does not apply the utilization of D6-Desaturase enzyme. The process of limited beta-oxidation in peroxisomes of DHA generating EPA and DPA is termed as retro-conversion [7]. With the increase in the degree of unsaturation in FA, there is a simultaneous decrease in melting point conferring to the fluidity on n-3 PUFA [8]. Thus, n-3 PUFA is comprised of ALA, EPA, SDA, DPA, and DHA. ALA is the main FA which undertakes various desaturation and elongation reactions to synthesize DHA or EPA. The human metabolic system is incompetent to synthesize all types of n-3 PUFA due to the absence of enzyme-involved insertion of a cis double bond.

5.3 Food Sources and Supplements

The n-3 PUFA are mostly present in marine organisms originating primarily from the liver of white lean fish, namely halibut and cod, oily fish such as salmon, menhaden, and mackerel, as well as fat from other mammals, namely whales and seals [4]. Fascinatingly, various marine cold-water fishes are enriched with n-3 PUFA, as most of these organisms feed on various zooplankton and phytoplankton, which are the richest source of n-3 PUFA. Thereby, the oil, namely krill oil acquired from Antarctica krill, contains a high amount of n-3 PUFA in the form of PL, has turned out to be the most prevalent form of n-3 PUFA supplement. In addition to supplements from food, nowadays various new conventional forms of foods have been explored including some formulas for infant feeding fortified using various fish DHA and microalgae [9]. The foremost type of n-3 PUFA obtained from aquatic sources includes DHA and EPA, with DHA present in low amounts in almost all oils obtained from the fish source (Table 5.1). Conversely, ALA plants are the prime source with the maximum amount acquired from nuts and some seeds, namely chia seed, flaxseed, echium seed, and walnut along with oil from some vegetable

Table 5.1 Food sources and supplements of omega-3 fatty acids.

Food source	n-3 PUFA			References
	DHA (%)	EPA (%)	ALA (%)	
Fish oil				
Herring oil	6.8	7.5	-	[4]
Cod liver oil	12.7	12.2	-	
Menhaden oil	9.6	18.3	-	
Butterfish oil	10.8	5.1	-	
Cod flesh oil	32.6	19.1	-	
Skipjack tuna oil	29.1	11.1	-	
Capelin oil	4.1	9.3	-	
Halibut oil	30.6	9.6	-	
Winter founder oil	20.1	14.4	-	
Mackerel oil	19.3	8	-	
Yellowtail founder oil	18.7	15	-	
Salmon oil	9.1	6.2	-	
Haddock oil	24.8	14.8	-	

(*Continued*)

Table 5.1 Food sources and supplements of omega-3 fatty acids. (*Continued*)

Food source	n-3 PUFA DHA (%)	EPA (%)	ALA (%)	References
Fish (raw)				
Trout	0.5	0.15	-	[4]
Herring	1.01	1.09	-	
Salmon	1.19	0.89	-	
Sardine	1.16	0.51	-	
Cod	0.62	0.02	-	
Marine mammals				
Harp seal oil	7.58	6.41	-	[4]
Grey seal oil	7.12	5.23	-	
Bearded seal oil	13.38	9.27	-	
Ringed seal oil	26.19	10.59	-	
Hooded seal oil	7.47	4.29	-	
Harbor seal oil	7.76	9.31	-	
Bivalves				
Blue mussel	13.2	19.6	-	[4]
Greenland cockle	16.5	22.6	-	
Icelandic scallop	25.9	26.9	-	
Surf calm	14.3	22.9	-	
Cephalopods				
Squid	16.9	13.9	-	[4]
Common octopus	20.6	16.1	-	
European squid	31.6	14.3	-	

(*Continued*)

Table 5.1 Food sources and supplements of omega-3 fatty acids. (*Continued*)

Food source	n-3 PUFA			References
	DHA (%)	EPA (%)	ALA (%)	
Green vegetables (raw)				
Green broccoli	-	-	0.11	[9]
Brussels sprouts	-	-	0.15	
Lettuce	-	-	0.17	
Meat				
Pork, fat	-	-	-	[9]
Beef, veal without fat	-	-	0.08	
Lamb, lean meat	-	-	0.11	
Seeds (dried)				
Almond	-	-	0.3	[9]
Walnut	-	-	6.64	
Chia	-	-	17.83	
Hazelnuts	-	-	0.11	
Fungi				
Mortierella	-	20.0	-	[10]
Pythium	-	12.0	-	
Thraustochytriumaureum	-	62.9	-	
Pythium irregulare	-	8.2	-	
Microalgae				
Nannochloropsisoceanica	-	23.4	-	[10]
***Thraustochytrium* sp.**	-	45.1	-	
***Nannochloropsis* sp.**	-	26.7	-	
Chlorella minutissima	-	39.9	-	

(*Continued*)

Table 5.1 Food sources and supplements of omega-3 fatty acids. (*Continued*)

Food source	n-3 PUFA			References
	DHA (%)	EPA (%)	ALA (%)	
Nannochloropsissalina	-	28.0	-	
Pinguiococcuspyrenoidosus	-	22.03	-	
Isocrysisgalbana	-	28.0	-	
Pavlova viridis	-	36.0	-	
Dunaliellasalina	-	21.4	-	
Pavlova lutheri	-	27.7	-	
Transgenic plants				
Brassica carinata	-	25.0	-	[10]
Soybean	-	20.0	-	
Nicotianabenthamiana	-	26.0	-	

sources [4]. The highest content of ALA, i.e., 49.2 g/100g is present in flaxseed oil including other sources such as canola oil, soybean oil, and walnuts. Thus, primarily ALA is attained from walnuts, green leafy vegetables, soybean, canola oils, and flaxseed. The EPA and DHA are the derivatives of ALA with a minimum amount of conversion and thus must be acquired from external sources such as breast milk and oils from different fish such as mackerel, salmon, anchovies, sardines, rainbow trout, and herring along with algae [9]. Additionally, the oil acquired from fish is one of the richest vitamins sources (fat-soluble). The liver obtained from haddocks, whales, cod, tuna, sharks, and halibut are the major supplement oil rich in vitamin A and D. Vitamin A content in cod liver oil is 1000 IU/g and vitamin D is 10 IU/g. However, oils obtained from various fish, namely halibut, skipjack tuna, and cod flesh exhibited the maximum amount of DHA, although haddock, cod flesh, and flounder species contain the maximum amount of EPA [4]. Apart from fish and various aquatic mammals, bivalves, cephalopods, and crustaceans are also known to contain n-3 PUFA.

The microalgae and various other fungi are the extran-3 PUFA source. Microalgae are the prominent producers of n-3 PUFA, e.g., DHA with

Schizochytrium and *Crypthecodiniumcohnii* species as the prominent algae source of DHA with total fatty acids levels of 40% and 55%, respectively [4]. The major n-3 PUFA, i.e., DHA and EPA, are produced by phytoplankton and algae, which in turn gets relocated from the web to aquatic system. The seeds acquired from the Boraginaceae family, namely Echium vulgare, Buglossoidesarvensis, and borage, including fish and hemp oil are efficient sources of SDA. Though SDA is not the chief constituent of diet, it has gained recent attention as supplementation of SDA enhances the EPA level more compared to ALA [4].

5.4 Intake and Safety of Fatty Acids

ALA, the basic n-3 PUFA is the antecedent molecule for the production of EPA and DHA. The ALA is measured as an indispensable FA as it is unable to get synthesized and therefore must be supplemented from external sources. The evidence which proved that ALA was crucial in human development was the reversing of abnormal neurological signs by supplementation of ALA to a six-year-old girl suffering from visual difficulties and sensory loss [2]. In subsequent consumption, mostly the ALA gets catabolized for energy generation through β-oxidation, with the small amount that undergoes the conversion for the synthesis of EPA and DHA. Thereby, according to the World Health Organization (WHO), the dietary intake of n-3 PUFA for adults must be within 0.5-2%. However, in the case of non-pregnant/non-lactating and adult males, the amount of EPA+DHA should be 0.25 g/day, with inadequate evidence for setting the minimum intake for DHA and EPA alone [11]. The recommended intake in accordance with the Food and Agriculture Organization (FAO) and on a daily basis for n-3 PUFA, especially EPA+DHA, is 100 and 250 mg. Children up to the age of 10 years must take in 100 and 250 mg of EPA+DHA, pregnant women up to 300 mg, and healthy adults up to 250 mg (with at least 200 mg of DHA) [12]. The EFSA has also estimated the recommended intake of 0.25 g of EPA and DHA on a daily basis [6]. In contrast, various reports in the literature have highlighted the adverse effects of increased uptake of n-3 PUFA. As reported by Weylandt *et al.* [2] induction or exacerbation of various neoplastic pathology was observed in case of increased supplementation of n-3 PUFA. In addition to this, various researchers also reported the enhancement in risk of developing various cancers [13]. The increased supplementation of cod liver oil along with polyunsaturated fatty acids was significantly associated with an increased risk of developing melanoma among women. Numerous studies have also recommended that augmented uptake of DHA might be the reason for an enhanced risk of prostate cancer. Though the stated correlations do not provide any decisive evidence about the consumption of fish oil as a supplement or whether fish whole can cause prostate cancer. However, supplements of n-3 PUFA may contain various other components such as vitamin E or other oxidative ingredients which might be associated with carcinogenesis [13]. Thereby, the potential adverse effects have been related to the supplements used as a source of n-3 PUFA, leading to serious health implications. Another major concern with n-3 PUFA which is imparting negative effects is the intake of oxidized lipids. As during storage, the n-3 PUFA gets easily oxidized, as a result, the n-3 PUFA supplements containing lipid peroxides and other secondary oxidative compounds get oxidized, whereas the level of other unoxidized FA progressively gets diminished. The oxidation of n-3 PUFA cannot be prevented, though it can be reduced with the addition of various anti-oxidants. Thereby, it is essential to identify such compounds that can delay the process of oxidation [13]. The oil present in n-3 PUFA supplements varies compared to the natural source as a supplement is a mixture of EPA, DHA, various other FA, and indefinite levels of possibly harmful lipid peroxides and other secondary oxidative compounds [13]. The health implications drawn are not sufficient enough to draw a decisive conclusion on its safety. However, it is essential to examine the health implications of oxidized n-3 PUFA along with the development of a safety limit for its consumption.

5.5 Health Benefits

Nutritional studies have revealed the impact of nutrient intake and deficiency in the prevention and development of diseased conditions, respectively. Recently, the role of omega-3 fatty acids has also been found useful in prevention of numerous illnesses and its deficiency has been found associated with advancements of numerous illnesses. Metabolic disorders associated with lifestyles as well as mental health illnesses are found to be influenced positively with the consumption of this beneficial nutrient. The brief of the health illnesses and the related beneficial health benefits on consumption of omega-3 fatty acids has been discussed in the sub-sections below.

5.5.1 Omega-3 Fatty Acid and CVDs

Lifestyle and dietary habits have been identified as a major reason for the development of heart-related disorders. The associated mechanism of these diseases is found to be impacted negatively with the intervention of omega-3 fatty acids. The consumption of fish oil or whole cooked fish providing rich supplementation of omega-3 fatty acids is also stated to be beneficial in coronary heart diseases [14]. The Food and Drug Administration (FDA) in 2004 suggested EPA, i.e., eicosapentaenoic acid and DHA, i.e., docosahexaenoic acid, omega-3 fatty acids consumption in diet for prevention of coronary heart diseases on the basis of supportive studies [15]. Although the defined amount has not been suggested yet for the consumption of these omega-3 fatty acids, the epidemiological studies supporting their consumption associated benefits and evidences from secondary prevention studies suggested the consumption range of 0.5-1.8 grams per day. This can be consumed in the form of dietary supplements or in the form of food such as fatty fish. Also, triglyceride level can also be reduced up to 20% to 40% by EPA+DHA supplementation. Reduced risk of arrhythmias and thrombosis is also associated with the consumption of these omega-3 fatty acids, thereby, reducing the associated effects of heart attacks and other CVDs (cardiovascular diseases) [16].

5.5.2 Omega-3 Fatty Acid and Mental Illnesses

As per studies, Omega-3 PUFA (Polyunsaturated fatty acids) is very beneficial for anxiety and mood disorders. DHA, a kind of omega-3 fatty acid, is associated with communication mechanism of brain neurons. They are also associated with modulation of these mechanisms. Studies based on animal models have also suggested that omega-3 fatty acid supplementation is associated with the upliftment of mood and controlling anxiety. Symptoms of another psychiatric disorder, ADHD (attention deficit hyperactivity disorder) are also observed associated in children with lack of certain fatty acid in blood plasma and membrane of erythrocytes [17]. In the PUFA replacement therapy, EPA is found associated with reduced impact of depression. Hippocampal neurogenesis responsible for fear memory is also found to be influenced positively with omega-3 fatty acid supplementation, thereby, providing relief from Post-traumatic stress disorder (PTSD). Though clear evidences are yet to be found on why and how much supplementation of PUFAs is essential for curing and preventing mental illnesses, the associated inverse relationship of PUFAs levels and mental illnesses clearly indicates the importance of consumption and supplementation of these omega-3 fatty acids in psychiatric disorders [18].

5.5.3 Omega-3 Fatty Acids and Diabetes

n-3 PUFA from both seafood and plant sources accounts for improvement in certain health disorders/illnesses. Although the reduction in triglycerides and VLDL level is found associated with PUFA supplementation, its impact on glycemic value is as yet unclear [19]. Improvement in *in-vivo* insulin sensitivity has been observed in a study by dietary supplementation of omega-3 fatty acids among non-insulin-dependent diabetic (NIDD) patients [20]. Though direct effect of omega-3 fatty acid supplementation in diabetes is not clear, the relief from associated health conditions of diabetes mellitus patients like reduction in blood pressure, enhanced platelet function and eicosanoid formation has been observed [21].

5.5.4 Omega-3 Fatty Acids and Cancer

The potential to exert anti-inflammatory activity of omega-3 fatty acids led to the investigation for their role in cancer illness improvement. During the process of active chemotherapy treatment, these fatty acids have been suggested to reduce inflammatory response and preserve muscle mass [22]. The modification in associated risk factors of breast and colon cancer have been claimed by dietary omega-3 fatty acid supplementation [22]. Although the direct impact of PUFA supplementation for cancer prevention is unlikely, the omega-3 fatty acids, arginine, and RNA supplementation in combination are found associated with reduced post-operational risk and reduction in hospital stay tenure for tumour patients [23]. According to recent studies, EPA and DHA are suggested to possess cancer protective effect. However, the underlying mechanisms to support these facts and observations are yet to be unveiled.

5.6 Conclusion

The n-3 PUFA are the active polyunsaturated fatty acids called as omega 3-FA. The foremost member in the family of omega 3-FA is ALA that is further convertible to various biologically active n-3 PUFA, namely DHA and EPA. For their production in human beings a series of reactions (desaturation and elongation) occur synthesizing EPA and DHA in a restricted amount. Thereby, it becomes necessary to supplement them from outside either from some food sources or through various supplements. The superlative sources of n-3 PUFA which includes ALA, EPA, and DHA, are aquatic (fish and other seafood like bacteria, fungi, algae, etc.), along with plant sources (flaxseed oil, soybean oil, and canola oil), nuts and seeds (flaxseed, chia seeds, and walnuts), fortified foods (eggs, yogurt, juices, milk, soy beverages, and infant formulas) as well as dietary supplements (fish oil). The uptake of n-3 PUFA either in the form of supplements or as a natural source should be taken within prescribed limits as higher doses are related to serious health implications in contradiction of its associated health benefits.

References

1. Cholewski, M., Tomczykowa, M. and Tomczyk, M., A comprehensive review of chemistry, sources and bioavailability of omega-3 fatty acids. *Nutrients, 10*, 11, 1662, 2018.
2. Weylandt, K.H., Serini, S., Chen, Y.Q., Su, H.M., Lim, K., Cittadini, A. and Calviello, G., Omega-3 polyunsaturated fatty acids: the way forward in times of mixed evidence. *BioMed Research International*. 2015, 143109, 2015. doi: 10.1155/2015/143109.
3. Umhau, J.C. and Dauphinais, K. M., Omega-3 polyunsaturated fatty acids and health. In *Low-Cost Approaches to Promote Physical and Mental Health*, 87-101, 2007. Springer, New York, NY.
4. Shahidi, F. and Ambigaipalan, P., Omega-3 polyunsaturated fatty acids and their health benefits. *Annual Review of Food Science and Technology*, 9, 345-381, 2018.
5. Troesch, B., Eggersdorfer, M., Laviano, A., Rolland, Y., Smith, A.D., Warnke, I., Weimann, A. and Calder, P. C., Expert opinion on benefits of long-chain omega-3 fatty acids (DHA and EPA) in aging and clinical nutrition. *Nutrients, 12*, 9, 2555, 2020.
6. Tur, J.A., Bibiloni, M. M., Sureda, A. and Pons, A., Dietary sources of omega 3 fatty acids: public health risks and benefits. *British Journal of Nutrition, 107*, S2, S23-S52, 2012.
7. Calder, P. C., Mechanisms of action of (n-3) fatty acids. *Journal of Nnutrition, 142*, 3, 592S-599S, 2012.
8. Ruxton, C.H.S., Reed, S.C., Simpson, M.J.A., and Millington, K. J., The health benefits of omega-3 polyunsaturated fatty acids: a review of the evidence. *Journal of Human Nutrition and Dietetics, 17*, 5, 449-459, 2004.
9. Balić, A., Vlašić, D., Žužul, K., Marinović, B. and BukvićMokos, Z., Omega-3 versus omega-6 polyunsaturated fatty acids in the prevention and treatment of inflammatory skin diseases. *International Journal of Molecular Sciences, 21*, 3, 741, 2020.

10. Adarme-Vega, T. C., Lim, D. K., Timmins, M., Vernen, F., Li, Y., & Schenk, P. M. (2012). Microalgal biofactories: a promising approach towards sustainable omega-3 fatty acid production. *Microbial Cell Factories, 11*(1), 1-10.
11. Alabdulkarim, B., Bakeet, Z.A. N. and Arzoo, S., Role of some functional lipids in preventing diseases and promoting health. *Journal of King Saud University-Science, 24*, 4, 319-329, 2012.
12. Rincón-Cervera, M.Á., González-Barriga, V., Romero, J., Rojas, R. and López-Arana, S., Quantification and distribution of omega-3 fatty acids in South Pacific fish and shellfish species. *Foods*, 9, 2, 233, 2020.
13. Lange, K.W., Nakamura, Y., Gosslau, A. M. and Li, S., Are there serious adverse effects of omega-3 polyunsaturated fatty acid supplements? *Journal of Food Bioactives, 7*, 2019. https://doi.org/10.31665/JFB.2019.7192
14. Din, J. N., Newby, D. E., & Flapan, A. D., Omega 3 fatty acids and cardiovascular disease—fishing for a natural treatment. *BMJ, 328*, 7430, 30-35, 2004. doi: https://doi.org/10.1136/bmj.328.7430.30
15. Bowen, K. J., Harris, W. S., & Kris-Etherton, P. M., Omega-3 fatty acids and cardiovascular disease: are there benefits?. *Current Treatment Options in Cardiovascular Medicine, 18*, 11, 1-16, 2016.
16. Kris-Etherton, P. M., Harris, W. S., & Appel, L. J., Omega-3 fatty acids and cardiovascular disease: new recommendations from the American Heart Association. *Arteriosclerosis, Thrombosis, and Vascular Biology, 23*, 2, 151-152, 2003.
17. Ross, B. M., Seguin, J., & Sieswerda, L. E., Omega-3 fatty acids as treatments for mental illness: which disorder and which fatty acid?. *Lipids in Health and Disease, 6*, 1, 1-19, 2007.
18. Yonezawa, K., Kusumoto, Y., Kanchi, N., Kinoshita, H., Kanegae, S., Yamaguchi, N., & Ozawa, H., Recent trends in mental illness and omega-3 fatty acids. *Journal of Neural Transmission, 127*, 11, 1491-1499, 2020.
19. Hartweg, J., Perera, R., Montori, V. M., Dinneen, S. F., Neil, A. H., & Farmer, A. J., Omega-3 polyunsaturated fatty acids (PUFA) for type 2 diabetes mellitus. *Cochrane Database of Systematic Reviews*, 1, 2008. https://doi.org/10.1002/14651858.CD003205.pub2
20. Popp-Snijders, C., Schouten, J. A., Heine, R. J., Van der Meer, J., & Van der Veen, E. A., Dietary supplementation of omega-3 polyunsaturated fatty acids improves insulin sensitivity in non-insulin-dependent diabetes. *Diabetes Research* (Edinburgh, Scotland), *4*(3), 141-147, 1987.
21. Axelrod, L., Omega-3 fatty acids in diabetes mellitus: Gift from the sea? *Diabetes, 38*, 5, 539-543, 1989.
22. Laviano, A., Rianda, S., Molfino, A., & Fanelli, F. R., Omega-3 fatty acids in cancer. *Current Opinion in Clinical Nutrition & Metabolic Care, 16*, 2, 156-161, 2013.
23. MacLean, C. H., Newberry, S. J., Mojica, W. A., Issa, A., Khanna, P., Lim, Y. W., ... & Shekelle, P. G., Effects of omega-3 fatty acids on cancer. *Evidence Report/Technology Assessment (Summary)*, 113, 1-4, 2005.

6

Role of Fermented Dairy Products in Enhancing the Immunity

Vaishali Dasriya[1]*, Soniya Ranveer[1], Rudrakshi Bajaj[2], Aakash Sharma[3], Yugal Dasriya[4] and Harmeet Singh Dhillon[1]

[1]*Dairy Microbiology Division, ICAR-NDRI, Karnal, Haryana, India*
[2]*Dairy Microbiology Division, ICAR-NDRI, Karnal, Haryana, India*
[3]*Food and Animal Sciences, College of Agriculture, Tennessee University, Nashville, USA*
[4]*Jayshri Gayatri Food Products Pvt Ltd., Sehore, India*

Abstract
Fermented dairy products are known for possessing numerous beneficial properties such as antimicrobial, anticarcinogenic, anti-allergic, and immunomodulatory effects. This chapter focuses on the immunomodulatory properties of fermented dairy products, which serve as a means for delivering lactic acid bacteria and their metabolites. Probiotics and lactic acid bacteria have already been studied for their immunomodulatory properties, as they stimulate immune cells and modulate gut microbiota to provide numerous benefits. Fermented dairy products can modulate the immune system through immunoregulation, increased disease resistance, and immunoenhancement. For example, yogurt has been found to decrease proinflammatory cytokines such as IL-1B and increase anti-inflammatory cytokines such as IL-10, which can help alleviate the harsh environment in epithelial cells and improve symptoms of colitis. However, the immunomodulatory properties of fermented dairy products are strain-dependent and influenced by the growth medium used. Function foods such as fermented dairy products could be the future of population food as an immunomodulatory product.

***Keywords*:** Fermentation, dairy, immunomodulation, probiotics, microbiota, pathogen, disease

**Corresponding author*: vaishalidasriya95@gmail.com

Tejpal Dhewa, Anil Kumar Puniya and Anil Panghal (eds.) Nutritional Science and Technology: Concept to Application, (117–134) © 2023 Scrivener Publishing LLC

6.1 Introduction

Human GIT contains huge variety of microorganism including bacteria, protozoa, virus and protists, which can directly or indirectly influence our immune system. Healthy gut microbiota aids in maintenance and development of immunity of our body (Pahwa, & Sharan, 2022). Fermented dairy products are those in which milk sugar gets converted into acid and/or alcohol with help of bacteria. These probiotic strains of bacteria exhibit beneficial effect on host, enhance immune function. Healthy humans possess balance between bad and good bacteria. Western lifestyles, infection, stress condition, antibiotic therapy can proliferate more pathogenic bacteria and thus disturbing the above said balance (Di *et al.*, 2022).

The health enhancement attributes by fermented products were first attribute of Metchnikoff. The adverse effect of bacteria could be suppressed by ingestion of probiotic strains of lactic acid bacteria contains fermented dairy products (Kim *et al.*, 2021). It is now well established that balanced microflora is vital for optimum gut microflora could be balanced by using probiotics as a result of which there is shift from pathogenic/harmful bacteria to health promoting microorganisms such as *Lactobacilli*, *Lactococcus* and *Bifidobacteria* (Bedi *et al.*, 2022). The fermentation of dairy product produces various metabolites like lactic acid, acetic acid, peptides, short chain fatty acid, neurotransmitter, bacteriocin and exopolysaccharide (Zheng *et al.*, 2020). Gopal & Dhanasekaran (2021) found other lactic acid bacteria to exhibit property of probiotics for e.g., *Enterococcus faecalis*, *E. faecium*, *Sporolactobacillus inulinus*, also non-lactic microorganisms like *Saccharomyces cerevisiae*, *Propionibacterium freudenreichii* could be employed as probiotics and is especially used in pharmaceutical and animal products.

It has described whether yoghurt culture *S. thermophilus* sand *L. delbrueckii ssp. bulgaricus* should be considered as probiotic or not, but they failed to satisfy the criteria laid for probiotics, although they are known to be associated lactose digestion, immune enhancement but unable to survive at high acidic conditions of gut (Paul *et al.*, 2021). There were having issues of safety concerns with use of other genera, such as *Escherichia* and *Enterococcus*, which were considered and marketed as probiotics (Surowiak *et al.*, 2022). They are found in non-dairy products with potential of drug resistant nosocomial in infections. In various studies, it is found that, fermented dairy product contain metabolites and probiotic culture provide the abandons of health benefits to reduce cancer, diabetes, cardiovascular (Peluzio *et al.*, 2021) and neurological disorder (Lekchand Dasriya *et al.*, 2022). In the current book chapter highlight the therapeutic effect of fermented dairy product in immunomodulation.

6.2 Immune Enhancing Potential of Fermented Foods: Mechanism

The Fermented dairy products tend to improve both innate and adaptive immune systems of human. The immunomodulatory response tends to cause activation of systematic and secretory immune through co-ordination of microbiota, epithelial cells and immune cells (Nongonierma *et al.*, 2016). Fermentation of milk and milk products results in release of short peptide possessing activity of bioactive peptide. These bioactive peptides are associated with proliferation of lymphocyte, synthesis of antibody and production of cytokine (Patil *et al.*, 2020). (Siddique *et al.*, 2022) reported that consumption of yoghurt triggers cytokine production, which can thus help in boosting immune system. Lactoferrin, a glycoprotein present in milk and milk product, acts as antioxidant, immunomodulatory, antimicrobial compound and impart beneficial function in human body showing multifunctional property (Bukowska-Ośko *et al.*, 2022).

The intestinal microbiota is the primary source of microbial stimulation and can have both beneficial and harmful effects on human health. The interaction between lactic acid bacteria, probiotics in fermented products, and enterocytes can initiate immunomodulation, which is strain-specific and can result in the production of cytokines (Kober *et al.*, 2022). Peyer's patches contain M cells that are critical in the human gut since they transport macromolecules, antigens, inert particles, and microorganisms from the lumen to lymphoid tissues through adsorptive endocytosis (Zhang *et al.*, 2021). When antigenic molecules cross the intestinal barrier, they can stimulate the innate and immune systems of the body (Ma *et al.*, 2021).

The immunomodulatory effect of probiotics is related to the release of cytokines, including Tumor Necrosis Factors, interleukins, Interferons, Transforming Growth Factor, and immune cell chemokines such as lymphocytes, granulocytes, macrophages, mast cells, epithelial cells, and dendritic cells (Figure 6.1) (Raheem *et al.*, 2021). Lipoteichoic acid, a cell wall component of *Bifidobacteria* and *Lactobacilli*, has been discovered to stimulate NO synthase, which has the potential to induce pathogen-infected cell death mechanisms by releasing TNF macrophages (Yousefi *et al.*, 2019).

NO increase rate level of Fc RIII surface phagocytosis receptors along with Toll Like Receptor (Palmieri *et al.*, 2020). It modulates pro- and/or

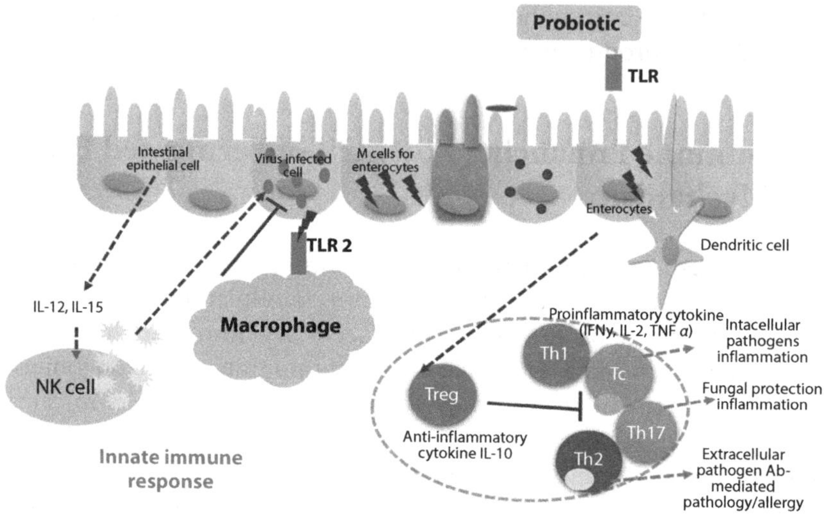

Figure 6.1 The potential role of fermented dairy product in immunomodulation.

anti-inflammatory action (adaptive immunity), according to probiotics interacts with enterocyte and dendritic, and Treg cell (Mahdavi-Roshan *et al.*, 2022). Probiotic strains have direct influence on the gut barrier it stimulates B cells, which is designed to produce IgA. It is found that probiotics have influence on the production of cytokine, which is responsible for adaptive responses. Cytokines is responsible for strengthening of the system of protection against invasion by bacterial, fungal, viral and any other pathogenic portion (Azad *et al.*, 2018). Numerous studies in field of immune system in animal models found that cytokines are crucial in binding to different cell membrane and for activation of cellular cascades for induction and development and also inhibit various cytokine gene in animal models. (Vodjgani *et al.*, 2020). The anti-inflammatory cytokine (IL-10) that inhibit the progression of inflammation in cell and it is generated by different receptors such as monocytes, B-cells, T-cells, macrophages, natural killer cells, and dendritic cells (Sun *et al.*, 2020). Immuno-stimulatory probiotics are act against various infections and carcinogenic cells. It also produces IL-12 and activates NK cells that can develops T1 cells. Those probiotics balancing T1/T2 cell and prevent allergy. Furthermore, immunoregulatory probiotics capable of producing IL-10 and Treg cells that leads to reduction in allergy, autoimmune disease, IBD, and progression of inflammatory response (Dargahi *et al.*, 2019).

6.3 Fermented Dairy Products Modulators of Intestinal Microbiota

The effects of the fermented products obtained from lactic acid bacteria found to be modulate in gut microbes and it was studied in healthy gut who administrated with concentrated whey from supernatant with *B. breve* for 1wk or diluted milk i.e., control group. After the feeding fermented product whey, it was found that their increase in *Bifidobacterium* in the faeces sample of that adult as compared to control group. From the above said it concluded that the fermented dairy product produces some bioactive compound which can modulate the gut microbiota (Granier *et al.*, 2013). Another study, show that oligosaccharide obtained by fermentation of lactose with *B. bifidum* NCIMB 41171 which can modulate the gut microbiota, so its confirming that lactic acid bacteria can use to produce the prebiotic can be food for microbes present in gut (Depeint *et al.*, 2008). Human milk oligosaccharide as prebiotic can modulate gut microbiota. The researchers conducted studies during the colonization period among new-born infants who administrated with standard non-fermented formula or FMBIF. They have compared composition of gut microbes and metabolism of new-born infants (Mullié *et al.*, 2004 & Paganini *et al.*, 2019). For four months, faeces of infants fed with fermented formula had a higher relative abundance of *bifidobacteria* with fewer adult species and the faeces were tested monthly for further confirmation. Breast feeder infant were easily colonizing the *B. longum infantis* and *B. breve*, species and can associated with the barrier function and immune modulation (Laursen *et al.*, 2021). Indeed, it is recognized that sufficient maturation of the infant immune system is associated with microbiota intestinal colonization (Calder *et al.*, 2006 & Di *et al.*, 2013) and in particular with some species of *bifidobacterium*, including those found in breast-fed infants (Sela and Mills DA, 2010).

6.4 Fermented Dairy Products Modulate Immune System

Our immune system may be modulated and protected against a variety of illnesses by the good microorganisms found in dairy products, as we previously mentioned. Fermented milk's live commensal microbiota is beneficial for the gut immune system's development and for building tolerance to dietary antigens (Wang *et al.*, 2022). When it comes to bacterial-immune interactions and the metabolites they produce, we're merely understanding exact mechanism behind all the functional properties. Antigen-presenting "Dendritic cells (DCs)" play an important part in this connection. Non-lymphoid tissues, particularly the gut mucosa, contain these first immature cells, which are responsible for capturing and processing antigens (Bošnjak *et al.*, 2022). After migrating into the lymphoid organs, DCs develop in the presence of "risk" signals, gaining the ability to present antigens to naïve T cells and inducing a particular immunogenic reaction in those cells. Additionally, DCs are capable of creating regulatory T cells, which may influence the immune system in a negative way (Carey *et al.*, 2021).

Tolerogenic and "immunogenic" DCs appear to be influenced by DC type, costimulatory molecule expression levels, and cytokine production profiles (Khan *et al.*, 2022). Integration of numerous intracellular signals, particularly "pathogen-associated molecular patterns" from distinct bacteria, contributes to the specialised DC functions to detect "pathogen-associated molecular patterns" (PAMP) in the microenvironment DCs express Toll-like receptors (TLRs) (Fernandes-Santos & Azeredo, 2022). TLR involvement may trigger a broad range of signal transduction pathways, such as mitogen-activated protein kinases or 3-kinase phosphatidylinositol, to affect the amplitude, and duration of immune responses and, hence, the immunological response. Immunogenic or tolerogenic DCs have the ability to induce DCs. There is a large spectrum of structures recognised by the human TLRs that are well conserved. Bacteria-specific transduction pathways may explain the different impacts on DC function that have been identified (Kordahi *et al.*, 2018). This makes DCs a good target for fermentation products with immunomodulatory qualities because of their diverse reactivity to stimuli and the wide variety of features they might produce.

6.4.1 Immune Enhancement

The immune system is boosted by consuming fermented dairy products containing live beneficial lactic acid bacteria.

- Enhancement of blood leucocyte phagocytic activity (PMN and mononuclear cells)
- Phosphatic cells secrete cytokines and produce reactive oxygen species to defend themselves against pathogens.
- Increasing the activation of NK cells

Lb. johnsonii La1 and *B. Lactic* Bb12 in fermented milk has been shown to increase the phagocytic activity of human participants' peripheral blood leucocytes (PMN and monocytes) for three weeks (Sterrett *et al.*, 2022). The phagocytic activity of the PMN was larger than that of the mononuclear cells. In comparison to mononuclear cells, the phagocytic activity of PMNs increased more than that of mononuclear cells. The expression of CR1, CR3, FcRI, and FcR by neutrophils has been shown to be considerably increased by *Lactobacillus* GG-containing milk consumption (Wu *et al.*, 2019). There is a rise in the phagocytic cell function and NK cell activity, as well as an increase in the proportion of NK cells in peripheral blood of human volunteers who consume yoghurt or probiotic-containing milk on a regular basis (Velikova *et al.*, 2020). We may conclude from our prior studies that some lactic acid bacteria strains may improve natural immunity and have a positive impact on human health when consumed in certain foods. In the following (Table 6.1), fermented food and its effect on the immune function are described.

Table 6.1 Listed the fermented dairy product and their role in immune enhancement.

Yoghurt/probiotic strain	**Study subjects**	**Effect**	**Reference**
Yoghurt enriched with *Lactobacillus brevis* KU200019 and fructo-oligosaccharides (FOS)		✓ High antioxidant ACE-inhibitory (70.37 ± 1.19%) ✓ Reactive oxygen species scavenging (2.52-fold) ✓ Immunomodulatory activities	Kariyawasam *et al.* 2021
Sheep and cow milk yoghurts		✓ Down-regulation of pro-inflammatory IL1B and IL12B ✓ Up-regulation of anti-inflammatory IL10 in neutrophils	Moschopoulou *et al.* 2018
Kefir	HCV- and HBV-infected cells	✓ Suppress the elevation of TNF-α ✓ Upregulate IL-10 and INF-γ	Ellatif *et al.* 2022

(*Continued*)

Table 6.1 Listed the fermented dairy product and their role in immune enhancement. (*Continued*)

Yoghurt/probiotic strain	**Study subjects**	**Effect**	**Reference**
Labaneh (High-acid and low-moisture yogurt)		✓ Cholesterol-lowering (>30%) ✓ Pathogen-inhibiting properties ✓ Hydrophobicity values of 18.0–87.3%, 7.1–86.0%, and 20.6–87.1% (for octane, xylene, and hexadecane, resp.)	Tarique *et al.* 2022
EPS (*Lactiplantibacillus plantarum* MM89)	Immuno-suppressed mouse model	✓ Increase in intestinal IgA levels ✓ Increase in serum cytokine levels ✓ Increase in lymphocyte proliferation	Rajoka *et al.* 2022
Yakult	40–65 years	✓ No effect on Ig levels ✓ NK cell activity ✓ Lymphocyte cytokine production	Spanhaak *et al.* (1998)
B. lactis Bb12	15–31-month-old children; all vaccinated with poliovirus by 12 months of age 23–62 years	✓ ↑ Anti-polio IgA levels in faeces	Fukushima *et al.* 1998
		✓ ↑ Phagocytic capacity of blood leucocytes (PMN and mononuclear cells)	Schiffrin *et al.* (1995)
Fermented milk containing *L. acidophilus* La1 and *B. bifidum* Bb12	19–45 years old	Slight increase in anti-*S. typhi* antibody response	Link-Amster *et al.* (1994)
		↑ Serum total IgA level	

6.4.2 Increased Disease Resistance

Numerous investigations conducted on humans have also shown a connection between LAB-induced immunostimulant and increased disease resistance (Ringø *et al.*, 2018). According to research done by Kaila *et al.*, 1992, *Lactobacillus* GG helps children suffering from acute rotavirus gastroenteritis recover more quickly from rotavirus diarrhoea by boosting the immunological responses of the host. Children who took *Lactobacillus* GG had shorter bouts of diarrhoea and a higher IgA response to anti-rotavirus treatment than those who did not. Majamaa *et al.* (1995) found a correlation between the protective benefits of *Lactobacillus* GG and the increase of certain immunological responses. This finding was consistent with previous research. After consuming certain dietary LAB, people showed an enhanced resistance to gastrointestinal infections and disorders that cause diarrhoea, according to the findings of many previous research (Ballini *et al.*, 2019). However, in many instances, there is still a lack of data that demonstrates beyond a reasonable doubt that an increased disease tolerance was the primary outcome of an elevated immune response to the consumption of LAB. LAB may help to host defence against pathogenic microbes through a variety of other pathways, some of which have been proposed. These include the rivalry for adhesion sites, the competition between different bacteria, and the competition to make antimicrobials (Wu *et al.*, 2019).

Yoghurt and probiotics may potentially lower the incidence of colon and bladder cancer, according to research (Parodi 1999). It's unclear whether these antitumor effects are the consequence of immune stimulation or of changes in the microbiota, which may play a role in cancer aetiology. However, studies conduction in tumor patients and experimental rats showed the probiotics activates immune system's and exhibited anti-cancer properties. It increased the response of ant-carcinogenic immune cell such as NK cell counts. The cytotoxicity linked with LAB ingestion (Al-Hilu and Al-Shujairi 2020). Oral treatment of *Lb. Casei* (Shirota) to mice following initial tumour excision boosted immune cell activity and, as a result, reduced secondary tumour development in comparison to mouse controls (Perillo *et al.*, 2020). Secondary tumours in bladder cancer patients have been shown to be effectively treated with *Lb. Casei* Shirota (Aso *et al.* 1995), and it may boost cellular immune responses to anti-tumor treatments (NK cell activity) (Nagao *et al.*, 2000).

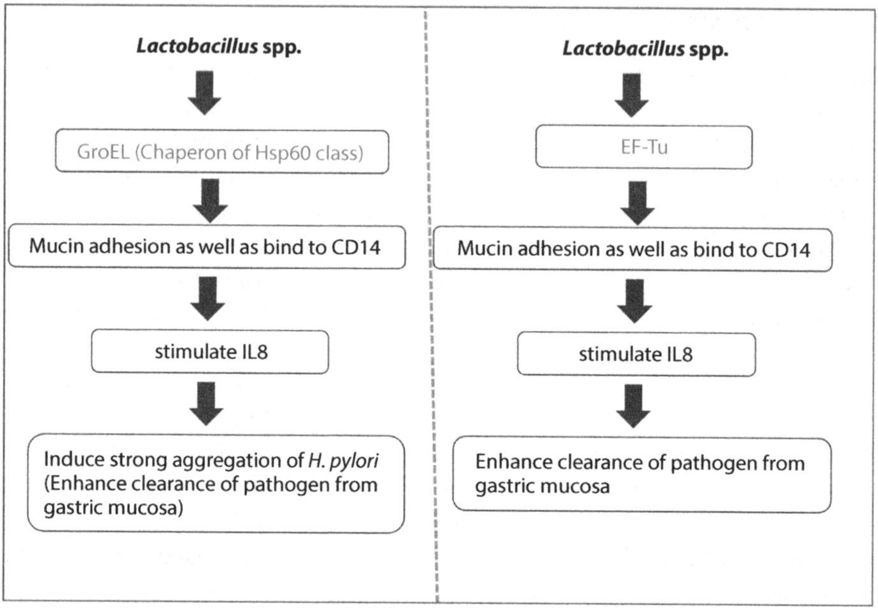

6.4.3 Immunoregulation

T helper 1 (Th1) and T helper 2 (Th2) responses must be in balance in order for the immune system to remain in a state of homeostasis. An imbalance (overactivation/downregulation) in the immunological response of Th1 vs Th2 is too responsible for many immune-mediated diseases, including allergies, autoimmune and immunoinflammatory disorders (Wang *et al.*, 2020). According to Shida *et al.* (1998) and Matsuzaki and Chin (2000), several *Lactobacillus* strains reduce IgE antibody responses in mice while reversing immunological reactivity following oral delivery. In human clinical research, it has been shown that long-term consumption of yoghurt or fermented milk products reduces serum IgE levels and alleviates some of the symptoms of nasal allergy in the elderly or atopic eczema in children (García-Burgos *et al.*, 2020). It is thought that probiotics may exert their anti-allergy effects by inducing the production of high levels of pro-Th1 cytokine expression (e.g., IL-12) and enhancing the intestinal mucosal barrier's ability to resist macromolecular sensitization (Murosaki *et al.*, 1998). When milk allergy sufferers consume casein hydrolysed with *Lb. rhamnosus*, the synthesis of the immunoregulatory peptide IL-4 by antigen-stimulated PBL was shown to be lower than in those given milk, suggesting that lactobacilli enzymes may be responsible for suppressing allergic responses (Rokka *et al.*, 1997; Azubuike-Osu *et al.*, 2021).

6.5 Future Trends

Integrating probiotic species into the production process for well-accepted conventional dairy products will be extensively investigated in order to create more probiotic products. The importance of producing convenience-based dry-cultured food containing viable bacteria using various techniques such as microencapsulation and fortification of these powders or concentrated products with minerals and vitamins is growing. The design of appropriate mixes containing healthy elements like dietary fibre, natural vitamins, and folic acid opens up new possibilities for this widely studied topic. The beverage sector would undoubtedly benefit from the development of various health drinks and dietetic preparations as value-added goods based on traditional fermented products. Apart from the overall nutrient increase, more work needs to be done to separate the therapeutic benefits of viable and non-viable species in diverse fermented products.

6.6 Conclusion

Probiotics, alone or in combination with prebiotics, have the capacity to modulate gut microbiota and immunological responses in the host, making fermented milk with probiotic treatment a promising therapeutic strategy. However, a lot of research papers on the impact of fermented milk with probiotics in avoiding obesity, inflammatory illnesses, and cancer are similar. The immunomodulatory benefits of probiotics in the treatment of degenerative and other diseases caused by pathogenic bacteria have received a lot of attention. Probiotics boost innate immunity and have a variety of antiviral activities. More research into the precise action of probiotics on inflammation may be recommended, as these findings will be crucial in the medical field and in enhancing human health.

References

Al-Hilu SA, Al-Shujairi WH. Dual role of bacteria in carcinoma: stimulation and inhibition. International Journal of Microbiology. 2020 Aug 24;2020.

Azubuike-Osu SO, Ekezie FG, Ibegbunam WE, Udenigwe CC. Allergic Effects of Bioactive Peptides Produced from Different Food Sources. In Bioactive Peptides 2021 Jun 14 (pp. 363-393). CRC Press.

Ballini A, Gnoni A, De Vito D, Dipalma G, Cantore S, Gargiulo Isacco C, Saini R, Santacroce L, Topi S, Scarano A, Scacco S. Effect of probiotics on the occurrence of nutrition absorption capacities in healthy children: A randomized double-blinded placebo-controlled pilot study. Eur. Rev. Med. Pharmacol. Sci. 2019 Oct 1;23(19):8645-57.

Bedi O, Chaudhary S, Singh TG. Protective Effect of Probiotic in Alcohol-Induced Liver Disorders: Role of Probiotics in Alcohol-Induced Liver Disorders. In Probiotic Research in Therapeutics: Volume 5: Metabolic Diseases and Gut Bacteria 2022 May 5 (pp. 143-160). Singapore: Springer Nature Singapore.

Bošnjak B, Do KT, Förster R, Hammerschmidt SI. Imaging dendritic cell functions. Immunological Reviews. 2022 Mar;306(1):137-63.

Bukowska-Ośko I, Sulejczak D, Kaczyńska K, Kleczkowska P, Kramkowski K, Popiel M, Wietrak E, Kowalczyk P. Lactoferrin as a Human Genome "Guardian"—An Overall Point of View. International Journal of Molecular Sciences. 2022 May 8;23(9):5248.

Calder PC. n−3 Polyunsaturated fatty acids, inflammation, and inflammatory diseases. The American journal of clinical nutrition. 2006 Jun 1;83(6):1505S-19S.

Carey ST, Gammon JM, Jewell CM. Biomaterial-enabled induction of pancreatic-specific regulatory T cells through distinct signal transduction pathways. Drug delivery and translational research. 2021 Dec;11:2468-81.

Dargahi N, Johnson J, Donkor O, Vasiljevic T, Apostolopoulos V. Immunomodulatory effects of probiotics: Can they be used to treat allergies and autoimmune diseases?. Maturitas. 2019 Jan 1;119:25-38.

Depeint F, Tzortzis G, Vulevic J, I'anson K, Gibson GR. Prebiotic evaluation of a novel galactooligosaccharide mixture produced by the enzymatic activity of Bifidobacterium bifidum NCIMB 41171, in healthy humans: a randomized, double-blind, crossover, placebo-controlled intervention study. The American journal of clinical nutrition. 2008 Mar 1;87(3):785-91.

Di Domenico M, Ballini A, Boccellino M, Scacco S, Lovero R, Charitos IA, Santacroce L. The intestinal microbiota may be a potential theranostic tool for personalized medicine. Journal of Personalized Medicine. 2022 Mar 24;12(4):523.

Di Mauro A, Neu J, Riezzo G, Raimondi F, Martinelli D, Francavilla R, Indrio F. Gastrointestinal function development and microbiota. Italian journal of pediatrics. 2013 Dec;39:1-7.

Ellatif SA, Abdel Razik ES, Abu-Serie MM, Mahfouz A, Shater AF, Saleh FM, Hassan MM, Alsanie WF, Altalhi A, Daigham GE, Mahfouz AY. Immunomodulatory Efficacy-Mediated Anti-HCV and Anti-HBV Potential of Kefir Grains; Unveiling the *In Vitro* Antibacterial, Antifungal, and Wound Healing Activities. Molecules. 2022 Mar 21;27(6):2016.

Fernandes-Santos C. Azeredo, ELd Innate Immune Response to Dengue Virus: Toll-like Receptors and Antiviral Response. Viruses 2022, 14, 992.

Fukushima Y, Kawata Y, Hara H, Terada A, Mitsuoka T. Effect of a probiotic formula on intestinal immunoglobulin A production in healthy children. International journal of food microbiology. 1998 Jun 30;42(1-2):39-44.

García-Burgos M, Moreno-Fernández J, Alférez MJ, Díaz-Castro J, López-Aliaga I. New perspectives in fermented dairy products and their health relevance. Journal of Functional Foods. 2020 Sep 1;72:104059.

Gopal V, Dhanasekaran D. Probiotics as a growth promotant for livestock and poultry production. InAdvances in Probiotics 2021 Jan 1 (pp. 349-364). Academic Press.

Granier A, Goulet O, Hoarau C. Fermentation products: immunological effects on human and animal models. Pediatric research. 2013 Aug;74(2):238-44.

Kariyawasam KM, Lee NK, Paik HD. Synbiotic yoghurt supplemented with novel probiotic Lactobacillus brevis KU200019 and fructooligosaccharides. Food Bioscience. 2021 Feb 1;39:100835.

Khan FU, Khongorzul P, Raki AA, Rajasekaran A, Gris D, Amrani A. Dendritic cells and their immunotherapeutic potential for treating type 1 diabetes. International Journal of Molecular Sciences. 2022 Apr 28;23(9):4885.

Kim H, Kim Y, Kang CH. *In vivo* confirmation of the antimicrobial effect of probiotic candidates against Gardnerella vaginalis. Microorganisms. 2021 Aug 9;9(8):1690.

Kober AH, Riaz Rajoka MS, Mehwish HM, Villena J, Kitazawa H. Immunomodulation potential of probiotics: a novel strategy for improving livestock health, immunity, and productivity. Microorganisms. 2022 Feb 7;10(2):388.

Kordahi MC, William DePaolo R. The influence of the microbiota on the etiology of colorectal cancer. Mechanisms underlying host-microbiome interactions in pathophysiology of human diseases. 2018:167-93.

Laursen MF, Sakanaka M, von Burg N, Mörbe U, Andersen D, Moll JM, Pekmez CT, Rivollier A, Michaelsen KF, Mølgaard C, Lind MV. Bifidobacterium species associated with breastfeeding produce aromatic lactic acids in the infant gut. Nature microbiology. 2021 Nov;6(11):1367-82.

Lekchand Dasriya V, Samtiya M, Dhewa T, Puniya M, Kumar S, Ranveer S, Chaudhary V, Vij S, Behare P, Singh N, Aluko RE. Etiology and management of Alzheimer's disease: Potential role of gut microbiota modulation with probiotics supplementation. Journal of Food Biochemistry. 2022 Jan;46(1):e14043.

Link-Amster H, Rochat F, Saudan KY, Mignot O, Aeschlimann JM. Modulation of a specific humoral immune response and changes in intestinal flora mediated through fermented milk intake. FEMS Immunology & Medical Microbiology. 1994 Nov 1;10(1):55-63.

Ma Y, Yin Z, Li L, Chen B, Dai H, Wu D, Cong J, Liao C, Li L, Ye Z, Huang Z. Food antigens exacerbate intestinal damage and inflammation following the disruption of the mucosal barrier. International Immunopharmacology. 2021 Jul 1;96:107670.

Mahdavi-Roshan M, Salari A, Kheirkhah J, Ghorbani Z. The effects of probiotics on inflammation, endothelial dysfunction, and atherosclerosis progression: a mechanistic overview. Heart, Lung and Circulation. 2022 Feb 10.

Matsuzaki T, Chin J. Modulating immune responses with probiotic bacteria. Immunology and cell biology. 2000 Feb;78(1):67-73.

Moschopoulou E, Sakkas L, Zoidou E, Theodorou G, Sgouridou E, Kalathaki C, Liarakou A, Chatzigeorgiou A, Politis I, Moatsou G. Effect of milk kind and storage on the biochemical, textural and biofunctional characteristics of set-type yoghurt. International dairy journal. 2018 Feb 1;77:47-55.

Mullié C, Yazourh A, Thibault H, Odou MF, Singer E, Kalach N, Kremp O, Romond MB. Increased poliovirus-specific intestinal antibody response coincides with promotion of Bifidobacterium longum-infantis and Bifidobacterium breve in infants: a randomized, double-blind, placebo-controlled trial. Pediatric research. 2004 Nov;56(5):791-5.

Murosaki S, Yamamoto Y, Ito K, Inokuchi T, Kusaka H, Ikeda H, Yoshikai Y. Heat-killed Lactobacillus plantarum L-137 suppresses naturally fed antigen–specific IgE production by stimulation of IL-12 production in mice. Journal of allergy and clinical immunology. 1998 Jul 1;102(1):57-64.

Paganini D, Uyoga MA, Kortman GA, Boekhorst J, Schneeberger S, Karanja S, Hennet T, Zimmermann MB. Maternal human milk oligosaccharide profile modulates the impact of an intervention with iron and galacto-oligosaccharides in Kenyan infants. Nutrients. 2019 Oct 29;11(11):2596.

Pahwa H, Sharan K. Food and nutrition as modifiers of the immune system: A mechanistic overview. Trends in Food Science & Technology. 2022 Mar 29.

Palmieri EM, McGinity C, Wink DA, McVicar DW. Nitric oxide in macrophage immunometabolism: hiding in plain sight. Metabolites. 2020 Oct 26;10(11):429.

Parodi PW. Conjugated linoleic acid and other anticarcinogenic agents of bovine milk fat. Journal of dairy science. 1999 Jun 1;82(6):1339-49.

Patil SM, Sujay S, Tejaswini M, Sushma PP, Prithvi S, Ramu R. Bioactive peptides: Its production and potential role on health. Innov. Food Sci. Emerg. Technol. 2020;7:167-82.

Paul MM, Anand N, Raj A, Raghavendra SN. Technological Challenges for Future Probiotic Foods.

Peluzio MD, Dias MD, Martinez JA, Milagro FI. Kefir and intestinal microbiota modulation: implications in human health. Frontiers in nutrition. 2021 Feb 22;8:638740.

Perillo F, Amoroso C, Strati F, Giuffrè MR, Díaz-Basabe A, Lattanzi G, Facciotti F. Gut microbiota manipulation as a tool for colorectal cancer management: recent advances in its use for therapeutic purposes. International Journal of Molecular Sciences. 2020 Jul 29;21(15):5389.

Raheem A, Liang L, Zhang G, Cui S. Modulatory effects of probiotics during pathogenic infections with emphasis on immune regulation. Frontiers in Immunology. 2021 Apr 8;12:616713.

Rajoka MS, Mehwish HM, Kitazawa H, Barba FJ, Berthelot L, Umair M, Zhu Q, He Z, Zhao L. Techno-functional properties and immunomodulatory potential of exopolysaccharide from Lactiplantibacillus plantarum MM89 isolated from human breast milk. Food Chemistry. 2022 May 30;377:131954.

Ringø E, Hoseinifar SH, Ghosh K, Doan HV, Beck BR, Song SK. Lactic acid bacteria in finfish—An update. Frontiers in microbiology. 2018 Aug 10;9:1818.

Rokka VM, Valkonen JP, Pehu E. Somatohaploid production by anther culture of interspecific somatic hybrids and their prospects in potato breeding. In Vitro Haploid Production in Higher Plants: Volume 5—Oil, Ornamental and Miscellaneous Plants. 1997:233-43.

Schiffrin EJ, Rochat F, Link-Amster H, Aeschlimann JM, Donnet-Hughes A. Immunomodulation of human blood cells following the ingestion of lactic acid bacteria. Journal of Dairy Science. 1995 Mar 1;78(3):491-7.

Sela DA, Mills DA. Nursing our microbiota: molecular linkages between bifidobacteria and milk oligosaccharides. Trends in microbiology. 2010 Jul 1;18(7):298-307.

Shida K, Makino K, Morishita A, Takamizawa K, Hachimura S, Ametani A, Sato T, Kumagai Y, Habu S, Kaminogawa S. Lactobacillus casei inhibits antigen-induced IgE secretion through regulation of cytokine production in murine splenocyte cultures. International archives of allergy and immunology. 1998;115(4):278-87.

Siddique F, Akram K, Alghamdi ES, Arshad Q, Siddique A. The Immunomodulatory Role of Probiotics. Rijeka, Croatia: IntechOpen; 2022 Apr 6.

Spanhaak S, Havenaar R, Schaafsma G. The effect of consumption of milk fermented by Lactobacillus casei strain Shirota on the intestinal microflora and immune parameters in humans. European Journal of Clinical Nutrition. 1998 Dec;52(12):899-907.

Sterrett JD, Andersen ND, Lowry CA. The influence of the microbiota on brain structure and function: implications for stress-related neuropsychiatric disorders. InEvolution, Biodiversity and a Reassessment of the Hygiene Hypothesis 2022 Mar 25 (pp. 267-337). Cham: Springer International Publishing.

Sun X, Huang Y, Zhang YL, Qiao D, Dai YC. Research advances of vasoactive intestinal peptide in the pathogenesis of ulcerative colitis by regulating interleukin-10 expression in regulatory B cells. World Journal of Gastroenterology. 2020 Dec 12;26(48):7593.

Surowiak AK, Sowała M, Talma M, Groborz K, Balcerzak L, Lochyński S, Strub DJ. Cytotoxicity, early safety screening, and antimicrobial potential of minor oxime constituents of essential oils and aromatic extracts. Scientific Reports. 2022 Mar 29;12(1):5319.

Tarique M, Abdalla A, Masad R, Al-Sbiei A, Kizhakkayil J, Osaili T, Olaimat A, Liu SQ, Fernandez-Cabezudo M, al-Ramadi B, Ayyash M. Potential probiotics and postbiotic characteristics including immunomodulatory effects of lactic acid bacteria isolated from traditional yogurt-like products. LWT. 2022 Apr 1;159:113207.

Velikova T, Tumangelova-Yuzeir K, Georgieva R, Ivanova-Todorova E, Karaivanova E, Nakov V, Nakov R, Kyurkchiev D. Lactobacilli supplemented with larch arabinogalactan and colostrum stimulates an immune response towards peripheral NK activation and gut tolerance. Nutrients. 2020 Jun 7;12(6):1706.

Vodjgani M, Salehi Z, Izad M. The influence of reactive oxygen species in the immune system and pathogenesis of multiple sclerosis. Autoimmune diseases. 2020 Oct;2020.

Wang C, Wei S, Liu B, Wang F, Lu Z, Jin M, Wang Y. Maternal consumption of a fermented diet protects offspring against intestinal inflammation by regulating the gut microbiota. Gut Microbes. 2022 Dec 31;14(1):2057779.

Wang W, Sung N, Gilman-Sachs A, Kwak-Kim J. T helper (Th) cell profiles in pregnancy and recurrent pregnancy losses: Th1/Th2/Th9/Th17/Th22/Tfh cells. Frontiers in immunology. 2020 Aug 18;11:2025.

Wu D, Lewis ED, Pae M, Meydani SN. Nutritional modulation of immune function: analysis of evidence, mechanisms, and clinical relevance. Frontiers in immunology. 2019:3160.

Yousefi B, Eslami M, Ghasemian A, Kokhaei P, Salek Farrokhi A, Darabi N. Probiotics importance and their immunomodulatory properties. Journal of cellular physiology. 2019 Jun;234(6):8008-18.

Zhang Y, Li M, Du G, Chen X, Sun X. Advanced oral vaccine delivery strategies for improving the immunity. Advanced Drug Delivery Reviews. 2021 Oct 1;177:113928.

Zheng D, Liu Z, Zhou Y, Hou N, Yan W, Qin Y, Ye Q, Cheng X, Xiao Q, Bao Y, Luo J. Urolithin B, a gut microbiota metabolite, protects against myocardial ischemia/reperfusion injury via p62/Keap1/Nrf2 signaling pathway. Pharmacological research. 2020 Mar 1;153:104655.

Azad M, Kalam A, Sarker M, Wan D. Immunomodulatory effects of probiotics on cytokine profiles. BioMed research international. 2018 Oct 23;2018.

Kaila M, Isolauri E, Soppi ES, Virtanen E, Laine S, Arvilommi H. Enhancement of the circulating antibody secreting cell response in human diarrhea by a human Lactobacillus strain. Pediatric research. 1992 Aug;32(2):141-4.

Majamaa H, Isolauri E, Saxelin M, Vesikari T. Lactic acid bacteria in the treatment of acute rotavirus gastroenteritis. Journal of pediatric gastroenterology and nutrition. 1995 Apr 1;20(3):333-8.

Nagao F, Nakayama M, Muto T, Okumura K. Effects of a fermented milk drink containing Lactobacillus casei strain Shirota on the immune system in healthy human subjects. Bioscience, biotechnology, and biochemistry. 2000;64(12):2706-8.

Nongonierma AB, O'keeffe MB, FitzGerald RJ. Milk protein hydrolysates and bioactive peptides. Advanced Dairy Chemistry: Volume 1B: Proteins: Applied Aspects. 2016:417-82.

Aso Y, Akaza H, Kotake T, Tsukamoto T, Imai K, Naito S, BLP Study Group. Preventive effect of a Lactobacillus casei preparation on the recurrence of superficial bladder cancer in a double-blind trial. European urology. 1995;27:104-9.

7

Potential Applications of Nanotechnology in Food Systems: An Overview

Sunny Dhiman*, Prachi Singh, Anu Kumar and Gunjan Mukherjee

University Institute of Biotechnology, Chandigarh University, Gharuan, Mohali, Punjab, India

Abstract

The increasing awareness amongst consumers regarding food safety, food quality, food safety, and health benefits are impelling researchers to look for ways to enhance food quality without compromising the product's nutritional value. Currently, approaches to minimizing the microbial hazards for ensuring food safety have severe drawbacks and inefficiencies. The severity and frequency of foodborne diseases as a result of food contamination has made it essential for research communities around the world to devise and execute advanced products for food safety. Considering the intense competition amongst food industries and the vitality of innovation, nanotechnology has emerged as an attractive potential to address the drawbacks of existing technologies from the food safety perspective. There are numerous nanotechnology applications in the food industry, including food processing, food packaging, shelf-life extension, functional food development, and food safety. Despite the mounting evidence of nanotoxicity's dangers, the commercial use of nanotechnology remains largely unregulated. This article aims to review the versatile applications of nanotechnology in food systems. Future trends of food nanotechnology and regulatory issues involved are also discussed.

***Keywords*:** Nanotechnology, food processing, nanoceutical, food safety, nanomaterials, nano food

**Corresponding author*: sdsunnydhiman473@gmail.com

Tejpal Dhewa, Anil Kumar Puniya and Anil Panghal (eds.) Nutritional Science and Technology: Concept to Application, (135–164) © 2023 Scrivener Publishing LLC

7.1 Introduction

The word "nano" means dwarf. A nanometer is denoted by 10^{-9} metres. Nanomolecules ranges from one-half the diameter of DNA up to 1/20 the size of a RBC [1]. Nanotech is the fabrication and integration of physiological, chemical, and biological systems at nanoscales [2]. Nanotech in conjunction with other sciences and technologies may increase the magnitude of its transformative potential. Nanotechnology can be applied in medicine, the food industry, manufacturing nanocoating materials, and wastewater treatment [3]. Enzymes could be used in several food processing techniques to modify food components to enhance taste, nutritional quality, medical benefits, etc. In comparison with standard macroscale support materials, nanomaterials assure advanced enzyme support systems (enhanced activity, storability, and expense) because of their substantial surface area/volume ratio [4]. Agro-food systems could benefit from nanotechnology. The nanometer-scale can impact food standards, efficiency, bioavailability, nutrient value, and the biochemical synthesis of recent products in the market [5]. Studies related to nanotechnology are advancing day by day. The food industry has developed novel products like nanoceuticals, nanoemulsions, and nanosensors, collaboratively with the help of nanotechnology [6].

7.2 Natural Self-Assembled Food Nanostructures

"Self-assembly is also known as self-organization and self-association. Self-assembly often happens when certain types of components under appropriate environment and preparation conditions undergo assembly, leading to well-defined structures to minimize the free energy of the system" [7]. Many natural foods contain nanoscale elements, and the structure of these elements specifies their properties. For generations, these have been consumed reliably. In fact, during standard food processing, several food raw materials like proteins, starches, etc., undergo structural reforms at the nanometric and micrometric dimensions [8]. Milk proteins are by far the most investigated animal proteins for manufacturing self-assembly configurations.

Self-assembled nanotubes made of lactalbumin have been suggested as a promising new carrier for the nanoencapsulation of fortified foods [9]. Caesins are a family of related phosphoproteins that lack disulfide bridges and have a large number of proline (Pro) molecules dispersed all across their primary structures [10]. Caseins have been found in abundance as self-assembled micelles (diameters ranging from 50 to 500 nm, on average is about 120nm) [11]. Casein micelles also can be used as nano vehicles for the entrapment, preservation, and delivery of susceptible nutrients [12]. "Whey protein isolate (WPI), on the other hand, is composed of approximately 80% lactoglobulin and 15% lactalbumin. It is widely used in the food industry because of its high functionality and low cost" [13]. The major determinants of WPI self-assembly are pH and temperature [14].

In case of fats, for example, monoglycerides, can self-assemble into many morphologies at the nanoscale level, such as di-nitrotriglycerides (10–100 nm), followed by arrangement into large clusters, flocs, and finally fat crystal networks [15].

7.3 Classification of Nanoparticles Applied in Food Industry

Nanoparticles mainly utilized in the food industry are classified into three categories: Organic, Inorganic, and Carbon Nanoparticles. Nanoparticles of relative dimensions have been highlighted in Figure 7.1.

7.3.1 Organic Nanoparticles

Organic nanoparticles (ONPs) are solid particles composed of proteins, carbohydrates, lipids, and other organic compounds [60]. Their size ranges from 100 nm to -1μm [61]. The most commonly used ONPs are liposomes, dendrimers, micelles, and carbon-based nanoparticles [62]. Starch and cellulose NPs are found to be non-toxic and biodegradable [63]. Micelles and liposomes have a hollow core termed nanocapsule sensitive to heat and electromagnetic radiations [64]. Organic nanoparticles are utilized in numerous fields, such as drug delivery systems, and they effectively deliver targeted drugs [65]. Colloidal and polymer science are derived from the physico-chemical characteristics of the source materials. Organic nanoparticles are manufactured by either a bottom-up approach or a top-down one, or a combination of both [16].

7.3.1.1 Liposomes

Liposomes are spherical synthetic vesicles formed from cholesterol and non-toxic phospholipids. Liposomes are found in a variety of sizes, from

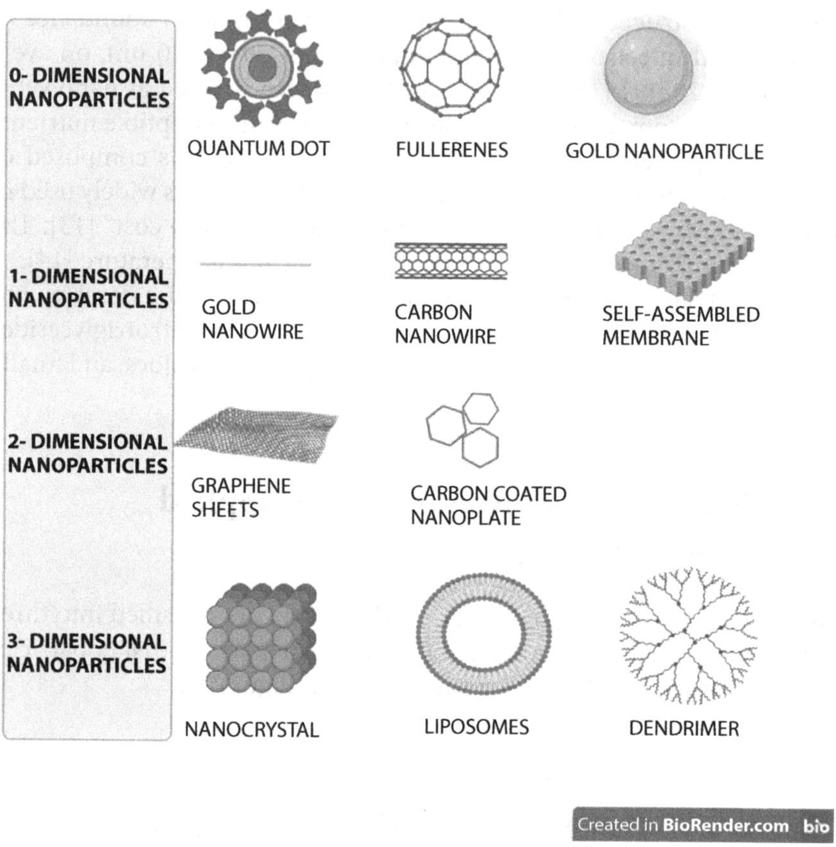

Figure 7.1 Schematic diagram illustrating nanoparticles of relative dimensions with examples of each category.

30 nanometers to several micrometers [66]. They are widely employed in the cosmetic and pharmaceutical industries as carriers for a variety of compounds. Additionally, the food and agriculture sectors have conducted extensive research towards applying liposome encapsulation to improve delivery systems that can entrap and secure volatile substances (such as anti-microbials, anti-oxidants, and flavouring agent). Liposomes entrap both hydrophobic and hydrophilic substances and release them at targeted locations, restricting the entrapped combinations from decomposing [67-69].

7.3.1.2 Dendrimers

Dendrimers are nanoscale molecules with homogeneous and monodispersed structures. Their structural attributes are similar to that of radially symmetric tree-like branches [70]. Their size varies from 1 to 15 nm [71]. Rather than a compound, dendrimer corresponds to an "architectural motif". Polyionic dendrimers do not retain their structure over time, and their size, structure, and versatility could change as the series of generations grows [72]. Dendrimers are a potential candidate for drug delivery systems due to their adjustable size. Recent studies involving dendrimers and drug delivery systems have primarily been in neoplastic diseases. Dendrimers are also being investigated in various therapeutic sectors, including anti-inflammatory, antiviral, antibacterial medicines, and cardiovascular illnesses [73].

7.3.1.3 Micelles

Micelles are described as spherical amphiphilic structures with a hydrophobic core and hydrophilic shell [74]. Micelles are produced when amphiphilic di- or triblock copolymers self-assemble into a sphere-resembling nanoshell framework in aqueous conditions [75].

Real-life applications of polymeric micelles are majorly found in ultrasonic-assisted micellar drug delivery [76].

7.3.1.4 Carbon Nanoparticles

Carbon nanoparticles (CNPs) are a novel class of nanomaterials. They are composed of an inner core of carbon (sp2) atoms. Analytical detection, bio-sensing, bio-imaging, and molecular interaction are just a few of the applications for such novel nano-dots. Their use as nanocatalysts is still in its formative development [77].

CNPs are carbon-nanotubes, fullerenes, graphene, and their derivatives, like graphene oxide, carbon quantum dots, and nanodiamonds [78].

7.3.2 Inorganic Nanoparticles

Inorganic nanoparticles are particles that are not composed of carbon [62]. Numerous inorganic NPs are currently being utilized in the food industry like silver, iron oxide, titanium dioxide, silicon dioxide, or zinc oxide. Inorganic NPs have a variety of surface properties and are available in a multitude of sizes determined by the source materials and preparation environment used in their manufacturing [79]. Inorganic NPs are widely used because of their ability to manipulate size and structure, crystallinity, and large surface-area. Inorganic particles possess unique optical, magnetic, catalytic, thermodynamic, and electrochemical properties with additional bioactivities [17]. Inorganic nanoparticles are categorized into metal and metal-oxide-based nanoparticles.

7.3.2.1 Metal Nanoparticles

Metal NPs are formed from nano-sized metallic materials using destructive or constructive methods. Nowadays, it is possible to manufacture nanoparticles from almost any metal [80]. The most common metal-based NPs are Al, Cu, Au, and Fe.

7.3.2.2 Metal-Oxide Nanoparticles

Metal oxide nanoparticles are synthesized to change the properties of their respective metal-based nanoparticles. For example, iron nanoparticles (Fe) quickly oxidize to iron oxide (Fe_2O_3) in the presence of oxygen at ambient temperature, increasing their reactivity. Metal oxide nanoparticles are synthesized because of their improved reactivity and efficiency [81]. Most commonly synthesized metal-oxide-based NPs are Al_2O_3, CeO_2, Fe_2O_3, Fe_3O_4, SiO_2, TiO_2, and ZnO.

7.4 Potential Applications: Nanotechnology in Food Industry

Nanotech is successful in various industries, including the food industry (Figure 7.2) [18]. Improved food shelf life, better monitoring of pathogens, enhanced food storage, and the inclusion of nutritional supplements for antibacterial agents in food have all been facilitated by nanotechnology [19]. Nanotechnology's numerous implementations and recent researches in food systems are discussed in greater detail. Food processing and packaging are thoroughly discussed.

7.4.1 Food Processing

Food processing is transforming the beverage and food industry to make the raw foodstuff of animal and plant origin suitable for consumption [20]. Irradiation, high hydrostatic heating, and ohmic heating are conventional food processing methods [21]. Food processing with nanomaterials involves the incorporation of nanoceuticals, viscosifying and deletion

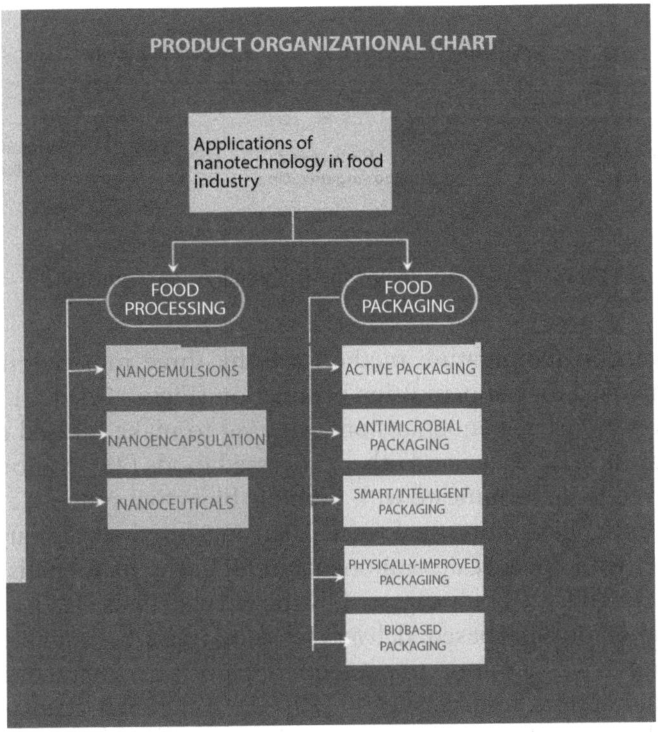

Figure 7.2 Flowchart demonstrating potential applications of nanotechnology.

agents, vitamin and mineral fortification, and flavoring nanoencapsulation [22]. Food processing is primarily done to preserve the integrity of the food and prolong the shelf life. One of the benefits of processed food is the year-round availability of various foods, particularly seasonal foods like peas and corn [23]. Along with processing fresh foods, healthy alternatives are becoming increasingly important, resulting in foods comprising micronutrients to satisfy consumers [24].

7.4.1.1 Nanoemulsions

"Emulsions are biphasic and liquid retentive formulations, which comprise two immiscible fluids: oil and water" [25]. Nanoemulsions are made up of oil droplets that are dispersed in a continuous aqueous state. Surfactant molecules surround each oil droplet. Their size ranges from 10-100 nanometers [26]. Nanoemulsions are thermodynamically unstable systems because some nanoemulsions require a significant amount of energy to form. High-pressure homogenization, microfluidization, and

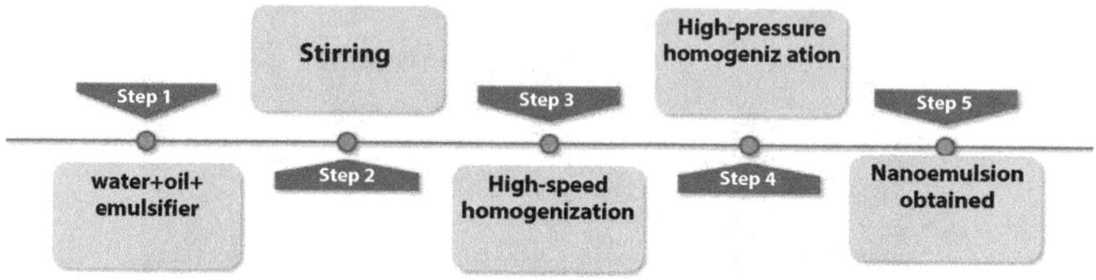

Figure 7.3 The general process to obtain kinetically stable nanoemulsion.

phase-inversion temperature method are the three most common processes employed for nanoemulsion production (Figure 7.3) [27].

In the food industry, nanoemulsions are used to make infused oils, fortified soft drinks and water, and other processed foods [28]. Nanoemulsions are observed to communicate with multiple bioactive systems, along with enzymes in the gastrointestinal tract. Their increased surface area aids in interaction with biological components of GIT. The nanoemulsion droplets found in GIT are easily digested by the enzyme lipase [29].

The nanoemulsion-based delivery system must be suitable to food matrix and have a minor impact on organoleptic characteristics like taste, visual appeal, and glossiness [30]. Antimicrobial agents found in nanoemulsions are known to be more effective against gram-positive bacteria. Nanoemulsions are also helpful in decontaminating food packaging materials [31]. Lemongrass oil nanoemulsion was used to assess the antimicrobial characteristics, as well as the physical and chemical changes in plums [32]. It was found capable of inhibiting *Salmonella* and *E. coli* population without changing the product's flavor, fracturability, and glossiness. Additionally, it was found to reduce the production of ethylene and delay the onset, retard the changes in lightness and concentration of phenolic compounds [33]. The growth of microbes has been assessed utilizing nanoemulsions formed from tributyl phosphate, soybean, or nonionic emulsifiers. As an outcome, food spoilage is reduced remarkably [34].

7.4.1.2 Nanoencapsulation

Nanoencapsulation is defined as the technology of packaging nanoparticles of solid, liquid, or gas, which are also called active or core, within a secondary material, named as shell or matrix, to form nanocapsules [35]. The active constituents (vitamins, prescription medications, fragrance, biotoxins, etc.) are contained in the core, whereas the shell segregates and safeguards the core from the external environment. This shielding can be temporary and permanent. The core is set to release by diffusion or in reaction to a stimulus, including enzymatic activity, pH, or shear stress, allowing for scheduled and controlled transfer to the targeted location [36]. Nanocapsules are found to have many advantages, including ease of processing, oxidation prevention, and enhanced mechanical stability. Its organoleptic impression lasts longer and has a higher absorbtivity rate in the body [37]. Nanocapsules guarantee food safety by capturing smell and undesired components in food while delivering the component to a specific location. Nanocapsules can transport their constituents through the gastrointestinal tract. This ability ensures bio-distribution and better absorption of nutritional supplements. Nanocapsules are also found to deliver lipid-soluble health products like minerals and multivitamins, boosting the food's nutrient content [38].

7.4.1.3 Nanoceuticals

Nanoceuticals are derived from the term "nutraceuticals", replacing the term "nutra" with "nano". As the name suggests, nanoceuticals are designed at the nano-scale size. It can be described as additives or natural ingredients, just differing in their size range, just like nutraceuticals. "Nutraceutical", a leading term for "nutrition" and "medicine", was invented in 1989 by Stephen L. DeFelice [39]. Converting "Nutra to nano" is a sensible technological advancement that is believed to enhance the integration and delivery to specific receptors, improve phytoconstituent availability, enhance flavor, incorporate antimicrobial activity nanoparticles into the food, and achieve durable storage life [40]. For efficient delivery of incorporated nutraceuticals, a variety of synthesis strategies such as nanoemulsions, micelles, nanoparticles, nanochealates, and nanocapsules have been used. The nanoformulation created allows for targeted distribution of encapsulated phytoconstituents along with the protracted release and improved bioavailability and therapeutic efficacy [41]. Nanoceuticals have the potential for preventive care based on their ability to provide therapeutic benefits with fewer side effects. In various ailments such as cancer [42], emotional and cognitive impairment [43], cardiovascular diseases, and even neurogenerative disorders [44], nanoceuticals have been shown to provide health-promoting and preventive effects.

Moreover, nanoceuticals are also found to combat severe medical conditions, like obesity, osteoporosis, arthritis, diabetes, and cholesterol [45]. Silver nanoparticles (AgNPs) are documented to be effective for wound healing [46]. AgNPs can stimulate keratinocyte differentiation and proliferation, leading to improved epidermal closure and re-epithelialization.

The secretion of anti-inflammatory cytokines is governed by neat silver nanoparticles (AgNPs), allowing faster non-hypertrophic scar-free wound repair [47]. According to research, novel polymeric nanoceuticals like chitosan are sustainable, strongly cationic, and get uniformly diffused with outstanding encapsulation efficiency for currently available antihypertensive drugs, like Calcium Chanel Blockers (CCB), Blockers (BB), and Angiotensin Converting Enzymes Inhibitors (ACEI) [48]. Nanoceuticals are a relatively new trend that is growing rapidly in a variety of supplement categories. Nanoparticles are incorporated to enhance the nutritional content of food [49]. Below are the examples of products related to nanoceuticals: Chinese nano-tea, which improves or stimulates selenium uptake; carotenoid particles are suspended in water and mixed with juice to enhance bio-availability; nutrients like vitamins and minerals are provided by canola oil-based nano-sized micelles [50, 51].

7.4.2 Food Packaging

Food packaging is the process of packaging food to protect it from microbial, physical, or chemical contamination or interference [52]. The current food economy revolves around food packaging. Good packaging reduces waste and ensures a longer shelf life [53]. The physical properties and chemical constituents used in packaging materials can impact the capacity to perform various functions. Transport, optical and mechanical characteristics, and chemical reactivity are the main properties to consider [54]. Food packaging can be optimized by the use of nanomaterials with Physico-chemical improvements like barrier properties, moisture and temperature resistance, durability, and mechanical toughness [55]. The packaging is usually improved by using nanomaterials with active functions, e.g., anti-oxidative, UV-resistant, and antimicrobial properties [56]. The prominent NMs utilized in food packaging are montmorillonite (MMT), zinc oxide (ZnO-NP), silver nanoparticles (AgNPs), and kaolinite [56]. Incorporating nanomaterials into packaging materials is feasible during the manufacturing process (e.g., extrusion or solvent casting) and through the formation of a layer (e.g., spray coating, electrospinning). This encapsulation process will protect active compounds such as antibacterial, antifungal, antioxidants and oxygen scavengers, allowing them to retain their properties for a longer duration [57]. This will enable the use of biologically active compounds that are generally ineligible for inclusion during packaging manufacture. This will also promote the usage of responsive materials, which will release the active component when prompted by extrinsic factors (e.g., pH and temperature [58]. One such example was produced utilizing smart poly (N-isopropyl acrylamide) nanohydrogels loaded with natamycin in polysaccharide films, where the natamycin release was temperature-responsive [59]. There are various types of food packaging currently in the application, such as active and antimicrobial-active packaging, improved food packaging, innovative packaging, and bio-based packaging.

7.4.2.1 Active Packaging

Active packaging integrates the active component applying nanotechnology into a food packaging material. Later, the carrier component is observed to facilitate interaction with internal or external stimuli. Active packaging is employed to prolong shelf life, nutritional content and ensure safe and fresh food [82]. Many food sectors require active packaging. This includes fresh case-ready meats, processed deli meats, bakeries, prepared foods, dairy goods, snack foods, tortillas, fresh pasta, and various beverages like beer and juice. Active Packaging systems are classified into active-releasing (emitters) systems that release chemicals to the headspace or packaged food and active-scavenging (absorbers) systems that eliminate unwanted materials from the food surface and area surrounding it. A few examples of active packaging are oxygen, moisture, ethylene absorbers, and antibacterial and carbon dioxide releasers. They are frequently packaged as a sachet that must be inserted into the food packaging. However, there are various disadvantages to such sachets, including the potential of unintentional breaking, which could result in the customer's involuntary intake of the content, incompatibility with beverages, and the need for an additional packaging operation step [83].

Furthermore, "customers in Europe are less accepting of sachet-based applications than in Asia or the United States" [84]. On the other hand, recent advances have precisely intended to incorporate active components into the polymer matrix. However, one of the most challenging aspects of developing such functional polymer films is ensuring that active agents have high activity and capacity while preserving the film's initial mechanical and physical qualities [85]. The transition of active components from the micro to nanometer scale offers many possibilities for decreasing the number of active components required while retaining the original attributes of the packaging material. Furthermore, compared to their microscaled counterparts, many active compounds perform better when applied at the nanoscale [86].

7.4.2.2 *Antimicrobial Packaging*

Antimicrobial packaging is an emerging technology designed to prevent food contamination in packaged food material [87]. Antimicrobial activity of a packaging system can be achieved in a variety of ways, including the use of antimicrobial volatile-containing pads or sachets, the application of packaging raw ingredients that are naturally antimicrobial, or the inclusion of antimicrobials into the packaging structure via co-extrusion, immobilization, or adsorption processes. There are three antimicrobial packaging mechanisms: (A) System dispersing antimicrobials: These systems need direct food contact. For example, organic acids are released directly via packing material into the food's surface. These systems necessitate intimate touch with the food. (B) Systems that release volatile antimicrobial agents from packaging materials, such as spice and herb extracts. There is no need for direct food interaction with these systems. (C) Antibacterial polymers that do not migrate and have an antimicrobial ingredient bonded to the polymer backbone. Food contact is essential in these systems since the antibacterial agent does not migrate out on its own [88]. According to a recent study, introducing zinc oxide nanoparticles (ZnO NPs) to pathogenic food-borne gram-positive (*B. cereus, S. aureus,* and *L. monocytogenes*) and gram-negative (*E. coli, S. typhimurium,* and *V. parahaemolyticus*) bacteria successfully inhibited their development. UV transmittance and oxygen permeability were lowered by 32% and 37%, respectively. Prepared films could be used to package meat items because of their considerable antibacterial action against the aforementioned microorganisms [89]. Recent research was conducted where polylactic (PLA) nanofibers containing Ethyly-Nα-dodecanoyl-L-arginate hydrochloride (LAE) were fabricated using the electrospinning method. Later on, their microstructure, thermal stability, and antimicrobial properties were assessed. Results indicated that active films with LAE could effectively extend the shelf life of strawberries at 25° Celsius. The active films showed remarkable antimicrobial properties against *E. coli O157:H7, Staphylococcus aureus*, and *Botrytis cinerea*. This study also suggested that PLA films incorporated with LAE could be used as an eco-friendly antimicrobial packaging material [90]. An investigation revealed that carbon nanotubes possess antibacterial properties. Direct contact with carbon nanotube granules is considered to be bactericidal for *E. coli*. This is due to the presence of thin and long CNTs, which penetrate inside microbial cells, causing irreversible harm [91].

7.4.2.3 *Physically Improved Food Packaging*

Physically improved food packaging aims to enhance physicomechanical characteristics like moisture stability, gas barrier, and heat resistance. Nanomaterials like nanoclays, metal oxide NPs and CNTs are commonly incorporated into the packaging [92]. Numerous nanomaterials have been constructed with reinforced polymer and nanostructures with NPs ranging from 5% to 5% (w/w), such as clay nanoparticle composites that improve barrier properties (80–90% reduction) in the making of certain beverages, and cooking oils [93]. Recent advances in nanolaminates and nanoclays are briefly discussed.

7.4.2.3.1 Nanolaminates

Nanolaminates are a type of composite material formed from two or more components with varying nanometric layers. Nanolaminates can have typical layer thicknesses ranging from a few atomic sheets to 100 nm [94]. The number of layers deposited, however, determines the final film thickness. Electrostatic bonding, hydrogen bonding, hydrophobic contacts, charge-transfer interactions, and other chemical interactions can be used to assemble the layers one by one [95]. In the future, power generation, logistics, and defense systems will require novel materials that can withstand temperature, stress, strain rate, and radiation extremes far beyond what today's most advancements can handle. Due to their excellent magnetic properties, electrical and material characteristics, nanolaminated materials may be able to meet future market expectations. Nanolaminated structures, unlike traditional materials, have a high density of interfaces, which gives them unique features. Interfaces serve as sources for nucleating plastic deformation carriers, barricades to prevent these carriers from spreading, and preferred sites for storing, reassembly, etc. [96]. For example, Cu–Nb-laminated composites have excellent heat resistance, high strength, shock resistance, and ion-irradiation-induced damage resistance [97].

Despite significant development in 2D-nanofiller-based composite nanolaminates, fabricating composite nanolaminates with improved ductility and toughness remains a challenge. A recent study obtained layered structure graphene oxide (GO)/gelatine powder (GP) composite nanolaminate with enhanced toughness and ductility by vacuum filtration of the aqueous dispersion number GO/GP composite solution containing 5wt% of GO. The composite film containing 5wt% of GO demonstrated an outstanding improvement of 200% in the stress at the break value; also, the toughness of the composite film significantly improved [98]. Thin films of nanolaminates are used to construct edible food-grade nanofilms, usually present on meat, chocolates, baked items, vegetables, etc. These nanolaminates mainly protect meals from moisture, lipids, and gases, improving food surface texture and acting as possible transporters of flavors, colors, antimicrobials, nutrients, and antioxidants. Recently, edible nanolaminates have been mostly synthesized from proteins, polysaccharides, and lipids [99].

7.4.2.3.2 Nanoclays

Nanoclays are defined as uniformly stacked silicates. Their single layer is estimated to have around 0.7 nm of thickness. Surfactants are usually utilized for the improvement of the plasticity and swelling capacity of the interlayer. The surface properties of nanoclays are determined by the charge of the nanosheet, the type of the surface atoms, and the presence of inter-layer exchangeable cations [100]. Nanoclays have a distinctive platelet shape [101]. To improve the properties of the polymers, various types of nanoclays like Montmorillonite (MMT) and organophilic MMT (organic modified MMT, OMMT) are incorporated. They have received significant attention owing to their high surface area with a relatively bigger aspect ratio (50–1000) and good compatibility with the majority of organic thermoplastics [102]. Recently, rice-flour gelatine nano clay film with catechin lysozyme was developed which induced mechanical improvement in film. The catechin-lysozyme nanocomposite film demonstrated high strength and increased shelf life. The developed film wrapped in pork had lower pH, and microbial growth [103]. A bio-nano composite was developed by mixing biopolymer chitosan with montmorillonite (MMK10) nanoclay and spirulina algae. The inclusion of MMT K10 and spirulina increased the surface color and opacity of chitosan bio nanocomposite films. The bio nanocomposite films had a high solubility and swelling property. The chitosan bio nanocomposite film with 0.5 g of spirulina demonstrated promising properties such as high tensile strength of 46.3 MPa, low oxygen and water vapor transmission rate of 996.79 cc/(m2.day.atm) and 7.12 g/m2/day, which outperformed other spirulina-based films of different concentrations [104].

7.4.2.4 Smart/Intelligent Packaging

Intelligent or smart packaging systems improve a package's communication aspect. This innovative packaging type identifies and reports any change in characteristics. Smart packaging helps to expand the global market. Such systems also help to comply with strict national and international food safety standards and also safeguard against food bioterrorism risks [105]. Nanosensors in packaging materials aid in sensing pathogens and chemical contaminants, resulting in a complete depiction of packaged food freshness [106]. Intelligent packaging with specialized preservatives is being devised, which releases preservatives as soon as the food spoilage occurs. A bio-switch is used in this type of "release on command" preservative packaging. Intelligent packaging consists of the following components: [107]

1. Time-Temperature Indicators (TT1)
2. Radio Frequency Identification Tags (RFID)
3. Sensors: Biosensors, gas sensors

Indicators make use of a distinguishing change, generally in coloration. Sensors identify and indicate the occurrence or absence of an analyte of interest as well as the intensity of chemical response [108]. Sensors vary from indicators in that they do not have receptor or transducer; instead, they communicate data through visual changes.

7.4.2.4.1 Time-Temperature Indicators

TTI (Time-Temperature Indicator) systems record the thermal behavior of food materials and clearly show the shelf life during packaging, delivery, and intake. This is achieved through an irreversible change in pigment brought on by the combined effects of time and temperature [109]. TTI gives consumers easily accessible visual information, allowing consumers to judge whether the food has deteriorated, ensuring food quality and safety. They also allow food producers to track their commodities throughout the logistics system. Timestrip has devised a refrigerated food system based on gold nanoparticles. When the device is beyond freezing temperature, it turns red, but the reddish pigment fades due to agglomeration of the gold NPs once it gets accidentally frozen [110]. A study devised a TTI based on a polyvinyl alcohol/Chitosan - anthocyanins complex (extracted from Red Cabbage) to indirectly indicate food quality changes by detecting changes in the pH when subjected to unsuited storage temperatures. Despite having a different modulus of elasticity than other food packaging polymers in the market, the developed Time-temperature indicator has appealing properties for usage in intelligent packaging [111]. These novel indicators are premised on Au-Ag Nanorods. They supervise the deterioration of pasteurized milk in the cold chain [112].

7.4.2.4.2 Radio Frequency Identification tags (RFIDs)

Radio Frequency Identification tags are inserted sensors in food packages. They could improve traceability and authentication (anti-counterfeiting of food material) and give instant and real-time reporting of conditions within the packaging headspace [113]. A reader transmits radio signals to collect data out of an RFID system. The recorded information is later transferred to a host device for tracking and assessment (that may be linked to the internet or nearby network). Single Wall Carbon-nanotube is a novel and feasible wireless-sensor [114]. A wireless H_2-sensing system can also detect H_2 in a sealed gas headspace by combining porous polymer-templated PdPt NPs (palladium-platinum nanoparticles) with a radio-frequency identification (RFID) system [115]. (Ru) integrated with (ZnO) nano-based RFID was documented to monitor ammonia gas of 100ppm concentration [116]. Moreover, technical difficulties associated with linking RFID sensing devices to the inner surface of packaged foods while reducing tag component exposure to food products, the full potential of nano-enabled RFID technology is still to be thoroughly investigated.

7.4.2.4.3 Gas Sensors

A gas sensor is an appliance that detects the presence of gases in a specific area. Gas sensors are frequently used in safety protocols. Air quality monitoring has now become a mandatory requirement in the contemporary age because of high pollution in the form of toxic gases (NO_2, NH_3, CO, CO_2, ethanol, methanol, etc.) [117]. Gas sensors are capable of monitoring and detecting these toxic gases. Gas sensors are divided into two categories based on the type of transducer: (1) optical and (2) electrochemical [118]. Optical oxygen sensors outperform conventional electro-chemical sensors because they do not fully exhaust oxygen and are not vulnerable to mixing. They have been employed in hazardous conditions, to perform remotely assessments, and to quantify gases and fluids [107]. A wireless gas sensor system based on a self-powered triboelectric nanogenerator (TENG) was newly innovated. It provided specific, sensitive, and wire-less food quality analysis [119]. Moreover, bilayer armchair graphene nanoribbon (B-AGNR) was also found to be a highly effective nanosensor for O_2 and CO gases [120].

7.4.2.5 Bio-Based Packaging

The term "bio-based food packaging materials" refers to "materials derived from renewable sources". Bio-based packaging materials are being developed on the assumption that they will have a lower carbon footprint than present petroleum-derived substances [121]. In general, bio-based packaging materials are pro-environmental because the decomposition of all products, like CO_2, water, and biomass, is entirely natural. The bio-based packaging adds renewable resources in packaging material for easy disposal. Nanotechnology improves barrier properties, adds bioactive components, senses important food information, changes the pervasion action of foils, and improves heat resistance and mechanical properties [122].

7.4.2.5.1 Starch and Derivatives

Starch is a prospective natural resource. It is generally obtained from plants. Mass commercial manufacturing of starch can fulfill present demands in lesser expenditure. Naturally, starch and its derivatives are unable to produce layered films with required physic-mechanical resilience. As an outcome, starch raw materials are firstly chemically treated [123]. It is transformed into a thermoplastic material after being altered with both thermo-mechanical energy in an extrusion process. Plasticizers are essential in producing thermoplastic starch granules because they reduce intra-molecular hydrogen bonds while ensuring the consistency of the product. As a result, there are numerous possibilities for thermoplastic starches to be used as packaging materials. It has been observed that date palm syrup (DPS), when added to the starch film, increases UV ray absorption properties and enhances durability [124]. Thermoplastic starch (TPS) can be fabricated to make foams and solid-shaped items at elevated temperatures and pressures. TPS are found to be brittle when stored for too long. Because of retrogradation, this brittleness worsens over time. Starch degrades in a wide range of surroundings, including soil and water/seawater. Environmental variables like microorganisms, abiotic components like heat, light, oxygen, moisture, and so on all have an impact on the biodegradability mechanism [125].

7.4.2.5.2 Polylactic Acid (PLA)

"Polylactic acid (PLA) is a recyclable and biodegradable polyester derived from renewable feedstock" [126]. Polylactic Acid is extensively electrospun to manipulate its functionality concerning end uses, including pure PLA, PLA mixtures, and nanostructured materials manufactured with metals, metal oxides, and CNT. Furthermore, drug compounds have already been incorporated in PLA electrospun mats for the targeted release of active components in the body. The pore size of electrospun PLA-nanofibers is compact, the porosity and the surface area are increased. Also, biodegradability is improved. As a result, PLA-nanofibers are being used in more manufacturing fields than ever before, including operational composites, diagnostic devices, filtration, synthetic biology, and high-tech drug delivery systems [127]. PLA has found applications in food packaging, textiles, and, more recently, engineering plastics.

7.4.2.5.3 Polybutylene Succinate (PBS)

PBS is a bio-polyester. PBS is synthesized via polycondensation of succinic acid and butanediol. The reaction is carried out in two stages. Firstly, esterification is observed between the di-acid and the diol. It is then followed by a polycondensation reaction at high temperatures. Thereafter, synthesis of high-molecular mass PBS takes place [128]. With many fascinating functionalities such as biodegradability, processing capability, and thermochemical resistance, PBS is a commonly available polyester [129].

7.4.2.5.4 Polycaprolactone (PCL)

"PCL is a semi-crystalline aliphatic polyester that is typically synthesized by ring-opening polymerization of caprolactone monomer using a diol, bivalent alcohol, and tin (II) or tin (10) salts as initiators" [130]. PCL is a highly functional bio-polymer for packaging products. Accessibility, eco-friendliness, and its physical characteristics make it unique and valuable. PCL is chemically and solvent resilient and tough. Because chain segments are mobile and intermolecular interactions are poor, the melting and glass transition temperatures are extremely low [131]. The disadvantage of PCL material is that it has poor antimicrobial properties. As a result, AgNPs are also mixed into the PCL composite to guarantee antimicrobial properties [132].

7.5 Nanotoxicity and Health Hazards

In the twenty-first century, fields ranging from medicine to the food industry have access to advanced nanotechnology. The way we utilize nanoparticles has revolutionized the traditional methods, but nothing great comes without a few drawbacks. Nanotoxicity is one of the major issues which is yet to be tackled. Nanoparticles are found to have a specific property of larger surface area/per unit volume. As a result, they behave entirely differently than their bulk counterparts [133]. For example, bulk gold is usually inert, but when converted into nanosize, it exhibits improved reactivity and other distinctive properties. Because of these distinct properties, gold NPs have large-scale applications ranging from being a drug carrier in the body to helping in diagnostic imaging. Nanoparticles are more likely to interact with bioactive molecules in cells like lipids and proteins. Nanoparticles can pass through cell membranes and enter different body parts, triggering inflammatory or immune responses [134]. AgNPs are known for being promising anti-cancer agents.

In contrast, a recent study showed that AgNPs were more toxic towards normal cell line HEK-293 (Human Embryonic Kidney 293 cells) than HeLa cells when used for cervical cancer therapy [135]. Nanoparticle toxicity is usually assessed using methods analogous to traditional toxicity analysis. A significant proportion of research has indicated that oxidative stress is an essential factor to consider when analyzing nanotoxicity. Growth in the number of oxidants in the cell is harmful. Much research has been done on air pollutants like carbon soot and other nanoscale effluents that generate ROS (Reactive Oxygen Species) and cause oxidative stress. This can gradually make an individual more prone to cardiovascular, neurodegenerative, and kidney-related diseases [136]. Nano-engineered materials pose a higher risk because of their increased reactivity and absorption in our bodies; they cause long-term clinical impacts and even genotoxicity. Several studies have monitored and assessed the health of factory floor laborers exposed to manufactured nanomaterials (MNMs). Lung function was observed to be reduced after exposure to a wide range of manufacturing nanomaterials (MNMs), including gold, silver, and CNT.

Furthermore, an increase in Clara cell protein 16 (a blood marker of airway damage), alterations in blood antioxidant and adhesion marker levels were observed in laborers exposed to MNMs [137]. Preclinical nanotoxicology is important for developing useful nanosafety systems and risk assessment tools. Monitored human exposure surveys can serve as a valuable addition to current approaches and a substantial upcoming move in nanotoxicological data analysis.

7.6 Nanotechnology in Food Industry: Regulatory Issues and Challenges

Nanotechnology has a variety of uses in food systems, from ingredients to packaging and product analysis. Besides the possible uses, their interplay with the food system elevates health concerns. Few nano-manufactured ingredients can be harmful for the ecosystem. There are no current legal laws limiting their use in agro-food industries. As a result, effective guidelines and rules are required for the safe use of NPs in food production. In the United States, the regulatory institution USFDA regulates production of nano-foods and packaging. Food Standards Australia and New Zealand (FSANZ), a regulatory body established under the Food Standards Code, regulate nano food additives and ingredients in Australia [138]. India stands among the top three for nanotech-based research and publication. The government invested Rs 10 billion in the Nano-Mission through the Department of Science and Technology. As a bio-diverse country, it may face hazards resulting from excess and unmonitored discharge of nanoparticles into the environment. As India shares land boundaries with nearby countries, it should regulate the exposure limits of nanoparticles. A statutory regulatory policy for nanoparticles and their uses must be outlined and executed [139]. Workplace safety and the environment are the significant implications of risks involved in nanomaterial applications. There are ethical, legislative, and sociocultural consequences to consider. Some countries have tried to partly control the amount or application of NPs due to increased global awareness.

7.7 Food Nanotech: Future Prospects and Conclusion

Nanotechnology's application in food systems is expected to grow in the coming years. Consumer acceptance is, of course, essential for success. Many nanostructured substances (NSMs) like inorganic metals, metal-oxide-based NPs, and bioactive nano-organic components have been applied to various food components. The individuals exposed to these nanomaterials have gradually observed negative health consequences. Nanomaterial exposure in the environment has become a matter of social interest and concern. The advancement of nanosensors to detect microbial and chemical pathogens can help control human health risksthrough food analysis. Nanosensors, nano-based food packaging, and nanoceuticals are anticipated to advance even more in the future. The inclusion of nano-science in a range of possible processes further enhances responsiveness, efficiency, reliability, and functionalization. At the same time, limited awareness regarding the consequences of nanoparticles on health and the environment is a severe impediment to the real-life application of nanotechnology. To optimize the current means of assessing health hazards, risk mitigation, and regulatory frameworks in nanotechnology it is necessary to establish effective governance and regulatory system in the field of food nanotechnology. This undoubtedly involves coordination between nanoparticle developers, risk assessors, regulatory bodies, and research scientists.

References

1. Dingman, Jim. "Guest commentary: nanotechnology: its impact on food safety." *Journal of Environmental Health* 70.6, 47-50, 2008.
2. Bhushan, Bharat, and Baumann. *Springer Handbook of Nanotechnology*. Ed. Bharat Bhushan. Vol. 2. Berlin: Springer, 2007.
3. Doyle, M. E. "Nanotechnology: A brief literature review. Food Research Institute briefings." 2006.
4. Xu, Xi-Jin, *et al.* "Preparation and formation mechanism of ZnS semiconductor nanowires made by the electrochemical deposition method." *Nanotechnology* 17.2, 426, 2005.
5. Aguilera, Jose Miguel. "Why food microstructure?" *Journal of Food Engineering* 67.1-2, 3-11, 2005.
6. Chaudhry, Qasim, *et al.* "Applications and implications of nanotechnologies for the food sector." *Food Additives and Contaminants* 25.3, 241-258, 2008.
7. McClements, David Julian, *et al.* "Structural design principles for delivery of bioactive components in nutraceuticals and functional foods." *Critical Reviews in Food Science and Nutrition* 49.6, 577-606, 2009.
8. Morris, V. J., and R. Parker. "Natural and designed self-assembled nanostructures in foods." *The World of Food Science: Food Nanotechnology* 4, 2008.
9. Bugusu, Betty, *et al.* "Global regulatory policies on food nanotechnology." *Food Technology* (Chicago) 63.5, 24-28, 2009.
10. Anema, Skelte G. "Spontaneous interaction of lactoferrin with casein micelles or individual caseins." *Journal of the Royal Society of New Zealand* 48.2-3, 89-110, 2018.
11. Fox, Patrick F., Paul LH McSweeney, and L. H. Paul. "Dairy chemistry and biochemistry", F6, 1998.
12. Semo, Efrat, *et al.* "Casein micelle as a natural nano-capsular vehicle for nutraceuticals." *Food Hydrocolloids* 21.5-6, 936-942, 2007.
13. Mohammadian, Mehdi, and Ashkan Madadlou. "Characterization of fibrillated antioxidant whey protein hydrolysate and comparison with fibrillated protein solution." *Food Hydrocolloids* 52, 221-230, 2016.
14. Nicolai, Taco. "Formation and functionality of self-assembled whey protein microgels." *Colloids and Surfaces B: Biointerfaces* 137, 32-38, 2016.
15. Yaktine, Ann, and Leslie Pray, eds. *Nanotechnology in Food Products: Workshop Summary*. National Academies Press, 2009.
16. Pan, Kang, and Qixin Zhong. "Organic nanoparticles in foods: fabrication, characterization, and utilization." *Annual Review of Food Science and Technology* 7, 245-266. 2016.

17. Meena, Jairam, *et al.* "Inorganic nanoparticles for natural product delivery: a review." *Environmental Chemistry Letters,* 1-12, 2020.
18. Singh, Trepti, *et al.* "Application of nanotechnology in food science: perception and overview." *Frontiers in Microbiology* 8, 1501, 2017.
19. Nile, Shivraj Hariram, *et al.* "Nanotechnologies in food science: applications, recent trends, and future perspectives." *Nano-Micro Letters* 12.1, 1-34, 2020.
20. Kuddus, Mohammed. "Enzymes in food biotechnology." 2019.
21. Varelis, Peter, Laurence Melton, and Fereidoon Shahidi. *Encyclopedia of Food Chemistry.* Elsevier, 2018.
22. Huang, Qingrong, Hailong Yu, and Qiaomei Ru. "Bioavailability and delivery of nutraceuticals using nanotechnology." *Journal of Food Science* 75.1, R50-R57, 2010.
23. Pradhan, Neha, *et al.* "Facets of nanotechnology as seen in food processing, packaging, and preservation industry." *BioMed Research International,* 2015.
24. Weiss, Jochen, Paul Takhistov, and D. Julian McClements. "Functional materials in food nanotechnology." *Journal of Food Science* 71.9, R107, 2006.
25. Shunmugaperumal, Tamilvanan, Raja SekharanThenrajan, and Ruchi Sharma. "Multifunctional nanosized emulsions for theragnosis of life threatening diseases." *Nanostructures for Drug Delivery.* Elsevier, 579-617, 2017.
26. Qian, Cheng, and David Julian McClements. "Formation of nanoemulsions stabilized by model food-grade emulsifiers using high-pressure homogenization: Factors affecting particle size." *Food Hydrocolloids* 25.5, 1000-1008, 2011.
27. Simonazzi, Analía, *et al.* "Nanotechnology applications in drug controlled release." In *Drug Targeting and Stimuli Sensitive Drug Delivery Systems.* William Andrew Publishing, 81-116, 2018.
28. Borthakur, Priyakshree, *et al.* "Nanoemulsion: Preparation and its application in food industry." *Emulsions.* Academic Press, 153-191, 2016.
29. Salvia-Trujillo, Laura, *et al.* "Influence of particle size on lipid digestion and β-carotene bioaccessibility in emulsions and nanoemulsions." *Food Chemistry* 141.2, 1472-1480, 2013.
30. Aswathanarayan, Jamuna Bai, and Ravishankar Rai Vittal. "Nanoemulsions and their potential applications in food industry." *Frontiers in Sustainable Food Systems* 3, 95, 2019.
31. Wang, Yan, *et al.* "Characterisation and cooperative antimicrobial properties of chitosan/nano-ZnO composite nanofibrous membranes." *Food Chemistry* 132.1, 419-427, 2012.
32. Kim, In-Hah, *et al.* "Plum coatings of lemongrass oil-incorporating carnauba wax-based nanoemulsion." *Journal of Food Science* 78.10, E1551-E1559, 2013.
33. Amaral, Daniel Mathias Fuzette, and Kanika Bhargava. "Essential oil nanoemulsions and food applications." *Adv Food TechnolNutr Sci Open J* 1.4, 84-87, 2015.

34. Sanguansri, Peerasak, and Mary Ann Augustin. "Nanoscale materials development–a food industry perspective." *Trends in Food Science & Technology* 17.10, 547-556, 2006.
35. Augustin, Mary Ann, and Yacine Hemar. "Nano- and micro-structured assemblies for encapsulation of food ingredients." *Chemical Society Reviews* 38.4, 902-912, 2009.
36. Jyothi, N. Venkata Naga, *et al.* "Microencapsulation techniques, factors influencing encapsulation efficiency." *Journal of Microencapsulation* 27.3,187-197, 2010.
37. Chaudhry, Qasim, *et al.* "Applications and implications of nanotechnologies for the food sector." *Food Additives and Contaminants* 25.3, 241-258, 2008.
38. Dreher, Kevin L. "Health and environmental impact of nanotechnology: toxicological assessment of manufactured nanoparticles." *Toxicological Sciences* 77.1, 3-5, 2004.
39. Kalra, Ekta K. "Nutraceutical-definition and introduction." *AapsPharmsci* 5.3, 27-28, 2003.
40. Chellaram, C., *et al.* "Significance of nanotechnology in food industry." *APCBEE Procedia* 8 109-113, 2014.
41. Kakkar, Vandita, Manoj Kumar, and Komal Saini. "Nanoceuticals governance and market review." *Environmental Chemistry Letters* 16.4,1293-1300, 2018.
42. Kakkar, Vandita, and Indu Pal Kaur. "Antidepressant activity of curcumin loaded solid lipid nanoparticles (C-SLNs) in mice." *Am J Pharm Res* 2.3, 729-736, 2012.
43. Kakkar, Vandita, *et al.* "Delivery of sesamol-loaded solid lipid nanoparticles to the brain for menopause-related emotional and cognitive central nervous system derangements." *Rejuvenation Research* 14.6, 597-604, 2011.
44. Kakkar, Vandita, *et al.* "Curcumin loaded solid lipid nanoparticles: an efficient formulation approach for cerebral ischemic reperfusion injury in rats." *European Journal of Pharmaceutics and Biopharmaceutics* 85.3, 339-345, 2013.
45. Mtewa, Andrew G. "Introduction to phytochemicals and the military." *Phytochemistry, the Military and Health.* Elsevier, 3-11, 2021.
46. Chakrabarti, Srijita, *et al.* "Aspects of nanomaterials in wound healing." *Current Drug Delivery* 16.1,26-41, 2019.
47. Parani, Madasamy, *et al.* "Engineered nanomaterials for infection control and healing acute and chronic wounds." *ACS Applied Materials & Interfaces* 8.16, 10049-10069, 2016.
48. Niaz, Taskeen, *et al.* "Antihypertensive nano-ceuticales based on chitosan biopolymer: Physico-chemical evaluation and release kinetics." *Carbohydrate Polymers* 142, 268-274, 2016.
49. Chen, Lingyun, Gabriel E. Remondetto, and Muriel Subirade. "Food protein-based materials as nutraceutical delivery systems." *Trends in Food Science & Technology* 17.5, 272-283, 2006.

50. Qureshi, M. A., *et al.* "Application of nanotechnology in food and dairy processing: an overview." *Pak J Food Sci* 22.1, 23-31, 2012.
51. Khan, M. I., *et al.* "Nanotechniques used in food industries: A mini review." *J. Food Biotechnol. Res* 2.1, 6, 2018.
52. Cachon, Remy, and Philippe Girardon, eds. *Gases in Agro-food Processes.* Academic Press, 2019.
53. Robertson, Gordon L., ed. *Food Packaging and Shelf Life: A Practical Guide.* CRC Press, 2009.
54. Berk, Zeki. *Food Process Engineering and Technology.* Academic Press, 2018.
55. Primožič, Mateja, Željko Knez, and Maja Leitgeb. "(Bio) Nanotechnology in Food Science—Food Packaging." *Nanomaterials* 11.2, 292, 2021.
56. Chaudhary, Pallavi, Faria Fatima, and Ankur Kumar. "Relevance of nanomaterials in food packaging and its advanced future prospects." *Journal of Inorganic and Organometallic Polymers and Materials* 30.12, 5180-5192, 2020.
57. Cerqueira, Miguel A., António A. Vicente, and Lorenzo M. Pastrana. "Nanotechnology in food packaging: opportunities and challenges." *Nanomaterials for Food Packaging* 1-11, 2018.
58. Brockgreitens, John, and Abdennour Abbas. "Responsive food packaging: Recent progress and technological prospects." *Comprehensive Reviews in Food Science and Food Safety* 15.1, 3-15, 2016.
59. Cerqueira, Miguel A., *et al.* "Development of active and nanotechnology-based smart edible packaging systems: physical–chemical characterization." *Food and Bioprocess Technology* 7.5, 1472-1482, 2014.
60. Pan, Kang, and Qixin Zhong. "Organic nanoparticles in foods: fabrication, characterization, and utilization." *Annual Review of Food Science and Technology* 7, 245-266, 2016.
61. Drexler, K. Eric. "Molecular engineering: An approach to the development of general capabilities for molecular manipulation." *Proceedings of the National Academy of Sciences* 78.9, 5275-5278, 1981.
62. Ealia, S. Anu Mary, and M. P. Saravanakumar. "A review on the classification, characterisation, synthesis of nanoparticles and their application." *IOP Conference Series: Materials Science and Engineering.* Vol. 263. No. 3. IOP Publishing, 2017.
63. Kümmerer, Klaus, *et al.* "Biodegradability of organic nanoparticles in the aqueous environment." *Chemosphere* 82.10, 1387-1392, 2011.
64. Tiwari, Dhermendra K., J. Behari, and Prasenjit Sen. "Application of nanoparticles in waste water treatment 1", 2008.
65. Virlan, Maria Justina Roxana, *et al.* "Organic nanomaterials and their applications in the treatment of oral diseases." *Molecules* 21.2. 207, 2016.
66. Akbarzadeh, Abolfazl, *et al.* "Liposome: classification, preparation, and applications." *Nanoscale Research Letters* 8.1, 1-9, 2013.

67. Atrooz, O. M. "Effects of alkylresorcinolic lipids obtained from acetonic extract of Jordanian wheat grains on liposome properties." *Int J Biol Chem* 5.5, 314-321, 2011.
68. Benech, R-O., *et al.* "Inhibition of Listeria innocua in cheddar cheese by addition of nisin Z in liposomes or by *in situ* production in mixed culture." *Applied and Environmental Microbiology* 68.8, 3683-3690, 2002.
69. Shehata, Tamer, *et al.* "Prolongation of residence time of liposome by surface-modification with mixture of hydrophilic polymers." *International Journal of Pharmaceutics* 359.1-2, 272-279, 2008.
70. Sampathkumar, Srinivasa-Gopalan, and Kevin J. Yarema. "Dendrimers in cancer treatment and diagnosis." *Nanotechnologies for the Life Sciences: Online*, 2007.
71. Mittal, Pooja, *et al.* "Dendrimers: A New Race of Pharmaceutical Nanocarriers." *BioMed Research International*, 2021.
72. Lyu, Zhenbin, *et al.* "Self-assembling supramolecular dendrimers for biomedical applications: Lessons learned from poly (amidoamine) dendrimers." *Accounts of Chemical Research* 53.12, 2936-2949, 2020.
73. Patel, Vruti, *et al.* "Dendrimers as novel drug-delivery system and its applications." *Drug Delivery Systems*. Academic Press, 333-392, 2020.
74. Aguilar, Zoraida. *Nanomaterials for Medical Applications*. Newnes, 2012.
75. Grumezescu, Alexandru Mihai, and Anton Ficai, eds. *Nanostructures for Cancer Therapy*. Elsevier, 2017.
76. Husseini, Ghaleb A., and William G. Pitt. "Micelles and nanoparticles for ultrasonic drug and gene delivery." *Advanced Drug Delivery Reviews* 60.10, 1137-1152, 2008.
77. Testa, Caterina, *et al.* "Catalysis with carbon nanoparticles." *RSC Advances* 9.47, 27659-27664, 2019.
78. Patel, Kapil D., Rajendra K. Singh, and Hae-Won Kim. "Carbon-based nanomaterials as an emerging platform for theranostics." *Materials Horizons* 6.3, 434-469, 2019.
79. McClements, David Julian, and Hang Xiao. "Is nano safe in foods? Establishing the factors impacting the gastrointestinal fate and toxicity of organic and inorganic food-grade nanoparticles." *npj Science of Food* 1.1, 1-13, 2017.
80. Salavati-Niasari, Masoud, Fatemeh Davar, and Noshin Mir. "Synthesis and characterization of metallic copper nanoparticles via thermal decomposition." *Polyhedron* 27.17, 3514-3518, 2008.
81. Tai, Clifford Y., *et al.* "Synthesis of magnesium hydroxide and oxide nanoparticles using a spinning disk reactor." *Industrial & Engineering Chemistry Research* 46.17, 5536-5541, 2007.
82. Lim, L. T. "Active and intelligent packaging materials." *Comprehensive Biotechnology*. 629-644, 2011.
83. Yildirim, Selçuk, and Bettina Röcker. "Active packaging." *Nanomaterials for Food Packaging*. Elsevier, 173-202, 2018.

84. Restuccia, Donatella, *et al.* "New EU regulation aspects and global market of active and intelligent packaging for food industry applications." *Food Control* 21.11, 1425-1435, 2010.
85. Silvestre, Clara, and Sossio Cimmino, eds. *Ecosustainable Polymer Nanomaterials for Food Packaging: Innovative Solutions, Characterization Needs, Safety and Environmental Issues*. CRC Press, 2013.
86. Imran, Muhammad, *et al.* "Active food packaging evolution: transformation from micro- to nanotechnology." *Critical Reviews in Food Science and Nutrition* 50.9, 799-821, 2010.
87. Yildirim, Selçuk, and Bettina Röcker. "Active packaging." *Nanomaterials for Food Packaging*. Elsevier, 173-202, 2018.
88. Lacroix, Christophe, ed. *Protective Cultures, Antimicrobial Metabolites and Bacteriophages for Food and Beverage Biopreservation*. Elsevier, 2010.
89. Saedi, Shahab, *et al.* "Semi-transparent regenerated cellulose/ZnONP nanocomposite film as a potential antimicrobial food packaging material." *Journal of Food Engineering* 307, 110665, 2021.
90. Li, Ting, *et al.* "Development of electrospun films enriched with ethyl lauroylarginate as novel antimicrobial food packaging materials for fresh strawberry preservation." *Food Control* 130, 108371, 2021.
91. Kang, Seoktae, *et al.* "Single-walled carbon nanotubes exhibit strong antimicrobial activity." *Langmuir* 23.17, 8670-8673, 2007.
92. Kuswandi, Bambang, and Mehran Moradi. "Improvement of food packaging based on functional nanomaterial." *Nanotechnology: Applications in Energy, Drug and Food*. Springer, Cham, 309-344, 2019.
93. Brody, Aaron L. "Case studies on nanotechnologies for food packaging." *Food Technology*, 2007.
94. Azadmanjiri, Jalal, *et al.* "Nanolaminated composite materials: structure, interface role and applications." *RSC Advances* 6.111, 109361-109385, 2016.
95. Borges, João, *et al.* "Layer-by-layer assembly of light-responsive polymeric multilayer systems." *Advanced Functional Materials* 24.36, 5624-5648, 2014.
96. Zheng, Shijian, *et al.* "High-strength and thermally stable bulk nanolayered composites due to twin-induced interfaces." *Nature Communications* 4.1, 1-8, 2013.
97. Demkowicz, M. J., R. G. Hoagland, and J. P. Hirth. "Interface structure and radiation damage resistance in Cu-Nb multilayer nanocomposites." *Physical Review Letters* 100.13, 136102, 2008.
98. Layek, Rama Kanta, *et al.* "Tailoring of Physical and Mechanical Property of Biocompatible Graphene Oxide/Gelatin Composite Nanolaminates via Altering of Crystal Structure and Morphology." *Materials Advances*, 2021.
99. Mandal, Abhishek, and Ena Ray Banerjee. "Introduction to Nanoscience, Nanotechnology and Nanoparticles." *Nanomaterials and Biomedicine*. Springer, Singapore, 1-39, 2020.
100. Kausar, Ayesha. "Flame retardant potential of clay nanoparticles." *Clay Nanoparticles*. Elsevier, 169-184, 2020.

101. Ganguly, Saheli, *et al.* "Organophilic nano clay: A comprehensive review." *Transactions of the Indian Ceramic Society* 70.4, 189-206, 2011.
102. Golebiewski, Jan, *et al.* "Low density polyethylene–montmorillonite nanocomposites for film blowing." *European Polymer Journal* 44.2, 270-286, 2008.
103. Pattarasiriroj, Kanokwan, Pimonpan Kaewprachu, and Saroat Rawdkuen. "Properties of rice flour-gelatine-nanoclay film with catechin-lysozyme and its use for pork belly wrapping." *Food Hydrocolloids* 107, 105951, 2020.
104. Ramji, V., and M. Vishnuvarthanan. "Chitosan Ternary Bio Nanocomposite Films Incorporated with MMT K10 Nanoclay and Spirulina." *Silicon*, 1-12, 2021.
105. Yam, Kit L., Paul T. Takhistov, and Joseph Miltz. "Intelligent packaging: concepts and applications." *Journal of Food Science* 70.1, R1-R10, 2005.
106. Liao, Frank, Christopher Chen, and Vivek Subramanian. "Organic TFTs as gas sensors for electronic nose applications." *Sensors and Actuators B: Chemical* 107.2, 849-855, 2005.
107. Kerry, J. P., M. N. O'grady, and S. A. Hogan. "Past, current and potential utilisation of active and intelligent packaging systems for meat and muscle-based products: A review." *Meat Science* 74.1, 113-130, 2006.
108. Lee, Seung Ju, and ATM Mijanur Rahman. "Intelligent packaging for food products." In *Innovations in Food Packaging*. Academic Press, 171-209, 2014.
109. Wang, Shaodong, *et al.* "Review of time temperature indicators as quality monitors in food packaging." *Packaging Technology and Science* 28.10, 839-867, 2015.
110. Robinson, D. K. R., and M. J. Morrison. "Nanotechnologies for food packaging: reporting the science and technology research trends: report for the Observatory NANO." *Phytotherapy Res* 15, 476-480, 2010.
111. Pereira Jr, Valdir Aniceto, Iza Natália Queiroz de Arruda, and Ricardo Stefani. "Active chitosan/PVA films with anthocyanins from Brassica oleraceae (Red Cabbage) as Time–Temperature Indicators for application in intelligent food packaging." *Food Hydrocolloids* 43, 180-188, 2015.
112. Gao, Tingting, *et al.* "Gold–silver core-shell nanorods based time-temperature indicator for quality monitoring of pasteurized milk in the cold chain." *Journal of Food Engineering* 306, 110624, 2021.
113. Wang, Yun, and Evangelyn C. Alocilja. "Sensor technologies for anti-counterfeiting." *International Journal of Comparative and Applied Criminal Justice* 36.4, 291-304, 2012.
114. Sharrock, K. R. "Advances in freshness and safety indicators in food and beverage packaging." *Emerging Food Packaging Technologies*. Woodhead Publishing, 175-197, 2012.
115. Koo, Won-Tae, *et al.* "Hydrogen sensors from composites of ultra-small bimetallic nanoparticles and porous ion-exchange polymers." *Chem* 6.10, 2746-2758, 2020.

116. Ali, Ibrahim, *et al.* "Ruthenium (Ru) doped zinc oxide nanostructure-based radio frequency identification (RFID) gas sensors for NH3 detection." *Journal of Materials Research and Technology* 9.6, 15693-15704, 2020.
117. Amrane, Abdeltif, *et al.*, eds. *Nanomaterials for Air Remediation.* Elsevier, 2020.
118. Puligundla, Pradeep, Junho Jung, and Sanghoon Ko. "Carbon dioxide sensors for intelligent food packaging applications." *Food Control* 25.1, 328-333, 2012.
119. Cai, Chenchen, *et al.* "Integration of a porous wood-based triboelectric nanogenerator and gas sensor for real-time wireless food-quality assessment." *Nano Energy* 83, 105833, 2021.
120. Kheirabadi, Sharieh Jamalzadeh, *et al.* "Selective gas sensor based on bilayer armchair graphene nanoribbon." *Physica E: Low-dimensional Systems and Nanostructures* 129, 114635, 2021.
121. Robertson, G. "State-of-the-art biobased food packaging materials." *Environmentally Compatible Food Packaging.* Woodhead Publishing, 3-28, 2008.
122. Berekaa, Mahmoud M. "Nanotechnology in food industry; advances in food processing, packaging and food safety." *Int J CurrMicrobiol App Sci* 4.5, 345-357, 2015.
123. Gonera, Antje, and Paul Cornillon. "Gelatinization of starch/gum/sugar systems studied by using DSC, NMR, and CSLM." *Starch-stärke* 54.11, 508-516, 2002.
124. Nikvarz, Navid, Gholam Reza Khayati, and Shahriar Sharafi. "Bio-based ultraviolet protective packaging film preparation using starch with incorporated date palm syrup." *Materials Chemistry and Physics*, 124794, 2021.
125. Bertolini, Andréa C. "Trends in starch applications." *Starches, Characterization, Properties, and Applications*, 1-19, 2009.
126. Avar, G., *et al.* "Polymer science: a comprehensive reference." *Polymer Science: A Comprehensive Reference* 10, 411-441, 2012.
127. Misra, M., *et al.* "Nanotechnologies for agricultural bioproducts." 111-119, 2011.
128. Takiyama, E., and T. Fujimaki. ""bionolle" biodegradable plastic through chemical synthesis." *Studies in Polymer Science.* Vol. 12. Elsevier, 150-174, 1994.
129. Ray, Suprakas Sinha, and MostoBousmina. "Biodegradable polymers and their layered silicate nanocomposites: in greening the 21st century materials world." *Progress in Materials Science* 50.8, 962-1079, 2005.
130. McKeen, Laurence W. *Permeability Properties of Plastics and Elastomers.* William Andrew, 2016.
131. Sikka, Monica Puri, and Vinay Kumar Midha. "The role of biopolymers and biodegradable polymeric dressings in managing chronic wounds." *Advanced Textiles for Wound Care.* Woodhead Publishing, 463-488, 2019.

132. Deshmukh, K., *et al.* "Biopolymer composites with high dielectric performance: interface engineering." *Biopolymer Composites in Electronics*. Elsevier, 27-128, 2017.
133. Daniel, Marie-Christine, and Didier Astruc. "Gold nanoparticles: assembly, supramolecular chemistry, quantum-size-related properties, and applications toward biology, catalysis, and nanotechnology." *Chemical Reviews* 104.1,293-346, 2004.
134. Gojova, Andrea, *et al.* "Induction of inflammation in vascular endothelial cells by metal oxide nanoparticles: effect of particle composition." *Environmental Health Perspectives* 115.3, 403-409, 2007.
135. Liu, Xiongwei, *et al.* "Nanotoxic effects of silver nanoparticles on normal HEK-293 cells in comparison to cancerous HeLa cell line." *International Journal of Nanomedicine* 16, 753, 2021.
136. Pizzino, Gabriele, *et al.* "Oxidative stress: harms and benefits for human health." *Oxidative Medicine and Cellular Longevity*, 2017.
137. Liao, Hui-Yi, *et al.* "Six-month follow-up study of health markers of nanomaterials among workers handling engineered nanomaterials." *Nanotoxicology* 8. sup1, 100-110, 2014.
138. Bowman, Diana M., and Graeme A. Hodge. "Nanotechnology: Mapping the wild regulatory frontier." *Futures* 38.9, 1060-1073, 2006.
139. Mishra, Mansi, *et al.* "Prospects, challenges and need for regulation of nanotechnology with special reference to India." *Ecotoxicology and Environmental Safety* 171, 677-682, 2019.

8

Nutritional Biomarkers in Metabolic Disorders

Komal Dagar[1], Mrinal Samtiya[1], Veda Krishnan[2], Sunil K. Srivastava[3], Ananya Rana[4*], Anita Kumari[1], Sanjeev Kumar[5] and Tejpal Dhewa[1]

[1]*Department of Nutrition Biology, Central University of Haryana, Mahendergarh, India*
[2]*Division of Biochemistry, ICAR-Indian Agricultural Research Institute, New Delhi, India*
[3]*Department of Microbiology, Swami Shraddhanand College (University of Delhi), New Delhi, India*
[4]*Department of Basic and Applied Sciences, National Institute of Food Technology Entrepreneurship and Management, Sonepat, New Delhi, India*
[5]*Department of Life Science and Bioinformatics, Assam University, Silchar, Assam, India*

Abstract

Metabolic syndrome is more than ever prevalent in the current scenario in both developing and developed countries. Most of them remain hidden, get complicated, get fatal over time, and have been well related to diet and lifestyle. Advancement in nutritional biomarkers have been in this direction and been majorly classified as various types of serum biomarkers including adipokines, neuropeptides, inflammatory cytokines, oxidized low-density lipoproteins, paraoxonase-1, and novel biomarkers like pentraxin-3, and hyocholic acid are closely related in earlier detection of various metabolic syndromes. Still, the concentration of these biomarkers may be increased or decreased in the patients depending upon the nature and type of metabolic syndrome. The level of biomarkers that usually increases in the metabolic syndrome includes adipokine (leptin), pro-inflammatory cytokine (IL-6), and oxidized low-density lipoproteins. In contrast to it, biomarkers whose concentration decreases with the metabolic syndrome are adipokine (adiponectin), neuropeptide (ghrelin), anti-inflammatory cytokine (IL-10), and antioxidant (paraoxonase-1). Novel biomarkers like HCA and PTX-3 are also positively

**Corresponding author*: ananya.rana.333@gmail.com

Tejpal Dhewa, Anil Kumar Puniya and Anil Panghal (eds.) Nutritional Science and Technology: Concept to Application, (165–198) © 2023 Scrivener Publishing LLC

related to metabolic syndromes and their severity. In subsequent sections, these serum biomarkers, their mechanism of action, and their interrelation with metabolic syndromes are discussed. Nutritional biomarkers have several limitations along with applications. These biomarkers are quite beneficial to the physicians for earlier treatment and in keeping a close watch on COVID-19 patients also, and their levels vary with the severity of COVID. The rising field of metabolomics aid in the advanced development of nutritional biomarkers, but progress in food metabolome databases is still required.

Keywords: Metabolic syndrome, nutritional biomarkers, cardiovascular disorders, metabolomics, serum biomarkers, adipokines

8.1 Introduction

An estimated 20–25% of the general population suffers from a metabolic condition known as the metabolic syndrome or disorder [1]. Mostly hidden at an earlier stage, this syndrome has been associated majorly with diet and lifestyle. Under/over and imbalanced nutrition over a period of time cause cascades of metabolic abnormalities resulting in metabolic syndromes. Mostly being associated with inflammatory cascades, these result in a considerable increase in Cardiovascular Disease (CVD) risk in an individual [2]. A person with metabolic syndrome is more likely to develop fatty liver, cholesterol gallstones, sleep difficulties, polycystic ovary syndrome, asthma, and several forms of cancer, including breast cancer, prostate cancer, colorectal cancer, and pancreatic cancer [3]. The condition of "insulin resistance" can also be seen in people with metabolic syndrome, which ultimately leads to an increase in the risk of type 2 diabetes. There is an excellent interrelation between the diet that an individual consumes and the risk of metabolic disorders and their severity in that particular individual. A group of researchers conducted a study on aged non-human primates (NHP) models fed with high fat/sugar diet (HFSD) and were able to successfully prove that the aged non-human primates (NHP) models fed with a high fat/sugar diet (HFSD) exhibit metabolic disorders and cardiac contractile dysfunction [4]. Undernourishment in terms of micronutrients, known as hidden hunger, has also been associated with metabolic syndromes as these micronutrients act as cofactors for various metabolic enzymes.

"Any material that is employed as an 'indicator' of a biological condition" is how a biomarker is defined. Biomarkers are "biological or chemical test findings in an investigated biological material that are associated to particular exposure, liability, or biological impacts," according to one definition [5].

These indicators are now highly useful in the early diagnosis and categorization of risk in individuals with metabolic syndrome (MeS) [1]. In terms of consumption and metabolism of necessary nutrients and dietary constituents, the nutritional biomarkers may be defined as any biological sample generally used to measure nutritional status (Potischman and Freudenheim, 2003) [6]. These are generally the hints left by the body that disclose the food we consume and indicate typically the actual amount of nutrients that reach the tissues after absorption and metabolism. For instance, in urine, nitrogen is used as a nutritional biomarker of protein intake in the diet.

An "ideal" dietary biomarker is one that is sensitive, specific, and relevant to a wide range of groups. It would also appropriately deliberate its food consumption amount [7]. These biomarkers are clinically helpful particularly in determining the insufficiency for aiding medical intervention and treatment (for example; Assessment of vitamins like vitamin B9, vitamin B12 and minerals like iron, copper, and zinc are beneficial to recognize the possible nutritional causes of anemia) (LN) [8] and nutritional adequacy (for example, the combination of serum ferritin and transferrin receptor are used to determine the deficiency of iron (DAPA Measurement toolkit). A single biomarker itself is sufficient and liable to identify the metabolic syndrome. It will be hard to tell. Instead of a single or individual biomarker, a group of different biomarkers is considered helpful because individual biomarkers have different roles and action mechanisms [2].

Jenab classified biomarkers on the basis of their intended use and biokinetics as replacement, recovery, predictive, and concentration biomarkers [7]. On the other hand, Lars O. said that by focusing only on their use, generally subdivides them as susceptibility, exposure, and effect biomarkers [9]. The biomarkers of nutritional status do not need to be only the nutrients. Some other compounds can also serve as dietary biomarkers. For example, an elevated level of Methylmalonic acid in urine or blood serves as early and sensitive indicator of deficiency of vitamin B12. The lower level of Erythrocyte glutathione reductase and RBC Riboflavin are the evidence of deficiency of Riboflavin. An elevated level of Homocysteine indicates the lack of vitamin B6, vitamin B12, and Folic acid.

The strength of nutritional biomarkers include that they avoid the recall and social desirability-related biases and do not increase the burden on participants. When compared to self-reported food consumption, the key benefit is that the odds of inaccuracies are quite low. The most significant downside of nutritional indicators is that they are more expensive than subjective techniques (DAPA Measurement toolkit). Serum biomarkers are often used to identify and monitor a variety of metabolic illnesses, including hypertension, hyperglycemia, obesity, and insulin resistance, and may aid in early identification and risk management [2].

We have explained various types of nutritional biomarkers, especially serum biomarkers and their interrelation with various types of metabolic disorders, and their limitations, challenges, and future perspective.

8.2 Metabolic Syndrome

Metabolic syndrome or disorders are a group of metabolic deformities which finally results in a considerable increase in Cardiovascular Disease (CVD) risk which include abnormal blood sugar levels, abdominal obesity, high blood pressure, increased triglyceride levels, and low high-density lipoprotein (HDL)–cholesterol levels in an individual [10]. The chances of various CVDs are twice as high as in the individual without the syndrome [2]. These cardiovascular risk factors or metabolic syndrome are prevalent in 20-25% of the general population [1]. WHO and the report of NCEP ATP 3 both considered cardiovascular disorders (CVDs) as the principal clinical outcome of metabolic syndrome. Insulin resistance can be seen in most individuals having metabolic syndrome,

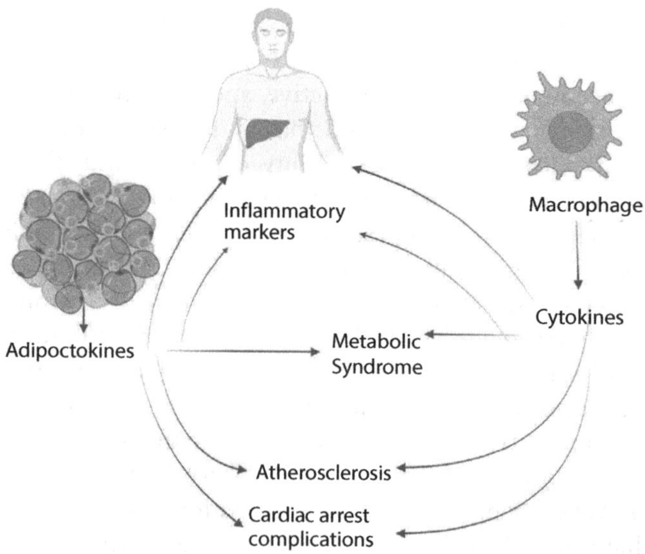

Figure 8.1 Interaction of adipokines, cytokines, and inflammatory markers that contribute to the development of the metabolic syndrome and its complications. (Created with BioRender.com)

and this condition finally leads to the high possibility for diabetes type 2 (Figure 8.1). Besides cardiovascular diseases, a person with metabolic syndrome is also more prone to various other diseases like fatty liver, cholesterol gallstones, sleep disturbances, polycystic ovary syndrome, asthma, and some types of cancers, including breast cancer, prostate, colorectal, and pancreatic cancer [3]. The criteria of a patient with metabolic syndrome are to have at least three of the five components: increased BP or hypertensive drugs treatment, drug treatment for low HDL or low HDL cholesterol, increased fasting glucose level, increased triglycerides level, increased waist circumference [11].

There is a great interrelation between the diet of individual consumers and the risk of metabolic disorders and their severity on that particular individual. A group of researchers studied aged non-human primates (NHP) models fed with a high fat/sugar diet (HFSD) for about eight months. After that, the metabolic disorders in NHP aged monkeys were diagnosed by serum biochemistry and H-NMR. On the other hand, cardiac function was assessed by echocardiography. Their results showed that after feeding with HFSD to the monkeys, their serum metabolic profiles were changed with contractile dysfunction, aortic tissue damage, and cardiac remodeling. Finally, they were able to successfully prove that the aged non-human primates (NHP) models fed with a high fat/sugar diet (HFSD) exhibit metabolic disorders and cardiac contractile dysfunction [4].

Different studies about the interrelationship of proteins w.r.t obesity and weight loss proved different results in the case of proteins. Some studies have shown that a high-protein diet can result in greater weight loss and upgrade the biomarkers of metabolic syndromes (MeS). For this purpose, a group of researchers did a study to know the effect of high protein intake in the diet on weight loss among Mexican individuals (age 20~60 years) with MeS. They divided these adults into two groups. One group was fed with a standard protein diet (SPD), and the other with a high protein diet (HPD) for approximately nine months. After that, they concluded that when the overall group was examined, there were no notable differences in weight loss and MeS biomarkers, but the participants of the high-protein diet group lost more weight than the participants in the standard protein diet group [12].

8.3 Nutritional Biomarkers (NB)

Generally, biomarkers may be defined as "Any of the substance that is used as an indicator of any biological state". Another definition of biomarker may be, "these can be more accurately defined as the biological or chemical test results associated to a certain susceptibility, exposure, or biological effects in an analyzed biological material" [5]. In terms of consumption and metabolism of necessary nutrients and dietary constituents, the nutritional biomarkers may be defined as any biological sample generally used to measure nutritional status [6]. These are generally the hints left by the body that disclose the food we have consumed and indicate typically the actual amount of nutrients that reached the tissues after absorption and metabolism. For instance, in urine, the amount of nitrogen is used as a nutritional biomarker of protein intake in the diet.

These indicators are now highly useful in the early diagnosis and categorization of risk in individuals with metabolic syndrome (MeS) [1]. Biomarkers are a better indicator of nutritional status than food intakes. Biomarkers, based on the nature of the sample used with respect to time period, is divided into short-term biomarkers (considering the intake in past hours or days, for example, Hydrogen breath test and lactose digestion test), medium-term biomarkers (assessing the intake in past weeks or months, for example, RBC for folate amount), and the long-term biomarkers (considering the intake in past months or years, e.g., hair and toenail sample [6]. These biomarkers are clinically helpful, particularly in determining the insufficiency for aiding medical intervention and treatment (for example, Assessment of vitamins like vitamin B9, vitamin B12 and minerals like iron, copper and zinc are beneficial to recognize the possible nutritional causes of anemia) [13] and nutritional adequacy (for example, the combination of serum ferritin and transferrin receptor are used to determine the deficiency of iron (DAPA Measurement toolkit). A single biomarker itself is sufficient and liable to identify the metabolic syndrome. It will be hard to tell. Instead of single or individual biomarkers, a group of different biomarkers is considered helpful because individual biomarkers have different roles and action mechanisms [2].

8.3.1 Classification of Nutritional Biomarkers (NB)

Nutritional biomarkers are classified based on various aspects and criteria. An individual NB can also be included in either of the categories listed below:

1. Dietary exposure biomarkers – This category evaluates the different dietary patterns, foods, non-nutritive components, different food groups, and intake of different nutrients.

Table 8.1 Different classes of dietary biomarkers and their intended uses in the dietary assessment methods.

S. no.	Class of nutritional biomarkers (N.B.)	Examples	Applications
1.	Recovery biomarkers	i. Urinary nitrogen ii. Doubly labeled water iii. Urinary potassium	Used as a reference measurement to evaluate the correctness of diet assessment methods.
2.	Predictive biomarkers	i. Urinary sucrose ii. Urinary fructose	Used as a reference measurement to evaluate the correctness of diet assessment methods. (The overall recovery in predictive biomarkers is quite lower than the recovery biomarkers.)
3.	Concentration biomarkers	i. Serum or plasma vitamins ii. serum carotenoids iii. blood lipids iv. urinary electrolytes	1. Evaluate the connection with estimates of dietary intake. 2. Used in the assessment of diet-disease risk interrelation.
4.	Replacement biomarkers	i. Phytoestrogens ii. Polyphenols iii. Sodium	1. Estimate compounds for which sufficient information is unavailable in food composition databases or nutrient databases. 2. Used as a substitute to dietary assessments.

2. Nutritional status biomarkers – Along with the evaluation of nutrient intake, this category can also assess the metabolism of nutrient or nutrients and their effects on particular diseases processes. Possibly these biomarkers throw light on the relationship and the interconnection between several nutrients rather than the nutritional status of a single nutrient.

Based on their intended use and biokinetics, they are divided as Replacement biomarkers, Recovery biomarkers, Predictive biomarkers, and Concentration biomarkers [7]. On the other hand, [9] and his colleagues divided NB by focusing on their use only into Susceptibility, Exposure, and Effect biomarkers (Table 8.1).

1. Recovery biomarkers – These are based on the idea of metabolic balance between the intake and excretion of nutrients within a fixed period so as to find the exact intake levels of nutrients [14]. Examples include doubly labeled water, urinary nitrogen, urinary potassium. We can say that the levels of excretion are highly associated with intake levels. Doubly labeled water is used to estimate the total energy expenditure and the metabolic rate. Total urinary nitrogen and potassium is used to measure total daily protein and potassium consumption, respectively [15];[16].
2. Predictive biomarkers – These are also sensitive, time-dependent biomarkers like the recovery biomarkers. But the major difference is that the overall recovery in predictive biomarkers is quite lower than the recovery biomarkers. Examples including urinary fructose and sucrose, which are closely correlated with intake of sugars [17]. They assess only some extent of dietary intake rather than evaluating the whole nutritional intake.
3. Concentration biomarkers – These are used to estimate the intake of food or nutrients and are not used to find the absolute intake because they depend upon various factors like metabolism, age, sex, the lifestyle of person, physical activity, etc.

   Examples include serum or plasma vitamins, serum carotenoids, blood lipids, urinary electrolytes [18].
4. Replacement biomarkers **are closely related to concentration biomarkers and especially** those compounds for which sufficient information is unavailable in food composition

databases or nutrient databases. Examples include some phytoestrogens [19], polyphenols, aflatoxins [20], and sodium.

8.3.2 Merits and Demerits of Nutritional Biomarkers

It should be kept in mind that all nutritional biomarkers have some weak points or demerits. But the level and category of limitation may change depending on the type of particular biomarker. The following are the strength and weaknesses of nutritional biomarkers.

Merits:

1) Avoid the recall and social desirability-related biases.
2) Errors associated with nutritional biomarkers are independent of the errors related to self-reported dietary intakes.
3) Some biomarkers are easy to obtain so do not increase the burden on participants.
4) Some biomarkers (concentration or predictive) can be used to rank individuals based on various factors.
5) Biomarkers are used as reference measurements to evaluate the validity and correctness of dietary assessment methods [15];[6];[17].
6) Some biomarkers (recovery biomarkers) provides an estimate of absolute intake levels [14].
7) Dietary biomarkers, including replacement and concentration, are used to assess the diet-disease risk relationship [6].

Demerits:

1) The dietary biomarkers evaluation is generally costlier than the subjective methods.
2) Some biomarkers (for ex. tissue biopsy) increase the burden or discomfort on participants because they are difficult to obtain.

8.4 Factors Affecting the Specificity and Utility of Nutritional Biomarkers

Certain components (Figure 8.2) may influence the utility and attentiveness of the nutritional biomarkers. These factors are as follows; [7]; (DAPA Measurement toolkit)

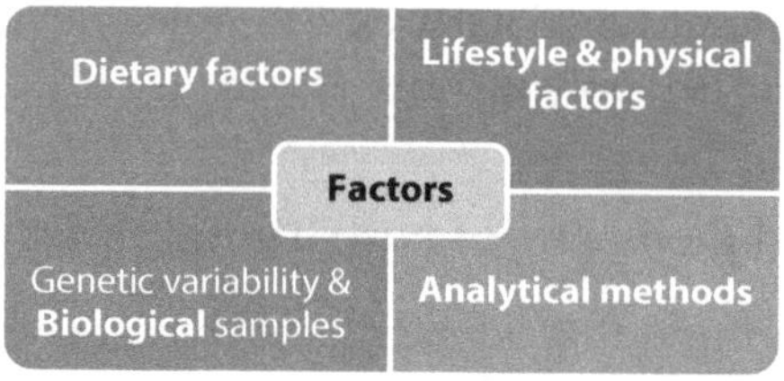

Figure 8.2 Factors affecting the specificity and utility of nutritional biomarkers.

1) Genetic Variability

   i. The genes affect the taste, attraction to a particular food or types of foods, dietary intake patterns, etc.
   ii. Biological difference in digestion, absorption, tissue turnover, metabolism and excretion of nutrients.

2) Lifestyle/Physical Factors

   i. Alcohol use, age, smoking, exercise, sex, body weight, height, socioeconomic status, etc.
   ii. Effect of gut microbiota (bioconversion and release of bioactive dietary compounds).
   iii. Enterohepatic circulation of nutrients (e.g., phyto-estrogens).
   iv. Inflammation and metabolic-related diseases, stress, etc.

3) Dietary Factors

   i. Intake frequency for a particular nutrient.
   ii. Nutrient-nutrient interrelations.
   iii. Nutrient bioavailability, the influence of food matrix.

4) Biological Sample

   i. Type of sample collected to examine biomarkers (e.g., urine, whole blood, plasma, serum).
   ii. Sample collection condition, transport, treatment, conditions of storage, the time span of storage.
   iii. Diurnal variation, day of the week, or season of sample collection.

5) Analytical Methodology

   i. Correctness, detection limits of the analytical technique.
   ii. Variations from method to method or laboratory to laboratory.

The biomarkers of nutritional status do not need to be only the nutrients. Some other compounds can also serve as nutritional biomarkers. For example, an elevated level of Methylmalonic acid in urine or blood serves as an early and sensitive indicator of deficiency of vitamin B12. The lower level of Erythrocyte glutathione reductase and RBC riboflavin are the evidence of deficiency of riboflavin. An elevated level of Homocysteine indicates the lack of vitamin B6, vitamin B12, and Folic acid.

8.5 Role of Different Nutritional Biomarkers (Serum Biomarkers) in Different Metabolic Syndromes

Nowadays, a nutritional biomarker plays a great role in early detection and diagnosing and risk grouping of metabolic syndrome in patients. Recent studies proved the important role of various types of serum biomarkers, for example, adipokines, inflammatory cytokines, prothrombotic factors, neuropeptides, etc., in the investigation and observation of metabolic disorders. These serum biomarkers are generally used to reveal and monitor various metabolic diseases, including hypertension, hyperglycemia, obesity, insulin resistance, etc., and are helpful in their early detection and risk management [2]. The concentration of these serum biomarkers may increase or decrease in the patients with MeS depending upon the nature and type of metabolic syndrome.

8.5.1 Adipokines

The adipokines or adipocytokines are cytokines that are produced by adipocytes or adipose tissue. A group of researchers has been shown that the accumulation of these adipocytes results in unbalanced adipokines formation, which are ultimately involved in the metabolic syndrome expansion [21]. The complete mechanism or process by which unbalanced production of adipokines takes place due to the result of the accumulation of these adipocytes is quite unclear at present, but it is partly as a result of the oxidative stress produced by obesity [22].

a) Leptin

Leptin was the first adipokine to be discovered, in 1994. It is a hormone primarily produced by white adipose tissue (WAT) or some also by non-adipose tissue (Matarese, 2005) [23] for example, stomach, placenta, mammary glands, immune system, and enterocytes in the small intestine. The main function of this hormone is to maintain energy balance by suppressing hunger, which results in a decrease in fat storage in adipocytes. Along with this function, leptin also enhances energy expenditure, improves sympathetic activity, ease in utilization of glucose, and upgrades glucose sensitivity [24]. The leptin level in the serum shows the amount of energy stored in the adipose tissues [24].

There are many factors on which the level and regulation of leptin in the plasma depends (Figure 8.3).

Mechanism of action: Leptin decreases caloric intake, and along with this, in non-adipose tissue, it regulates fatty acid oxidation to reduce abnormal fatty acid spillover. Leptin damages non-adipose tissues, which ultimately results in leptin resistance among patients with impaired non-adipose tissue [25]. The primary leptin receptor is located in the hypothalamus, where its main function is to enhance energy expenditure and decrease appetite. Elevated levels of leptin can cause selective or global leptin resistance. Thus high levels of leptin are connected with metabolic syndromes [2] and [1]. After binding leptin to its functional receptor located in the hypothalamus, it activates JAK/STAT pathways, which is leptin's main pathway to impose its consequences. Apart from JAK/STAT pathway, leptin signaling also

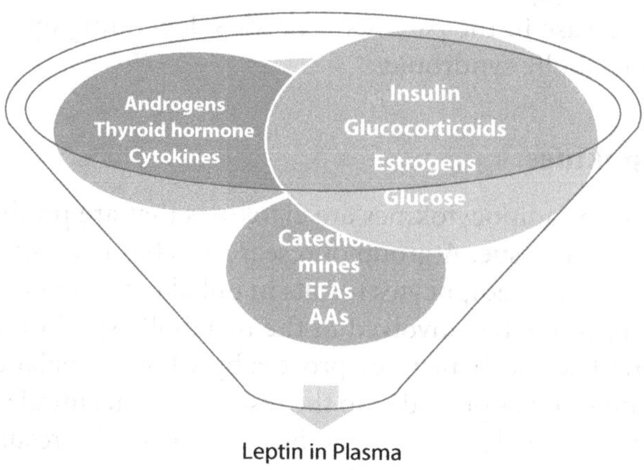

Figure 8.3 Factor that regulates level of leptin in the plasma [1].

Table 8.2 Interrelationship between leptin and MeS.

S. no.	Year of research	Subjects (number & category)	Conclusions	Reference
1.	2012	153, women	1. Elevated leptin in postmenopausal women with MeS. 2. Also found positive interaction between leptin and abdominal obesity.	[27]
2.	2008	321, elementary schoolchildren	Leptin was the most liable biomarker for CVDs and MeS in primary school children.	[28]
3.	2010	9,995, Korean population	1. With the increase in MeS components, serum leptin levels increase also. 2. Thus, reduction of leptin levels may be protective.	[29]
4.	2012	103, adult population	1. Positive relation of obesity, hyperinsulinemia and insulin resistance with leptin. 2. Weak relation of leptin to other MeS components.	[30]
5.	2006	153, healthy non-diabetic men	Leptin was related with waist size strongly. Weakly related with lipid profile, which vanished by the adjustment in BMI.	[31]
6.	2015	204, Mexican University workers	Concentration of leptin in plasma is proportional to the degree of central obesity causing leptin resistance. Leptin is strongly associated with BMI.	[32]

includes some other pathways, such as AMPK and MAPK pathways. Insulin signaling is also interrelated with leptin [26]. Numerous studies confirm the correlation of leptin with various MeS, which has been shown in Table 8.2.

b) Adiponectin

Adiponectin was firstly discovered in 1995 in differentiating adipocytes. It is the most abundant plasma protein secreted by adipocytes. Earlier it was thought that adipocytes only secreted adiponectin in addition to adipocytes, other cells, for example, Cardiac and skeletal myocytes, and endothelial cells also produce adiponectin [33]. In females, the circulating levels of adiponectin are higher than in males due to the stimulating activity of testosterone on the secretion of adiponectin [34]. Adiponectin has an immense physiological function that helps in metabolic balance in normal individuals (Figure 8.4) and its lower levels are correlated with an increased metabolic risk and cardiac and vascular abnormalities. For example, lower level plays an important role in diseases related to obesity, including insulin resistance or type 2 diabetes and CVDs.

Mechanism of action: Various studies for MeS like obesity, atherosclerosis, and diabetes on human and animal models highlights the potential roles of adiponectin and adiponectin receptors in these metabolic disorders, mainly in insulin-sensitizing, anti-atherogenic, and anti-inflammatory effects, and

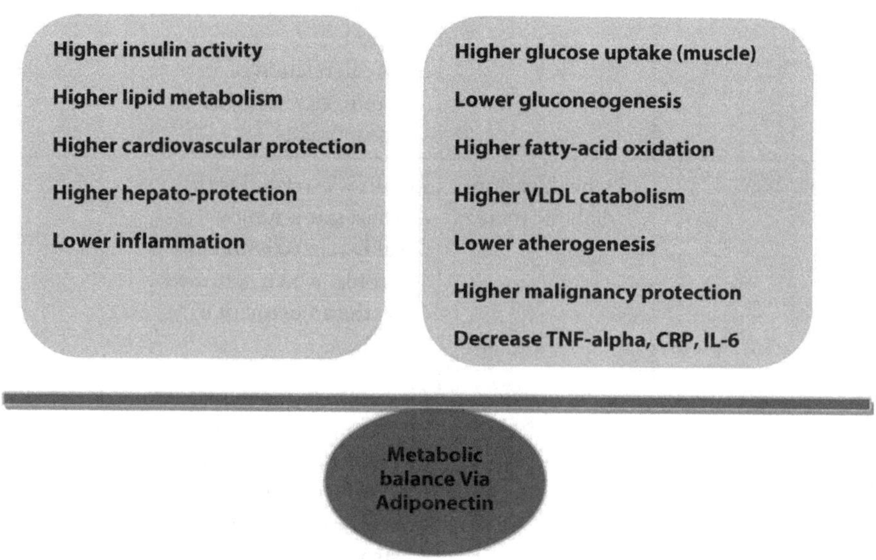

Figure 8.4 Metabolic balance connected with adiponectin [1].

Table 8.3 Adiponectin and MeS interrelationship.

S. no.	Year of research	Subjects (number & category)	Conclusions	Reference
1.	2004	661 Japanese adults	Men had lower levels of adiponectin than women, which is the reason why women have a lower risk of CAD.	[21]
2.	2006	153, healthy non-diabetic men	Adiponectin is inversely correlated with MeS, independent of BMI.	[31]
3.	2006	1041 Finnish subjects	Low levels of adiponectin correlated with an increasing number of MeS components in both sexes, independent of BMI.	[35]
4.	2013	-	HMW adiponectin should be the best biomarkers for MeS.	[36]
5.	2006	298 patients admitted for diabetes treatment	HMW adiponectin to plasma adiponect in ratio was better detector of insulin resistance and MeS.	[37]
6.	2020	144 subjects of middle age	HOMA-AD could be a better predictor of MeS than HOMA-IR among middle-aged Indians.	[38]

also in reducing body weight. Thus, adiponectin replacement therapy in humans is suggested to treat diabetes, insulin resistance, atherosclerosis, and obesity [33]. Numerous scientific studies show the interrelationship between adiponectin and various metabolic syndrome, as shown in Table 8.3.

Table 8.4 LAR and the metabolic syndrome.

S. no.	Year of research	Subjects (number & category)	Conclusions	Reference
1.	2010	460 people with incident Type 2 diabetes and 1,474 non-diabetic people	Leptin and adiponectin both interact with each other for modulating the risk of diabetes.	[39]
2.	2009	2,097 non-diabetic individuals	In non-diabetic adults, LAR is a useful biomarker for insulin resistance.	[40]
3.	2014	–	LAR is used in the assessment of insulin resistance.	[41]
4.	2011	678 Japanese people (208 men and 470 women)	LAR in subjects with MeS was notably higher than that in subjects without MeS, regardless of gender.	[42]
5.	2013	–	High LAR, for the diagnosis of MeS, is a preferable biomarker than leptin or adiponectin alone.	[36]
6.	2012	199 adult healthy subjects	Men without MeS had significantly lower LAR than men with MeS. Women without MeS had a lower leptin level and LAR than women with MeS, but had significantly higher adiponectin.	[43]

c) Leptin:adiponectin ratio (LAR)
Leptin and adiponectin, both molecules, have an antagonistic effect. But still, many studies have shown that the ratio of leptin and adiponectin, i.e., LAR is beneficial rather than alone, as shown in Table 8.4.

8.5.2 Neuropeptide

8.5.2.1 Ghrelin

Ghrelin, sometimes called "lenomorelin", is a hormone produced by enteroendocrine cells of the GI tract, especially the stomach, and is mainly called a "hunger hormone" because it enhances food intake and appetite (main function). This hormone travels to the brain through the bloodstream, conveying a message to the brain to become hungry and search out food. The level of ghrelin in the blood is highest before meals when we are hungry, and this range comes back to normal or lower levels after the food intake. Besides its effects in increasing hunger or appetite, ghrelin also has some major effects on other functions, including-energy homeostasis, muscular atrophy, glucose homeostasis, bone metabolism, and modulation of CV function [44]. Ghrelin goes through various posttranslational modifications, resulting in unacylated and acylated ghrelin, i.e., UAG and AG, respectively, the two circulating forms of ghrelin [24]. Ghrelin plays various functions in the GI system, including gastric acid secretion, GI motility, and pancreatic protein output. Most of these functions have been assigned to the activity of AG. The balance between secretion, deterioration, and clearance rate of ghrelin decides the circulating level of ghrelin [45].

Ghrelin and MeS relationships: the low level of ghrelin is inversely related to the components of MeS, including obesity, high BP, atherosclerosis, etc., but continuously lowering the ghrelin levels may be associated with the severity of metabolic disorders. A group of researchers highlights the interconnections between obese people with MeS with the low ghrelin levels [46]. The fasting ghrelin concentrations in patients with diabetes of type 2 are lower in obese people than in lean people [47]. In people with MeS, the administration of ghrelin upgrades the endothelial function by inhibiting the factors that promote atherogenic activity [48], also reduces BP, and finally results in results. The relation of ghrelin with MeS is mainly predicted based on obesity. Along with this, the level of ghrelin is quite low in patients with obesity or MeS as compared to non-obese patients with MeS [49].

8.5.3 Pro-Inflammatory Ctytokine

IL-6 – Interleukin-6 (IL-6) is a pro-inflammatory cytokine that plays an important role in acute and chronic inflammatory responses, immune cell development, and the etiology of autoimmune disorders. M1 macrophages generally produce it as part of the normal inflammatory response in opposition to an infection and injury [50]. IL-6 shows its pro-inflammatory effects or properties by different pathways and is linked with different components of inner immunity.

Interrelation of IL-6 with MeS: various studies confirm the interrelation of IL-6 with all components of MeS. The elevated activity of IL-6 gene is related to an increased risk of developing diabetes mellitus and other MeS [51] Another group of researchers [52] found that insulin resistance was connected with elevated IL-6 levels. A study was conducted on postmenopausal women; it was found that the increased levels of IL-6 were associated with low HDL, central obesity, and high triglycerides levels [53]. In the case of animal studies on mice [54] the administration of IL-6 into mice resulted in imperfect insulin signaling in liver and muscle tissues, which ultimately leads to insulin resistance and hyperglycemia condition in mice. Two different groups of researchers show that IL-6 is correlated with hypertension, hypertriglyceridemia, and fasting plasma glucose [55] & [56]. In postmenopausal women with MeS, there are higher levels of IL-6, and the condition of endothelial dysfunction was found along with metabolic and lipid abnormalities [53].

8.5.4 Anti-Inflammatory Cytokine

IL-10 – Interleukin-10 is an effective anti-inflammatory cytokine that regulates the immune response to avoid excessive activation and auto-damage [57]. A study was conducted by a group of researchers on children and young adolescents, it was proved that IL-10 levels in plasma were low in obese or overweight children. So, they concluded that IL-10 is a biomarker of metabolic risk factors [8]. In contrast to this, other researchers found that the high circulating levels of IL-10 can be seen in obese women, and the low IL-10 levels are related to MeS [10]. The level of IL-10 in plasma of patients with diabetes mellitus and MeS is found to be reduced [58]. In the case of study on an animal model, mice, it was found that when IL-6 was served to mice to influence the condition of insulin resistance, *in vivo* use of IL-10 shows defensive mechanism from the disabled insulin signaling which was due to the administration of IL-6, thus renovate the insulin sensitivity and normal glucose concentration into muscle tissue and liver [54].

8.5.5 Oxidized Low-Density Lipoproteins (OxLDL)

Low-density lipoprotein (LDL) particles in humans go through a chain of oxidation processes, which results in OxLDL particles and the generation of reactive oxygen species (ROS). This OxLDL particle acts as an oxidative stress biomarker. It should be noted that, in low concentrations, these two lipid oxidation products act as signaling compounds for cellular antioxidants pathways. But in case of dysfunctioning in cellular antioxidant capacity (like in MeS), both of these compounds will take part in an oxidative cascade that finally results in apoptosis and cellular damage [59]. This malfunctioning of OxLDL molecules lead to various CVDs and MeS. At the time of oxidation of LDL, along with the generation of OxLDL, different other products are also formed related to lipids, proteins, and fatty acids, as shown in Figure 8.5 [60]. The OxLDL particles get associated with LOX-1 receptors found on the surface of endothelial cells, regulate their activation [61], and cause the dysfunction of endothelial cells.

Interrelation of OxLDL with MeS: Various studies confirmed the interrelation of OxLDL with MeS and various types of CVDs. A group of researchers in their research established the relationship of high OxLDL level with all the components of MeS [62]. The OxLDL molecules can cause atherosclerosis due to damaging blood vessels by capturing them [63]. In a study with 3,987 subjects, a group of researchers proved the direct relationship of OxLDL with MeS and its components, especially with triglycerides levels, independently of central obesity and insulin resistance [64]. In adults along with CVDs the high levels of OxLDL compounds are also related to insulin resistance and obesity,

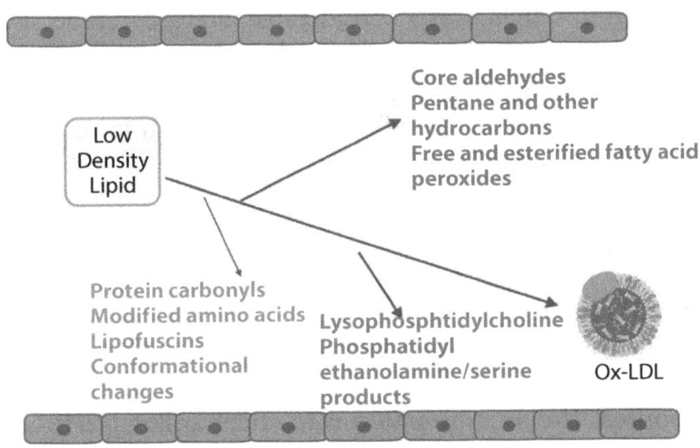

Figure 8.5 Products formed during oxidation of LDL [1] (Created with BioRender.com).

and a group of obese and overweight children resulted that OxLDL is directly and positively related with BMI, abdominal fat, % body fat, waist circumference, % trunk fat [65]. Studies by two different groups of scientists show that in MeS patients, the level of OxLDL gets increased, which is related to poor arterial elasticity, which is the main cause of CAD development [63] & [66].

8.5.6 Paraoxonase-1 (PON-1)

PON-1 is an enzyme with antioxidant and antitoxic properties, produced mainly by the liver, and defends against lipid oxidation and various extrinsic toxins. It is primarily transported by HDL, and it can break lipid peroxides into lipoproteins, thus reducing oxidative stress. The low level of PON-1 is closely found in various oxidative stress conditions, including DM, smoking, aging, dyslipidemia, etc. [67]. Along with antioxidant and antitoxic properties, PON-1 is also showing some anti-inflammatory effects. For proving this, various studies on the mice model were performed and concluded that PON-1 interconnects with macrophage scavenger receptor class B type I (SR-BI). This slows down the production of TNF-alpha and IL-6. In this way, PON-1 promotes anti-inflammatory properties.

Interrelation of PON-1 with MeS: PON-1 is confirmed related with all the five components of MeS, and its level and activity are generally lower in patients having all the MeS components, and there is an inverse relationship between MeS severity and levels of PON-1 [68]. The anti-inflammatory effect of PON-1 is mainly confirmed due to its anti-atherogenic property [69]. In a study on 130 participants for about seven months, PON-1 activity was measured by ELISA technique and showed that the concentration of PON-1 enzyme serum biomarkers as comparatively lower in MeS patients than the normal healthy ones. Along with this, there was a considerable increase in PON-1 activity in female patients with waist circumference 140-160 cm compared to males of the same waist circumference group [70]. It is concluded that PON-1 plays an important role in decreasing the effect of MeS through its anti-atherogenic, antioxidant, and anti-inflammatory properties.

8.6 Novel Biomarkers

Pentraxin-3 (PTX-3)

PTX or pentraxins are the families of pattern recognition proteins of similar structures found in clusters and are involved in the acute immunological responses. The main function is to attach cellular debris/microbial pathogens at the time of infection or inflammation. Thus, complement activations help in the expulsion of necrotic cells and pathogens [71].

PTX-3 is an acute-phase protein that is encoded by PTX-3 gene in humans. In normal conditions, the level of PTX-3 in blood or serum is quite low. The level of PTX-3 increases in the extreme infectious condition in a human. *In vitro,* the PTX-3 protein expression into endothelial cells is suppressed to a higher range by pitavastatin than other genes, suggesting that PTX-3 can be a novel biomarker for inflammatory CV diseases [72].

Interrelation of PTX-3 with MeS: The long-term PTX-3, as compared to the short-term PTX-3, is interrelated with the severity of infections and disease. The level of PTX-3 generally gets higher in these critical conditions. Thus, PTX-3 is a reliable and direct indicator of infections and inflammation [73]. In a study by a group of researchers on obese patients with MeS, it was found that the concentration of PTX-3 is high in obese patients with MeS as compared to the participants without any MeS. They also concluded that the higher PTX-3 level is also interconnected with high triglycerides and low HDL cholesterol levels [74]. Ohbayashi *et al.*, from their study for the very first time, showed that PTX-3 is a useful blood biomarker for the observation of atherosclerosis directly [75].

Hyocholic Acid (HCA)

HCA is a kind of bile acid that is mainly found in the bile acid pool of pigs in high concentration, but its minor or trace concentrations can also be found in other species like a human. Small amounts of HCA can be found in the urine and blood of human beings. The composition of bile acid profile, mainly hyocholic acid species, differ remarkably within mammalian species [76].

Interrelation of HCA with MeS

Zheng *et al.*, in their study, concluded that HCA could be a better and novel biomarker for detecting and predicting MeS in the population. For this purpose, they did about five different case studies on humans. They show the interrelation between diabetes, obesity, and HCA, and the lower level of HCA in serum is associated with both obesity and diabetes. But further deep studies are required in this area [4].

8.7 Various Analytical Techniques Related to Different Nutritional Biomarkers

The static and functional basic forms of laboratory analysis are taken into consideration. Static assays determine the amount of a nutrient, bioactive, or biomarker that is currently present in a biological sample. The analysis of serum iron, blood glucose, cholesterol, and other parameters are examples of this type. Circulating levels don't always correspond to a substance's bioavailability or amount in bodily reserves. Depending on the biomarker,

Table 8.5 Nutritional biomarkers and their corresponding analytical techniques.

S. no.	Biomarker	Sample type	Test	Analytical method	Reference
1.	Protein	Urine	Urinary Nitrogen	Kjeltec method	[77]
2.	Omega-3 fatty acids	Blood	Fatty acids concentration	Folch method	[78]
3.	Total energy	Urine/saliva blood	Doubly labeled water	Mass spectrometry	[79]
4.	Sodium	Urine/serum	Sodium-potassium	Flame photometer	[80]
5.	Potassium	Urine/serum	Urinary potassium	Flame photometer	[80]
6.	Iron	Serum/blood	Serum ferritin & Hemo-globin	ELISA & Cyanmethemo-globin method	(DAPA Measurement Toolkit)
7.	Iodine	Urine	Urinary iodine	Sandell – Kolthoff method	[81]
8.	Calcium	Serum	Serum ionized calcium	Ion-specific electrodes	[82]
9.	Vitamin-E	Serum	Serum tocopherol to serum cholesterol ratio	Reverse phase HPLC	[83]

(Continued)

Table 8.5 Nutritional biomarkers and their corresponding analytical techniques. (*Continued*)

S. no.	Biomarker	Sample type	Test	Analytical method	Reference
10.	Vitamin-K1	Plasma	Plasma Vit-K	Mass spectrometry and HPLC	[84]
11.	Vitamin-B3	Blood	Ratio of NAD to NADP	HPLC	[85]
12.	Vitamin-C	Serum/blood	Serum/leucocytes ascorbic acid	HPLC	[86]
13.	Vitamin-B9	Serum/blood	Erythrocytes and serum folate & Serum homocysteine	Microbiological assay using *L. casei* & Reverse phase HPLC	(DAPA Measurement Toolkit)
14.	Vitamin-B12	Serum	Serum methyl malonic acid & Serum Vit-B12	Mass spectrometry & Radioimmunoassay	(DAPA Measurement Toolkit)

recent consumption can affect the amount of the biomarker in plasma, serum, or any other fluid or tissue sample. However, this limitation can be at least partially overcome by taking the sample while fasting. Functional analysis gauges a reaction as opposed to static analysis. The search for health biomarkers has helped functional evaluation of a biological function, which enables a dynamic assessment to which a biomarker corresponds, even though it has been known and utilised for years. Functional analysis reflects the level of the person's homeostatic robustness and allows for the quantification of their phenotypic flexibility.

Functional analyses include tests like the oral glucose tolerance test for determining insulin sensitivity and pre-diabetic status, as well as the measurement of triglycerides after an oral lipid load to determine dynamic lidipaemia as a biomarker of cardiovascular risk and early metabolic syndrome detection.

8.8 Level of Biomarkers in Extreme Coronavirus Infection

Different biomarkers play an important role in the pathogenesis of coronavirus disease (Table 8.5). The level of biomarkers may rise or fall depending on the severity or stage of the COVID-19 disease. Estimating the level of these biomarkers is quite beneficial to the physicians for earlier treatment and keeping a close watch on COVID-19 patients, so that physicians can easily predict the disease development rate [87]. The change in the levels of some biomarkers in the course of extreme infection of COVID-19 are shown in Table 8.6.

Table 8.6 Change in biomarker levels during extreme infection of COVID-19 [87].

S. no.	Name of biomarker	Level in extreme COVID-19 infection
1.	C-reactive protein	increases
2.	Serum amyloid	increases
3.	Interleukin- 6	increases
4.	White cell count	Neutrophil to Lymphocyte Ratio (NLR) - increases Lymphocytes - decreases
5.	Platelet count	decreases
6.	Renal biomarkers	Creatinine and urea increase

8.9 Health Biomarkers

The main purpose of a diet is to enhance and support the optimal health of an individual. Therefore, it is necessary to have those biomarkers that promote physical health and functions. Such types of biomarkers are generally regarded as health biomarkers. According to the concept of nutrition, these health biomarkers are the keystone to research that sets up the functional consequences of nutrition or diet on health-disease associations. Health biomarkers can direct various food policies, nutrition, and health [88].

8.10 Concept of Omics in Development of New and Integrative Nutritional Biomarkers

During that time, the postgenomic era and the establishment of knowledge had a great impression in the omics science applications in studying the interrelation between diet, health, and lifestyle. The omics science includes transcriptomics, microbiomics, genomics, metabolomics, epigenomics, proteomics, etc. [89].

The development of novel integrative nutritional biomarkers helps find the close correlation between nutrition and metabolism. These biomarkers

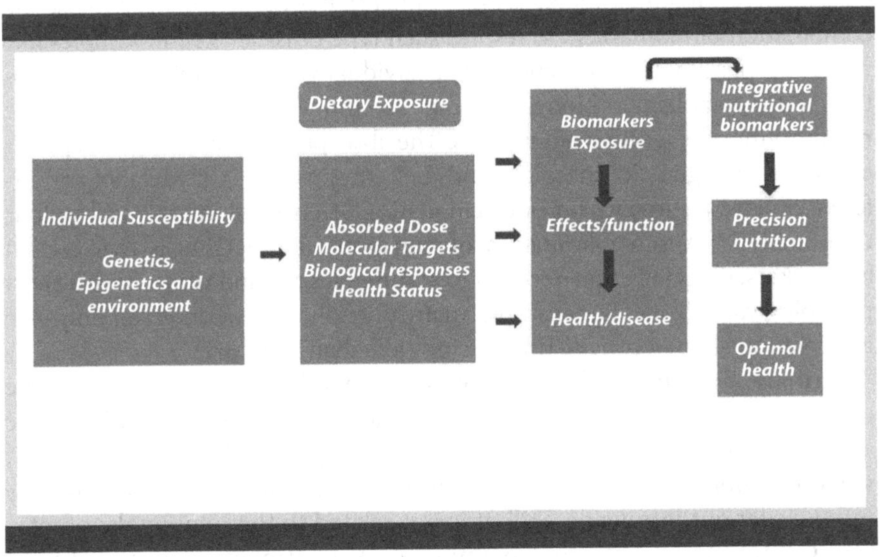

Figure 8.6 Involvement of integrative nutritional biomarkers and omics technologies in precision nutrition [88].

can be expressive for both the intake of nutrients or diet and their consequences in the body because they work on the fact that both diet and metabolism are closely interlinked. Along with this, these integrative nutritional biomarkers also put together the concept of precision nutrition [88].

Various factors (genetic factors, environmental conditions, lifestyle, epigenetics, etc.) can influence an individual's reaction to dietary exposure and its impacts on their health. The new integrative nutritional biomarkers can be developed due to various omics techniques and tools, as shown in Figure 8.6.

8.11 Limitations and Challenges in the Field of Nutritional Biomarkers

Nutritional biomarkers play a very important role in nutritional epidemiology. They permit the dietary exposure assessment either directly or indirectly. The direct assessment is done by the use of recovery biomarkers and the indirect assessment is by their union with self-reported intake [90].

Nutritional biomarkers also have several limitations. The insufficiency of detailed data in food composition and a great irregularity in foods make it hard to evaluate the dietary exposure. It is the main drawback of nutritional biomarkers [91]. Evaluation for the intake of various food groups is very difficult. For example, vegetables and fruits consumption is related to health benefits, but assessing their intake is quite difficult. It is quite beneficial to use the combination of biomarkers to resolve such type of issue [92]. In samples like urine and plasma, the inter-and intra-individual variations related to absorption and metabolism of biomarkers can also influence the measured and the real amount of respective biomarkers. The absorption of biomarkers depends on factors like genetic variability and diet composition [7]. Metabolism differences have an impact on both circulating levels of biomarkers as well as their analysis. Another limitation of nutritional biomarkers is that certain biomarkers, especially those related to serum, plasma, and urine, have a short physiological half-life. For example, many phenolic compounds can only give information about a concise time period (less than 24 hours).

Urinary biomarkers for nutrients are generally not very useful as the excretion is a control point. For instance, no calcium can be found in urine as all the calcium is reabsorbed by the kidneys [93]. In terms of vitamin C, when the glomerular filtration rate becomes saturated, the excess of vitamin C is let out into urine. Still, it is not related to the total intake of vitamin C [94]. Along with this, many medical treatments, drugs, and food additives can also cause nutrient metabolism and excretion.

8.12 Future Directions and Perspectives

Biomarkers should objectively estimate nutrient status, which is the major drawback of subjective dietary assessment methods. Future research related to the field of NB should highlight their development to evaluate the recommendations for certain food groups, including fruits, vegetables, saturated fat foods, high-fat foods, sugar-sweetened beverages, etc. [95]. In human nutrition, the rising field into metabolomics and the development of well-founded Food Frequency Questionnaire, and the development of Food metabolome databases will allow the recognition of specific dietary components in food. Thus, it will produce more suitable and reasonable biomarkers of exposure to particular food along with the advanced experimentation in nutritional sciences whose objective is to assess the relationship between diet and disease.

8.13 Conclusions

A metabolic syndrome is a group of metabolic imperfection which results in increasing the risk of CVDs among individuals. These metabolic syndromes can be detected by using serum biomarkers. The use of these biomarkers in a panel or group would be more effective in avoiding the complications of these metabolic disorders. Nowadays, nutritional biomarkers play a great role in early detection and specific treatment along with the risk grouping of metabolic syndrome in patients. Various recent studies resulted in the importance of different types of serum biomarkers, including adipokines (leptin, adiponectin, and their ratios), inflammatory cytokines (IL-6, IL-10), prothrombotic factor (PAI-1), neuropeptide (ghrelin), oxidized low-density lipoproteins, paraoxonase-1 and other novel biomarkers including pentraxin-3, hyocholic acid, etc., in the investigation and observation of metabolic disorders. The concentration of biomarkers may be increased or decreased in patients with MeS and severe COVID-19 infection, depending upon the nature and severity of the disease. Using these nutritional serum biomarkers in predicting various metabolic syndrome reduces the burden forced on patients. Further deep studies and interventions are required in this area. The nutritional biomarkers are predominant for future research that is related to the interlink between health and diet. The rising field of metabolomics and new techniques are quite helpful in the advance development of unique nutritional biomarkers.

References

1. Zaman G. *Ultimate Guide to Insulin*. BoD – Books on Demand; 2019. 132 p.
2. Srikanthan K, Feyh A, Visweshwar H, Shapiro JI, Sodhi K. Systematic Review of Metabolic Syndrome Biomarkers: A Panel for Early Detection, Management, and Risk Stratification in the West Virginian Population. *Int J Med Sci*. 2016;13(1):25–38.
3. Grundy SM, Brewer HB, Cleeman JI, Smith SC, Lenfant C. Definition of Metabolic Syndrome: Report of the National Heart, Lung, and Blood Institute/American Heart Association Conference on Scientific Issues Related to Definition. *Circulation*. 2004 Jan 27;109(3):433–8.
4. Zheng S, Tan W, Li X, Li B, Gong B, Pyle WG, *et al.* Aged Monkeys Fed a High-Fat/High-Sugar Diet Recapitulate Metabolic Disorders and Cardiac Contractile Dysfunction. *J Cardiovasc Transl Res*. 2021 Oct;14(5):799–815.
5. Biesalski HK, Dragsted LO, Elmadfa I, Grossklaus R, Müller M, Schrenk D, *et al.* Bioactive compounds: Definition and assessment of activity. *Nutrition*. 2009 Nov;25(11–12):1202–5.
6. Potischman N, Freudenheim JL. Biomarkers of Nutritional Exposure and Nutritional Status: An Overview. *J Nutr.* 2003 Mar 1;133(3):873S-874S.
7. Jenab M, Slimani N, Bictash M, Ferrari P, Bingham SA. Biomarkers in nutritional epidemiology: applications, needs and new horizons. *Hum Genet*. 2009 Jun;125(5–6):507–25.
8. Chang JS, Bai CH, Huang ZC, Owaga E, Chao KC, Chang CC, *et al.* Interleukin 10 and clustering of metabolic syndrome components in pediatrics. *Eur J Clin Invest*. 2014 Apr;44(4):384–94.
9. Gao Q, Praticò G, Scalbert A, Vergères G, Kolehmainen M, Manach C, *et al.* A scheme for a flexible classification of dietary and health biomarkers. Genes Nutr. 2017 Dec;12(1):34.
10. Esposito K, Chiodini P, Capuano A, Bellastella G, Maiorino MI, Rafaniello C, *et al.* Colorectal cancer association with metabolic syndrome and its components: a systematic review with meta-analysis. *Endocrine*. 2013 Dec;44(3):634–47.
11. Park S, Lee S, Kim Y, Lee Y, Kang MW, Han K, *et al.* Altered Risk for Cardiovascular Events With Changes in the Metabolic Syndrome Status: A Nationwide Population-Based Study of Approximately 10 Million Persons. *Ann Intern Med*. 2019 Dec 17;171(12):875.
12. Campos-Nonato I, Hernandez L, Barquera S. Effect of a High-Protein Diet versus Standard-Protein Diet on Weight Loss and Biomarkers of Metabolic Syndrome: A Randomized Clinical Trial. *Obes Facts*. 2017;10(3):238–51.
13. Chen L, Ma Q, Chen Y, Wang B, Pei D. Identification of major walnut cultivars grown in China based on nut phenotypes and SSR markers. Sci Hortic. 2014 Mar;168:240–8.
14. Kaaks R, Riboli E, Sinha R. Biochemical markers of dietary intake. *IARC Sci Publ*. 1997;(142):103–26.

15. Bingham SA. Urine Nitrogen as a Biomarker for the Validation of Dietary Protein Intake. *J Nutr.* 2003 Mar 1;133(3):921S-924S.
16. Day N. Epidemiological assessment of diet: a comparison of a 7-day diary with a food frequency questionnaire using urinary markers of nitrogen, potassium and sodium. *Int J Epidemiol.* 2001 Apr 1;30(2):309–17.
17. Tasevska N. Urinary Sucrose and Fructose as Biomarkers for Sugar Consumption. *Cancer Epidemiol Biomarkers Prev.* 2005 May 1;14(5):1287–94.
18. Bingham S, Luben R, Welch A, Low YL, Khaw KT, Wareham N, *et al.* Associations between dietary methods and biomarkers, and between fruits and vegetables and risk of ischaemic heart disease, in the EPIC Norfolk Cohort Study. *Int J Epidemiol.* 2008 Oct 1;37(5):978–87.
19. Grace PB, Taylor JI, Low YL, Luben RN, Mulligan AA, Botting NP, *et al.* Phytoestrogen Concentrations in Serum and Spot Urine as Biomarkers for Dietary Phytoestrogen Intake and Their Relation to Breast Cancer Risk in European Prospective Investigation of Cancer and Nutrition-Norfolk. *Cancer Epidemiol Biomark Prev.* 2004;11.
20. Qian GS, Ross RK, Yu MC, Yuan JM, Gao YT, Henderson BE, *et al.* A follow-up study of urinary markers of aflatoxin exposure and liver cancer risk in Shanghai, People's Republic of China. *Cancer Epidemiol Biomarkers Prev.* 1994 Jan 1;3(1):3–10.
21. Ryo M, Nakamura T, Kihara S, Kumada M, Shibazaki S, Takahashi M, *et al.* Adiponectin as a Biomarker of the Metabolic Syndrome. *Circ J.* 2004;68(11):975–81.
22. Furukawa S, Fujita T, Shimabukuro M, Iwaki M, Yamada Y, Nakajima Y, *et al.* Increased oxidative stress in obesity and its impact on metabolic syndrome. *J Clin Invest.* 2004 Dec 15;114(12):1752–61.
23. Matarese G, Moschos S, Mantzoros CS. Leptin in Immunology. *J Immunol.* 2005 Mar 15;174(6):3137–42.
24. Mantzoros CS. The Role of Leptin in Human Obesity and Disease: A Review of Current Evidence. *Ann Intern Med.* 1999 Apr 20;130(8):671.
25. Unger RH, Clark GO, Scherer PE, Orci L. Lipid homeostasis, lipotoxicity and the metabolic syndrome. *Biochim Biophys Acta BBA - Mol Cell Biol Lipids.* 2010 Mar;1801(3):209–14.
26. Hegyi K. Leptin-induced signal transduction pathways. *Cell Biol Int.* 2004;28(3):159–69.
27. Lee SW, Jo HH, Kim MR, You YO, Kim JH. Association between metabolic syndrome and serum leptin levels in postmenopausal women. *J Obstet Gynaecol.* 2012 Jan;32(1):73–7.
28. Yoshinaga M, Sameshima K, Tanaka Y, Wada A, Hashiguchi J, Tahara H, *et al.* Adipokines and the Prediction of the Accumulation of Cardiovascular Risk Factors or the Presence of Metabolic Syndrome in Elementary School Children. *Circ J.* 2008;72(11):1874–8.

29. Yun JE, Kimm H, Jo J, Jee SH. Serum leptin is associated with metabolic syndrome in obese and nonobese Korean populations. *Metabolism.* 2010 Mar;59(3):424–9.
30. Martins M do C, Lima Faleiro L, Fonseca A. Relationship between leptin and body mass and metabolic syndrome in an adult population. *Rev Port Cardiol Engl Ed.* 2012 Nov;31(11):711–9.
31. Gannagé-Yared MH, Khalife S, Semaan M, Fares F, Jambart S, Halaby G. Serum adiponectin and leptin levels in relation to the metabolic syndrome, androgenic profile and somatotropic axis in healthy non-diabetic elderly men. *Eur J Endocrinol.* 2006 Jul;155(1):167–76.
32. García-Jiménez S, Bernal Fernández G, Martínez Salazar MF, Monroy Noyola A, Toledano Jaimes C, Meneses Acosta A, *et al.* Serum Leptin is Associated With Metabolic Syndrome in Obese Mexican Subjects: Leptin, Obesity, and MetS in Mexican Subjects. *J Clin Lab Anal.* 2015 Jan;29(1):5–9.
33. Achari AE, Jain SK. Adiponectin, a Therapeutic Target for Obesity, Diabetes, and Endothelial Dysfunction. *Int J Mol Sci.* 2017 Jun;18(6):1321.
34. Robinson K, Prins J, Venkatesh B. Clinical review: Adiponectin biology and its role in inflammation and critical illness. *Crit Care.* 2011;15(2):221.
35. Santaniemi M, Kesäniemi YA, Ukkola O. Low plasma adiponectin concentration is an indicator of the metabolic syndrome. *Eur J Endocrinol.* 2006 Nov;155(5):745–50.
36. Falahi E, Khalkhali Rad AH, Roosta S. What is the best biomarker for metabolic syndrome diagnosis? *Diabetes Metab Syndr Clin Res Rev.* 2015 Oct;9(4):366–72.
37. Hara K, Horikoshi M, Yamauchi T, Yago H, Miyazaki O, Ebinuma H, *et al.* Measurement of the High–Molecular Weight Form of Adiponectin in Plasma Is Useful for the Prediction of Insulin Resistance and Metabolic Syndrome. *Diabetes Care.* 2006 Jun 1;29(6):1357–62.
38. Banerjee J, Dhas Y, Mishra N. HOMA-Adiponectin Closely Associates with Cardiometabolic Risk Markers in Middle-Aged Indians with Metabolic Syndrome. *Exp Clin Endocrinol Diabetes.* 2021 Jun;129(06):449–56.
39. Thorand B, Zierer A, Baumert J, Meisinger C, Herder C, Koenig W. Associations between leptin and the leptin / adiponectin ratio and incident Type 2 diabetes in middle-aged men and women: results from the MONICA / KORA Augsburg Study 1984-2002: Adipokines and incident Type 2 diabetes. *Diabet Med.* 2010 Sep;27(9):1004–11.
40. (on behalf of the European Group for the Study of Insulin Resistance: Relationship between Insulin Sensitivity and Cardiovascular Disease Risk Study Group), Finucane FM, Luan J, Wareham NJ, Sharp SJ, O'Rahilly S, *et al.* Correlation of the leptin:adiponectin ratio with measures of insulin resistance in non-diabetic individuals. *Diabetologia.* 2009 Nov;52(11):2345–9.
41. López-Jaramillo P, Gómez-Arbeláez D, López-López J, López-López C, Martínez-Ortega J, Gómez-Rodríguez A, *et al.* The role of leptin/adiponectin

ratio in metabolic syndrome and diabetes. *Horm Mol Biol Clin Investig.* 2014 Apr 1;18(1):37–45.

42. Kotani K, Sakane N. Leptin:Adiponectin Ratio and Metabolic Syndrome in the General Japanese Population. *Ann Lab Med.* 2011 Jul 1;31(3):162–6.
43. Cicero AFG, Magni P, Moré M, Ruscica M, Borghi C, Strollo F. Metabolic Syndrome, Adipokines and Hormonal Factors in Pharmacologically Untreated Adult Elderly Subjects from the Brisighella Heart Study Historical Cohort. *Obes Facts.* 2012;5(3):319–26.
44. Pradhan G, Samson SL, Sun Y. Ghrelin: much more than a hunger hormone. *Curr Opin Clin Nutr Metab Care.* 2013 Nov;16(6):619–24.
45. Ibrahim Abdalla MM, Senior Lecturer, Physiology Department, Faculty of Medicine, MAHSA University, Malaysia. Ghrelin – Physiological Functions and Regulation. *Eur Endocrinol.* 2015;11(2):90.
46. Ukkola O. Ghrelin and Metabolic Disorders. *Curr Protein Pept Sci.* 2009 Feb 1;10(1):2–7.
47. Shiiya T, Nakazato M, Mizuta M, Date Y, Mondal MS, Tanaka M, *et al.* Plasma Ghrelin Levels in Lean and Obese Humans and the Effect of Glucose on Ghrelin Secretion. *J Clin Endocrinol Metab.* 2002 Jan;87(1):240–4.
48. Tesauro M, Schinzari F, Rovella V, Di Daniele N, Lauro D, Mores N, *et al.* Ghrelin Restores the Endothelin 1/Nitric Oxide Balance in Patients With Obesity-Related Metabolic Syndrome. *Hypertension.* 2009 Nov;54(5):995–1000.
49. McLaughlin T, Abbasi F, Lamendola C, Frayo RS, Cummings DE. Plasma Ghrelin Concentrations Are Decreased in Insulin-Resistant Obese Adults Relative to Equally Obese Insulin-Sensitive Controls. *J Clin Endocrinol Metab.* 2004 Apr;89(4):1630–5.
50. Aroor AR, McKarns S, DeMarco VG, Jia G, Sowers JR. Maladaptive immune and inflammatory pathways lead to cardiovascular insulin resistance. *Metabolism.* 2013 Nov;62(11):1543–52.
51. Qu D, Liu J, Lau CW, Huang Y. IL-6 in diabetes and cardiovascular complications: IL-6 in diabetes and cardiovascular complications. *Br J Pharmacol.* 2014 Aug;171(15):3595–603.
52. Indulekha K, Surendar J, Mohan V. High Sensitivity C-Reactive Protein, Tumor Necrosis Factor-α, Interleukin-6, and Vascular Cell Adhesion Molecule-1 Levels in Asian Indians with Metabolic Syndrome and Insulin Resistance (CURES-105). *J Diabetes Sci Technol.* 2011 Jul;5(4):982–8.
53. Chedraui P, Escobar GS, Pérez-López FR, Palla G, Montt-Guevara M, Cecchi E, *et al.* Angiogenesis, inflammation and endothelial function in postmenopausal women screened for the metabolic syndrome. *Maturitas.* 2014 Apr;77(4):370–4.
54. Kim HJ, Higashimori T, Park SY, Choi H, Dong J, Kim YJ, *et al.* Differential Effects of Interleukin-6 and -10 on Skeletal Muscle and Liver Insulin Action *In Vivo. Diabetes.* 2004 Apr 1;53(4):1060–7.

55. Weiss TW, Arnesen H, Seljeflot I. Components of the Interleukin-6 trans-signalling system are associated with the metabolic syndrome, endothelial dysfunction and arterial stiffness. *Metabolism*. 2013 Jul;62(7):1008–13.
56. Sarbijani HM, Khoshnia M, Marjani A. The association between Metabolic Syndrome and serum levels of lipid peroxidation and interleukin-6 in Gorgan. *Diabetes Metab Syndr Clin Res Rev*. 2016 Jan;10(1):S86–9.
57. Iyer SS, Cheng G. Role of Interleukin 10 Transcriptional Regulation in Inflammation and Autoimmune Disease. *Crit Rev Immunol*. 2012;32(1):23–63.
58. van Exel E, Gussekloo J, de Craen AJM, Frölich M, Bootsma-van der Wiel A, Westendorp RGJ. Low Production Capacity of Interleukin-10 Associates With the Metabolic Syndrome and Type 2 Diabetes. *Diabetes*. 2002 Apr 1;51(4):1088–92.
59. Lamb J, Crawford ED, Peck D, Modell JW, Blat IC, Wrobel MJ, *et al.* The Connectivity Map: Using Gene-Expression Signatures to Connect Small Molecules, Genes, and Disease. *Science*. 2006 Sep 29;313(5795):1929–35.
60. Parthasarathy S, Raghavamenon A, Garelnabi MO, Santanam N. Oxidized Low-Density Lipoprotein. In: Uppu RM, Murthy SN, Pryor WA, Parinandi NL, editors. *Free Radicals and Antioxidant Protocols* [Internet]. Totowa, NJ: Humana Press; 2010 [cited 2022 Jun 9]. p. 403–17. (*Methods in Molecular Biology*; vol. 610). Available from: http://link.springer.com/10.1007/978-1-60327-029-8_24
61. Frostegard J, Haegerstrand A, Gidlund M, Nilsson J. Biologically modified LDL increases the adhesive properties of endothelial cells. *Atherosclerosis*. 1991 Oct;90(2–3):119–26.
62. Holvoet P. Association Between Circulating Oxidized Low-Density Lipoprotein and Incidence of the Metabolic Syndrome. *JAMA*. 2008 May 21;299(19):2287.
63. Pohjantähti-Maaroos H, Palomäki A, Kankkunen P, Laitinen R, Husgafvel S, Oksanen K. Circulating oxidized low-density lipoproteins and arterial elasticity: comparison between men with metabolic syndrome and physically active counterparts. *Cardiovasc Diabetol*. 2010;9(1):41.
64. Hurtado-Roca Y, Bueno H, Fernandez-Ortiz A, Ordovas JM, Ibañez B, Fuster V, *et al.* Oxidized LDL Is Associated With Metabolic Syndrome Traits Independently of Central Obesity and Insulin Resistance. *Diabetes*. 2017 Feb 1;66(2):474–82.
65. Kelly AS, Jacobs DR, Sinaiko AR, Moran A, Steffen LM, Steinberger J. Relation of circulating oxidized LDL to obesity and insulin resistance in children. *Pediatr Diabetes*. 2010 Dec;11(8):552–5.
66. Sigurdardottir V, Fagerberg B, Hulthe J. Circulating oxidized low-density lipoprotein (LDL) is associated with risk factors of the metabolic syndrome and LDL size in clinically healthy 58-year-old men (AIR study). *J Intern Med*. 2002 Nov;252(5):440–7.

67. Sentí M, Tomás M, Fitó M, Weinbrenner T, Covas MI, Sala J, *et al.* Antioxidant Paraoxonase 1 Activity in the Metabolic Syndrome. *J Clin Endocrinol Metab.* 2003 Nov 1;88(11):5422–6.
68. Martinelli N, Micaglio R, Consoli L, Guarini P, Grison E, Pizzolo F, *et al.* Low Levels of Serum Paraoxonase Activities are Characteristic of Metabolic Syndrome and May Influence the Metabolic-Syndrome-Related Risk of Coronary Artery Disease. *Exp Diabetes Res.* 2012;2012:1–9.
69. Garelnabi M, Litvinov D, Mahini H. Antioxidant and anti-inflammatory role of paraoxonase 1: Implication in arteriosclerosis diseases. *North Am J Med Sci.* 2012;4(11):523.
70. Adhe-Rojekar A, Mogarekar MR, Rojekar MV. Paraoxonase activity in metabolic syndrome in children and adolescents. *Casp J Intern Med.* 2018;9(2):116–20.
71. Lu J, Marjon KD, Mold C, Du Clos TW, Sun PD. Pentraxins and Fc receptors. *Immunol Rev.* 2012 Nov;250(1):230–8.
72. Inoue K, Kodama T, Daida H. Pentraxin 3: A Novel Biomarker for Inflammatory Cardiovascular Disease. *Int J Vasc Med.* 2012;2012:1–6.
73. Muller B, Peri G, Doni A, Torri V, Landmann R, Bottazzi B, *et al.* Circulating levels of the long pentraxin PTX3 correlate with severity of infection in critically ill patients: *Crit Care Med.* 2001 Jul;29(7):1404–7.
74. Kardas F, Akın L, Kurtoglu S, Kendirci M, Kardas Z. Plasma Pentraxin 3 as a Biomarker of Metabolic Syndrome. *Indian J Pediatr.* 2015 Jan;82(1):35–8.
75. Ohbayashi H, Miyazawa C, Miyamoto K, Sagara M, Yamashita T, Onda R. Pitavastatin Improves Plasma Pentraxin 3 and Arterial Stiffness in Atherosclerotic Patients with Hypercholesterolemia. *J Atheroscler Thromb.* 2009;16(4):490–500.
76. Zheng X, Chen T, Jiang R, Zhao A, Huang F, Zhang Y, *et al.* Hyocholic acid species and the risk of type 2 diabetes [Internet]. bioRxiv; 2018 [cited 2022 Jun 10]. p. 503532. Available from: https://www.biorxiv.org/content/10.1101/503532v1
77. Watkins KL, Veum TL, Krause GF. Total Nitrogen Determination of Various Sample Types: A Comparison of the Hach, Kjeltec, and Kjeldahl Methods. *J AOAC Int.* 1987 May 1;70(3):410–2.
78. Zhang X lei, Wu Y feng, Wang Y shan, Wang X zhe, Piao C hong, Liu J mei, *et al.* The protective effects of probiotic-fermented soymilk on high-fat diet-induced hyperlipidemia and liver injury. *J Funct Foods.* 2017 Mar;30:220–7.
79. Kalhan SC, Guo L, Edmison J, Dasarathy S, McCullough AJ, Hanson RW, *et al.* Plasma metabolomic profile in nonalcoholic fatty liver disease. *Metabolism.* 2011 Mar;60(3):404–13.
80. Garcia RA, Vanelli CP, Pereira Junior O dos S, Corrêa JO do A. Comparative analysis for strength serum sodium and potassium in three different methods: Flame photometry, ion-selective electrode (ISE) and colorimetric enzymatic. *J Clin Lab Anal* [Internet]. 2018 Nov [cited 2022 Jun 9];32(9). Available from: https://onlinelibrary.wiley.com/doi/10.1002/jcla.22594

81. Zhang YJ, Gan RY, Li S, Zhou Y, Li AN, Xu DP, *et al.* Antioxidant Phytochemicals for the Prevention and Treatment of Chronic Diseases. *Molecules*. 2015 Nov 27;20(12):21138–56.
82. Jafri L, Khan AH, Azeem S. Ionized Calcium Measurement in Serum and Plasma by Ion Selective Electrodes: Comparison of Measured and Calculated Parameters. *Indian J Clin Biochem*. 2014 Jul;29(3):327–32.
83. Zaman Z, Fielden P, Frost PG. Simultaneous determination of vitamins A and E and carotenoids in plasma by reversed-phase HPLC in elderly and younger subjects. *Clin Chem*. 1993 Nov 1;39(11):2229–34.
84. Fu X, Peterson JW, Hdeib M, Booth SL, Grusak MA, Lichtenstein AH, *et al.* Measurement of Deuterium-Labeled Phylloquinone in Plasma by High-Performance Liquid Chromatography/Mass Spectrometry. *Anal Chem*. 2009 Jul 1;81(13):5421–5.
85. Çat J, Yaman M. Determination of Nicotinic Acid and Nicotinamide Forms of Vitamin B3 (Niacin) in Fruits and Vegetables by HPLC Using Postcolumn Derivatization System. *Pak J Nutr.* 2019 May 15;18(6):563–70.
86. Robitaille L, Hoffer LJ. A simple method for plasma total vitamin C analysis suitable for routine clinical laboratory use. *Nutr J.* 2015 Dec;15(1):40.
87. Kermali M, Khalsa RK, Pillai K, Ismail Z, Harky A. The role of biomarkers in diagnosis of COVID-19 – A systematic review. *Life Sci.* 2020 Aug;254:117788.
88. Picó C, Serra F, Rodríguez AM, Keijer J, Palou A. Biomarkers of Nutrition and Health: New Tools for New Approaches. *Nutrients*. 2019 May 16;11(5):1092.
89. De Caterina R, Martínez Hernández JA, Kohlmeier M, editors. *Principles of nutrigenetics and nutrigenomics: fundamentals of individualized nutrition.* London: Academic Press; 2020. 566 p.
90. Ocké MC, Kaaks RJ. Biochemical markers as additional measurements in dietary validity studies: application of the method of triads with examples from the European Prospective Investigation into Cancer and Nutrition. *Am J Clin Nutr.* 1997 Apr 1;65(4):1240S-1245S.
91. Kuhnle GG. Nutritional biomarkers for objective dietary assessment. *J Sci Food Agric*. 2012 Apr;92(6):1145–9.
92. Baldrick FR, Woodside JV, Elborn JS, Young IS, McKinley MC. Biomarkers of Fruit and Vegetable Intake in Human Intervention Studies: A Systematic Review. *Crit Rev Food Sci Nutr*. 2011 Oct;51(9):795–815.
93. Johnston E, Pye H, Bonet-Carne E, Panagiotaki E, Patel D, Galazi M, *et al.* INNOVATE: A prospective cohort study combining serum and urinary biomarkers with novel diffusion-weighted magnetic resonance imaging for the prediction and characterization of prostate cancer. *BMC Cancer*. 2016 Dec;16(1):816.
94. Shils ME, Olson JA, Shike M. *Modern nutrition in health and disease.* Philadelphia: Lea & Febiger; 1994.
95. McGuire S. U.S. Department of Agriculture and U.S. Department of Health and Human Services, *Dietary Guidelines for Americans*, 2010. 7th Edition, Washington, DC: U.S. Government Printing Office, January 2011. Adv Nutr. 2011 May 1;2(3):293–4.

9

Food Safety and Quality Assurance in the Food Chain: Focus on Foodborne Outbreaks

Pierina Visciano* **and Maria Schirone**

Department of Bioscience and Technology for Food, Agriculture and Environment, University of Teramo, Teramo, Italy

Abstract
This chapter describes the foodborne outbreaks reported in the European Union during the last years, and number and type of notifications for different food categories contaminated with pathogenic bacteria, viruses, and parasites. Some food safety hazards are associated with the primary production, i.e., on field or at farm level for food of plant or animal origin respectively, but also during the production process, foods can be improperly handled, processed, or stored, contributing to the incidence of consumer illnesses. The most representative pathogens in the reported data are *Salmonella* spp., *Campylobacter* spp., *Escherichia coli* and *Listeria monocytogenes*. The high percentage of hospitalization in some situations underlines the importance of both preventive and control measures to guarantee the public health.

Keywords: Food, outbreaks, pathogens, zoonoses, notifications, alert

9.1 Introduction

Foodborne diseases constitute a significant public health problem as large numbers of persons are affected each year. Nevertheless, they are sometimes underestimated when people do not present to the healthcare system or require hospitalization. A foodborne outbreak is

**Corresponding author*: pvisciano@unite.it

Tejpal Dhewa, Anil Kumar Puniya and Anil Panghal (eds.) Nutritional Science and Technology: Concept to Application, (199–222) © 2023 Scrivener Publishing LLC

characterized by the appearance of the same symptoms in two or more individuals after ingestion of the same water or food or, if that cannot be demonstrated, sharing a common meal or food facility (Hajikarim *et al.*, 2020; Wu *et al.*, 2018). Many factors contribute to their appearance, as a consequence of contamination of raw matter and/or inappropriate practices during production and preparation, i.e., undercooking, incorrect chilling or storage, cross-contamination, and use of contaminated ingredients (Figure 9.1).

Conversely, their prevention is based above all on the general principles of food hygiene in all steps of the food production chain, from production to consumption. Some studies indicate that foodborne outbreaks are also associated with home-prepared food, and the most common risk factors are the purchase of foodstuffs from an unsafe source, their improper cooking, and cross-contamination inside domestic refrigerators or during meal preparation (Saeed *et al.*, 2021). The European Union (EU) annual report revealed that 41.3% of foodborne outbreaks described in 2019 occurred at home (EFSA-ECDC, 2021).

The different food matrices can influence the bacterial growth and survival, their virulence, and biofilm formation, as well as antibiotic resistance. They particularly regulate specific genes and pathways of microorganisms through integrated global metabolic responses, as demonstrated in many phenotypic and transcriptomic studies. The transcriptomic sequencing of *Salmonella* Typhimurium and *Cronobacter sakazakii*, developed respectively in low water activity foods and cured vegetables with high osmotic

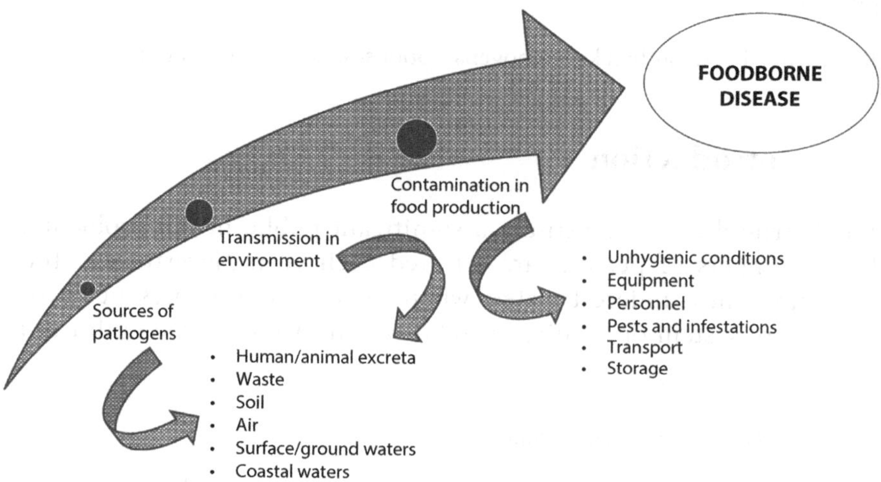

Figure 9.1 Potential transmission pathways of foodborne illnesses.

potential, revealed some characteristics facilitating desiccation and osmoregulation for the microbial cell survival (Chen *et al.*, 2021).

Foodborne illnesses can be caused by a great variety of bacteria, viruses, and parasites. Other forms of intoxication are associated with the consumption of food contaminated by harmful toxins or undesirable chemical substances. Most of them are characterized by acute gastroenteritis from pathogenic species belonging to the genera *Salmonella*, *Escherichia*, *Campylobacter*, *Listeria* and *Vibrio*, or viruses (e.g., Norovirus, Rotavirus, Adenovirus, and Astrovirus) and parasites, i.e., *Giardia* spp., *Cryptosporidium* spp., and *Entamoeba histolytica* (Hu *et al.*, 2020). However, some foodborne pathogenic bacteria can cause health complications, such as recurring intestinal inflammation, chronic renal malfunction, central nervous system (CNS) disorders, reactive arthritis, blindness, and even death (Qiao *et al.*, 2020).

9.2 Classification of Foodborne Diseases by Symptomatology

Foodborne diseases can be distinguished based on the mainly involved apparatus of the human body in:

- gastrointestinal: salmonellosis, campylobacteriosis, shigellosis, yersiniosis, vibriosis, cryptosporidiosis, etc.;
- reproductive: listeriosis, toxoplasmosis, dioxin or further endocrine disruptor exposure;
- cardiovascular: scombroid poisoning or histamine intoxication;
- muscle tissue: trichinellosis, sarcosporidiosis;
- CNS: botulism, neurotoxic or amnesic shellfish poisoning, methylmercury intoxication.

Campylobacteriosis is one of the leading foodborne infections worldwide, with over 400-500 million cases each year, caused mainly by *Campylobacter jejuni* and *Campylobacter coli*. It is characterized by gastrointestinal disorders (i.e., vomiting, stomach cramps, watery or bloody diarrhea, and fever) occurring 2-5 days after the consumption of contaminated foods (raw and/or improperly cooked chicken meat, but also milk products or vegetables) and predominantly observed in children, and in elderly and immuno-compromised individuals. However, severe complications are also described, such as inflammatory bowel disease, reactive arthritis, autoimmune disorders, and the Guillain-Barré syndrome (Igwaran and Okoh 2019; Sher *et al.*, 2021). The latter is a common cause of acute flaccid paralysis, starting distally with a symmetrical pattern from the limbs, up to hyporeflexia or areflexia. The immune response to an antecedent infection, for instance from *C. jejuni*, can generate cross-reactive antibodies against gangliosides, causing an acute inflammatory demyelinating polyneuropathy or acute motor axonal neuropathy (van den Berg *et al.*, 2014).

As foodborne disease, salmonellosis is mostly caused by the two main serovars of *Salmonella enterica*, i.e., *Salmonella* Enteritidis and *Salmonella* Typhimurium, from poultry meat and eggs, but also pork and beef, milk, seafood, and fruits and vegetables. The most common symptoms appear 8-48 h after the ingestion of contaminated food or water and consist of fever (38-39 °C), headache, nausea, vomiting, abdominal cramp, and non-bloody diarrhea. They are generally self-limiting, even if stool culture can remain positive for 4-5 weeks up to 1 year after the infection. Recurrent or chronic forms, such as the aseptic reactive arthritis and Reiter's syndrome, are also described (Di Febo *et al.*, 2019; Nadi *et al.*, 2020).

Even if *Escherichia coli* has a mutual relationship with the human body, some strains cause serious complications, such as bloody diarrhea, urinary tract infections, neonatal meningitis, and even kidney failure. Enterohemorrhagic *E. coli* (EHEC) are strains producing Shiga toxin, and therefore they are also called Shiga toxin-producing *E. coli* (STEC). *Escherichia coli* O157:H7 is a common serotype of EHEC responsible for multiple foodborne outbreaks worldwide (Glowacki *et al.*, 2019).

Listeriosis can be asymptomatic or present different clinical signs varying from a mild fever to gastroenteritis in immuno-compromised individuals, or abortion, preterm or stillbirth in pregnant women. The most severe complications are meningitis and encephalitis, showing even a fatal outcome (Iannetti *et al.*, 2020).

The symptoms of yersiniosis vary from self-limiting gastroenteritis to acute enteritis with diarrhea and abdominal pain. Moreover, mesenteric lymphadenitis and pseudo-appendicular syndromes seemly to acute appendicitis can occur in children over 5 years, and septicemia in elderly and immuno-compromised individuals. Further complications can be pneumonia, endocarditis, meningitis, and osteomyelitis (Luciani *et al.*, 2018).

Bacillus cereus causes two gastrointestinal forms of illness, i.e., the emetic syndrome within 5 h from the ingestion of the contaminated food and due to a heat-stable and pre-formed toxin named cerulide, and the diarrheal syndrome, occurring after 8-16 h and associated with two enterotoxins (i.e., hemolysin and non-hemolytic) produced by viable cells in the small intestine (Osimani *et al.*, 2018). Also, *Vibrio parahaemolyticus* produces a thermostable hemolysin, and gastroenteritis, septicemia, and wound infection are the most common symptoms (Li *et al.*, 2019).

Staphylococcal food poisoning is caused by the consumption of food containing staphylococcal enterotoxins produced by enterotoxigenic strains of *Staphylococcus aureus.* Such microorganism is reported in 30-80% of the human population, and therefore the personnel working in a food industry represent the first source on food contamination by unhygienic food handling and manufacturing practices. The typical symptoms occur rapidly (2-4 h) and include nausea, vomiting, abdominal pain, and diarrhea (Iannetti *et al.*, 2019; Umeda *et al.*, 2017).

Scombroid poisoning is due to high concentrations of histamine in fish and fish products subjected to temperature abuse, which can act as neurotransmitter and vasodilator, causing bronchospasm, respiratory distress, hypotension, heart palpitations, and vasodilatory shock in the worst situations, but also gastro-intestinal or cutaneous effects that can be treated with antihistamine drugs (Visciano *et al.*, 2020).

Botulism is caused by the botulinum neurotoxin produced by *Clostridium botulinum* and formed by a heavy chain for the transport into the nervous cell, and a light chain responsible for the inhibition of the release of acetylcholine at the neuromuscular junctions, causing the muscle paralysis. It is generally already present in food, but it can be produced also inside the body, e.g., in the intestine of infants fed with food containing the spores of the microorganism (Åberg *et al.*, 2021).

Foodborne viruses differ from pathogenic bacteria for a long persistence (for months) in the food products or the environment, as well as a higher resistance to the most common control measures during food production. The main foods contaminated with viruses are fruits, salads of raw vegetables, mussels, oysters, and raw fish. The hepatitis A virus is the most common causative agent of viral foodborne outbreaks worldwide, while the hepatitis E virus derives mainly from uncooked or undercooked pork or game meat, as these animal species represent a reservoir. The first causes an acute self-limiting hepatitis, that can be fulminant in 0.6% of the cases, while the second is mostly dangerous for pregnant women and individuals with other underlying liver diseases (Di Cola *et al.*, 2021). Another common viral source of illness is represented by Norovirus from contaminated live bivalve mollusks, that cause symptoms within 10-51 h, such as nausea, sudden vomiting and/or watery non-bloody diarrhea, abdominal or general muscle pain, headache, and mild fever. Dehydration, hospitalization, and even death are described only in vulnerable people, including children and elderly (Berti *et al.*, 2020). Besides food, the transmission of both Norovirus and hepatitis A virus occurs via the fecal-oral route, including direct person-to-person contact, or contact with contaminated environmental surfaces (Hennechart-Collette *et al.*, 2020).

Among the foodborne illnesses originating from parasites, trichinellosis must be carefully monitored as it can lead to death when *Trichinella spiralis* larvae arrive to some vital organs, such as heart or brain. It recognizes three consequent forms based on the life cycle of the nematode, i.e., an intestinal phase, with diarrhea and abdominal pain, then pyrexia, periorbital or facial oedema, and myalgia arise during the larvae migration (second phase) in the whole host body and, only in some cases, complications such as myocarditis, meningitis or encephalopathy, or pneumonia and obstructive bronchitis can occur. In the last phase, the larvae form a nurse cell-parasite complex into myocytes, where they can survive up to 10-15 years (Schirone *et al.*, 2019). Another common parasitic zoonosis is due to the migration of *Anisakis* spp. larvae, that normally infects fish and crustaceans as intermediate and marine mammals as definitive hosts. The human ingestion of the third stage larvae of the parasite, by consumption of raw or undercooked fish, causes a fish-borne zoonosis known as anisakiasis, that is characterized by digestive disorders, but also IgE-mediated hypersensitivity, angioedema, urticaria or anaphylaxis (Sánchez-Alonso *et al.*, 2021).

9.3 The EFSA-ECDC Reports on Zoonoses and Foodborne Outbreaks

Many countries have surveillance systems to monitor outbreaks of foodborne diseases. The EU system is based on the Zoonoses Directive 2003/99/EC, which requires Member States to monitor and collect relevant data on zoonoses, zoonotic agents, antimicrobial resistance, and foodborne outbreaks. Each Member State submits an annual report to the EU Commission, which assembles and sends all of them to the European Food Safety Authority (EFSA) for publishing – in collaboration with the European Centre for Disease Prevention and Control (ECDC) – the EU Annual Summary Report, that illustrates and identifies the etiological agents that cause the most common zoonotic infections in humans, but includes also some information regarding animals, food, and feeding. By such way, the trend of most zoonoses during the years can be observed and discussed by EU Commission and Member States, with the objective to find potential strategies to combat them.

In Figure 9.2, the foodborne outbreaks with the main symptoms obtained from the EFSA-ECDC Annual Reports from 2010 to 2019 are presented as a word cloud, while the numbers of confirmed human cases of the most common zoonoses are shown in Figure 9.3.

Figure 9.2 Analysis of relevant keywords for foodborne diseases and symptoms from the EFSA-ECDC Reports with a word cloud generator.

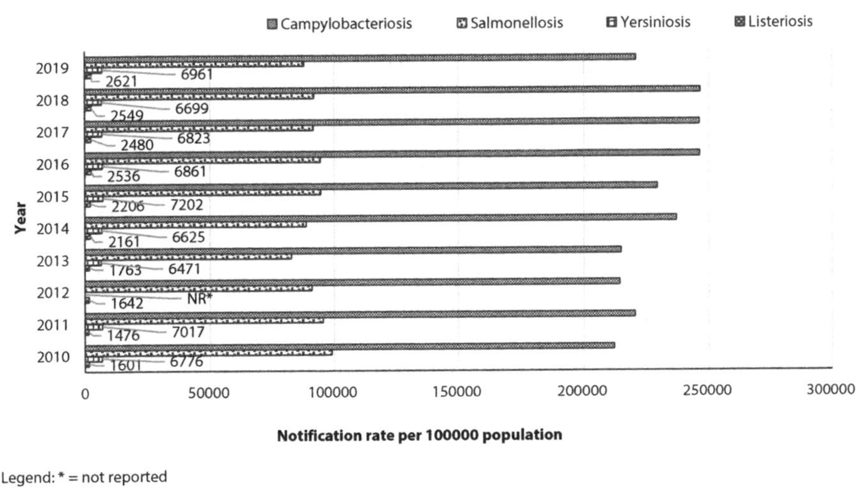

Legend: * = not reported

Figure 9.3 Number of confirmed human cases of major zoonoses in the EFSA-ECDC Reports in the years 2010-2019.

Campylobacteriosis has been first since 2005 and represents 50% of all the reported cases in 2019, followed by other bacterial diseases, i.e., salmonellosis, STEC infections, and yersiniosis. Hospitalizations and case fatalities correspond respectively to 20,432 and 47 for campylobacteriosis, and 16,628 and 140 for salmonellosis. The highest number of deaths is caused by listeriosis, up to 300 in 2019 (EFSA-ECDC 2021).

According to the technical specifications for the harmonization among the EU Member States when reporting the foodborne outbreaks as required by the Directive 2003/99/EC, such events must be categorized as having strong or weak evidence based on the strength of evidence implicating a suspected food vehicle (i.e., the food suspected of causing human cases). For strong-evidence outbreaks, a detailed data set describing the implicated food vehicle, contributory factors, and source is presented, whereas for weak-evidence outbreaks this reporting is not compulsory. However, the minimum required information regards the total number of outbreaks over a year, the number of human illnesses and deaths, the causative agent or foodstuff implicated in the outbreak, the place where the incriminated

Table 9.1 *Salmonella* foodborne outbreaks data collected from the EFSA-ECDC Reports related to the years 2010-2019.

	Outbreak		Strong evidence outbreak				Weak evidence outbreak
Year	N	%	N	Cases	Hospitalized	Deaths	(N)
2010	1604	30.5	341	5212	994	9	1263
2011	1501	26.6	283	4662	815	6	1218
2012	1533	28.6	347	5787	985	10	1186
2013	1168	22.5	315	4371	1134	3	853
2014	1049	20.0	226	3677	890	11	823
2015	953	21.9	184	6616	1719	3	769
2016	1067	22.3	215	9061	1766	10	852
2017	1241	24.4	269	9600	2227	11	972
2018	1581	30.7	297	11581	2298	8	1284
2019	926	17.9	265	9169	1915	7	661

Table 9.2 *Campylobacter* foodborne outbreaks data collected from the EFSA-ECDC Reports related to the years 2010-2019.

	Outbreak		Strong evidence outbreak				Weak evidence outbreak (N)
Year	**N**	**%**	**N**	**Cases**	**Hospitalized**	**Deaths**	
2010	470	8.9	27	398	10	0	443
2011	598	10.6	39	720	17	1	559
2012	501	9.3	25	198	17	0	476
2013	414	8.0	32	478	15	0	382
2014	446	8.5	31	525	40	0	415
2015	387	8.9	25	1440	129	1	362
2016	461	9.6	24	4606	140	0	437
2017	395	7.8	33	1445	207	1	362
2018	524	10.2	29	2335	135	0	495
2019	319	6.2	18	1254	125	0	301

foodstuff was produced, purchased, or consumed, and some contributory factors, as for instance, the deficiencies in food processing hygiene (EFSA 2014).

The data collected from the EFSA-ECDC Reports related to 2010-2019 for *Salmonella* and *Campylobacter* are exhibited in Tables 9.1 and 9.2 respectively, while Table 9.3 shows such information for some bacterial toxins including *Bacillus*, *Clostridium*, and staphylococcal enterotoxins. As before 2015 *Listeria* belonged to the category "other bacterial agents" the detailed trend for such pathogen can be presented only after that date, i.e., from 2015 to 2019 (Table 9.4).

9.4 The Notifications from the Rapid Alert System for Food and Feed Portal

The Rapid Alert System for Food and Feed (RASFF) guarantees the efficient and real-time sharing of information regarding food safety among

Table 9.3 Bacterial toxins including *Bacillus*, *Clostridium* and staphylococcal enterotoxins foodborne outbreaks data collected from the EFSA-ECDC Reports related to the years 2010-2019.

Year	Outbreak		Strong evidence outbreak				Weak evidence outbreak (N)
	N	%	N	Cases	Hospitalized	Deaths	
2010	461	8.8	87	2297	215	1	374
2011	730	12.9	119	2102	165	1	611
2012	777	14.5	127	2938	121	2	650
2013	834	16.1	208	4006	163	1	626
2014	843	16.1	109	3026	187	3	734
2015	849	19.5	102	8847	497	3	747
2016	848	17.7	113	8967	401	1	735
2017	818	16.1	115	8468	583	7	703
2018	950	18.5	120	9726	534	6	830
2019	997	19.3	104	10555	361	14	893

Table 9.4 *Listeria monocytogenes* foodborne outbreaks data collected from the EFSA-ECDC Reports related to the years 2015-2019.

Year	Outbreak		Strong evidence outbreak				Weak evidence outbreak (N)
	N	%	N	Cases	Hospitalized	Deaths	
2015	14	0.3	5	230	25	4	9
2016	5	0.1	2	25	14	2	3
2017	10	0.2	4	39	22	2	6
2018	14	0.3	7	158	98	21	7
2019	21	0.4	9	349	236	31	12

the EU Member States, EFSA and the Commission, leading to prompt actions to deal with risks in the food chain that can affect consumer health. The RASFF Portal has a public access and provides an information summary about the most recently transmitted notifications. According to the Commission Implementing Regulation (EU) 2019/1715 laying down rules for the functioning of the information management system for official controls and its system components, they are distinguished as follows: i) alert, when there is a serious direct or indirect risk deriving from food and therefore it must be recalled from the market; ii) border rejection, for foods that have been tested and rejected at borders of the EU Member States or the other States participating to the RASFF (i.e., Norway, Iceland, Liechtenstein, and Switzerland); iii) information for attention, when the nature of the risk does not require rapid action, and the product has not reached the market of all Member States, but it is present only in the notifying country, or it is not present on the market anymore; iv) information for follow up, related to a product that is or may be placed on the market of another RASFF network member country (Caldeira *et al.*, 2021).

The notifications collected from the RASFF Portal for some pathogens in 2020 are presented in Table 9.5.

The highest number (n = 670) is related to *Salmonella* spp. or *S. enterica*, followed by *L. monocytogenes* (n = 139). With regard to the food categories, the presence of *Salmonella* spp. was observed mainly in poultry meat and poultry meat products, herbs and spices, and nuts and nut products. Recent foodborne outbreaks have been associated with atypical food vehicles, that are characterized by conditions not favorable to the microbial development, such as a low moisture content. For instance, *Salmonella* infections have been connected to the consumption of tree nuts (pistachios, hazelnuts, pine nuts, and almonds), peanut butter, spices, or powdered infant formula (Kase *et al.*, 2017). Fish and fish products, meat other than poultry, and milk and milk products were the food categories more contaminated by *L. monocytogenes*, whereas *E. coli* and Shiga toxin-producing *E. coli* have been detected mainly in meat and meat products and bivalve mollusks. In all these notifications, for *Salmonella* spp. the only presence in 25 g was reported, whereas for *L. monocytogenes* also the detected values were given, up to 85,000 and 22,000 CFU/g in frozen chicken and chilled pork confit, respectively.

The most notified country of origin for the presence of *Salmonella* spp. in food was Poland (40.7%), followed by Brazil and France (Figure 9.4), while the most reported notifying countries for such pathogen were Poland, Germany, Lithuania, and France (Figure 9.5). Instead, in Figures 9.6 and 9.7

Table 9.5 Number of notifications distinguished for pathogens and food categories deriving from the RASFF Portal in the year 2020.

	Product categories																		
Pathogens	**A**	**B**	**C**	**D**	**E**	**F**	**G**	**H**	**I**	**J**	**K**	**L**	**M**	**N**	**O**	**P**	**Q**	**R**	**Total**
Bacillus cereus						5				2			1		1		2		11
Campylobacter jejuni																1			1
Clostridium botulinum								1					1						2
Clostridium perfringens									1										1
*Clostridium**												3							3
*Escherichia coli***	20								10	5		29	14	1		1			80
Listeria monocytogenes		2	1		2			36	6		1	32	26		2	22	8	1	139
*Salmonella****	5		1	2	2	2	14	3	21	85		55	4	65	3	406	2		670
Staphylococcus aureus			1																1
Vibrio parahaemolyticus					1														1
Vibrio cholerae					1			1											2
Yersinia enterocolitica																			-
Total	25	2	3	2	6	7	14	41	38	92	1	119	46	66	6	430	12	1	911

Legend: *= *Clostridium* sulphite reducer or *Clostridium sporogenes*; **= *Escherichia coli* or Shigatoxin-producing *Escherichia coli*; ***= *Salmonella* spp. or *Salmonella enterica*

A = bivalve mollusks and products thereof; B = Cephalopods and products thereof; C = Cereals and bakery products; D = Confectionery; E = Crustaceans and products thereof; F = Dietetic foods, food, supplements, fortified foods; G = Eggs and egg product; H = Fish and fish products; I = Fruits and vegetables; J = Herbs and spices; K = Ice and desserts; L = Meat and meat products (other than poultry); M = milk and milk products; N = Nuts, nut products and seeds; O = Other food product/mixed; P = Poultry meat and poultry meat products; Q = Prepared dishes and snacks; R = Soups, broths, sauces and condiments.

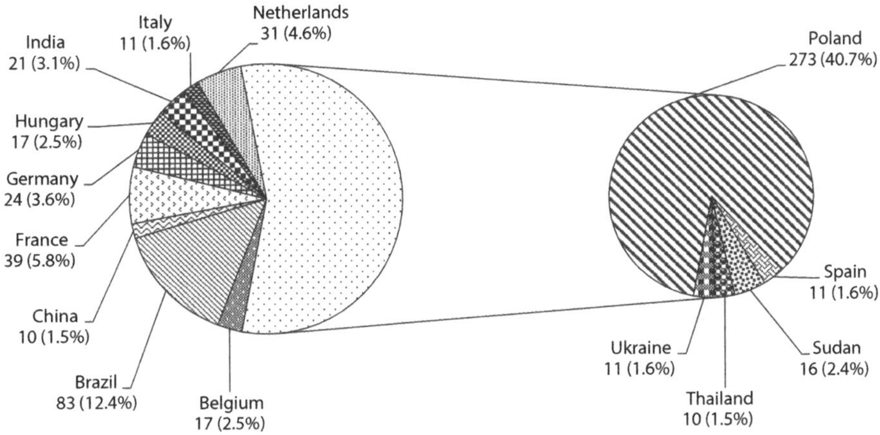

Figure 9.4 Countries of origin and number of notifications (equal or major than 10) for *Salmonella* spp. or *S. enterica* from the RASFF Portal in the year 2020.

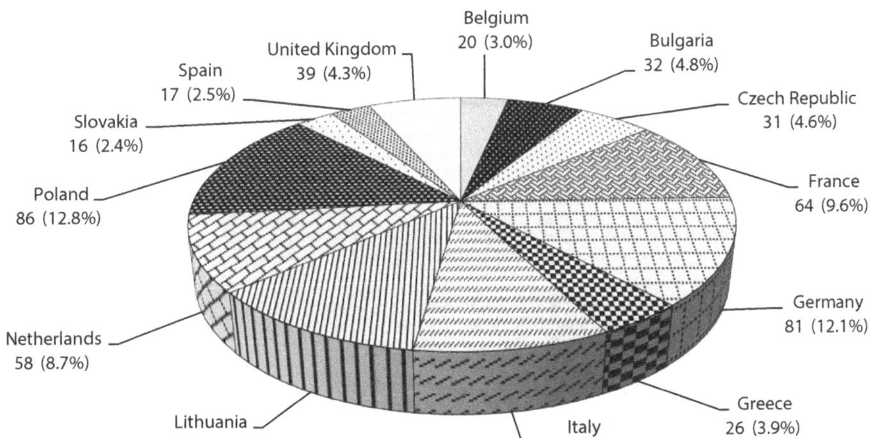

Figure 9.5 Notifying countries and number of notifications (equal or major than 10) for *Salmonella* spp. or *S. enterica* from the RASFF Portal in the year 2020.

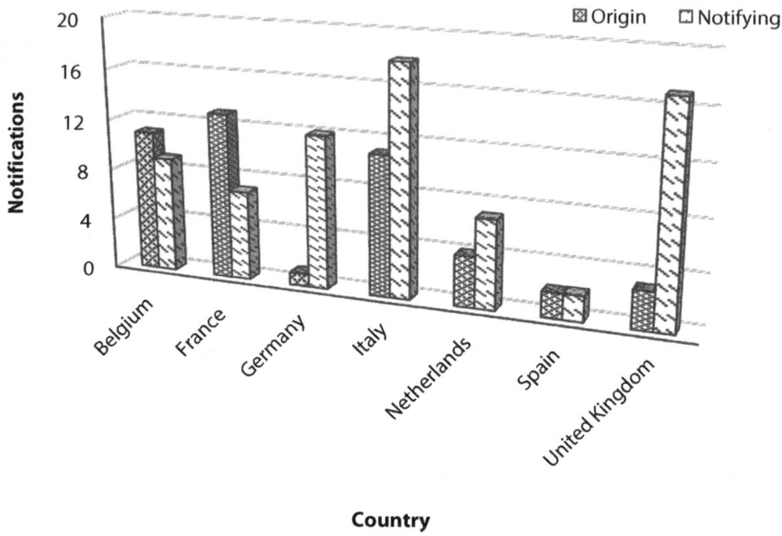

Figure 9.6 Number of notifications for countries resulting both countries of origin and notifying countries for *E. coli* in the year 2020.

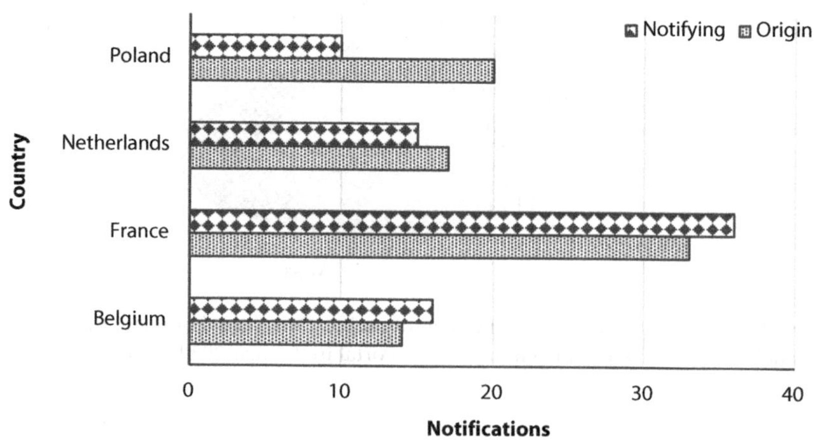

Figure 9.7 Number of notifications for countries resulting both countries of origin and notifying countries for *L. monocytogenes* in the year 2020.

some countries resulting both countries of origin and notifying countries for *E. coli* and *L. monocytogenes* respectively, are shown.

In Table 9.6 the different types of notifications for the most representative pathogens in 2020 were reported. Moreover, the number of foodborne outbreaks was also presented, corresponding to 7 and 3 episodes for *Salmonella* spp. and *L. monocytogenes*, respectively. Regarding the first pathogen, different serovars of *S. enterica* were involved, and the contaminated food categories were meat and meat products, poultry meat, eggs and egg products, and fruits and vegetables, including nuts and seeds. Such notifications were above all alert, i.e., the food was recalled from consumer or withdrawn from recipient, and only two were information for attention, but the product was no longer on the market. The outbreaks associated with *L. monocytogenes* were due to chilled smoked trout fillets from Denmark (68,000 CFU/g), and the remaining two were only suspected to be caused by the pathogen and referred to chilled mozzarella from Italy and chilled smoked trout fillets from the Netherlands. The only food poisoning suspected to be caused by *B. cereus* was linked to food supplement from the US and dispatched by the United Kingdom.

Salmonella spp. and *L. monocytogenes* were the most reported microorganisms in the RASFF Portal also during the period 2010-19. The numbers of notifications (n = 4139) and outbreaks (n = 39) associated with *Salmonella* spp. and/or different serovars of *S. enterica* are shown in Figure 9.8. The highest number (n = 706) was related to 2019, while in 2010 and 2011,

Table 9.6 Classification of notifications for the most represented pathogens by the RASFF Portal in the year 2020.

	Notifications				**Foodborne outbreaks**
Pathogens	**Alert**	**IA**[b]	**IF**[c]	**BR**[d]	
Bacillus cereus	7	1	2	0	1
Escherichia coli	17	22	3	1	0
Shiga toxin-producing *E. coli*	25	10	0	2	0
Listeria monocytogenes	86	34	16	0	3
Salmonella[a]	206	233	72	152	7

Legend: [a]= *Salmonella* spp. or *Salmonella enterica*; [b]= Information for attention; [c]= Information for follow up; [d]= Border rejection.

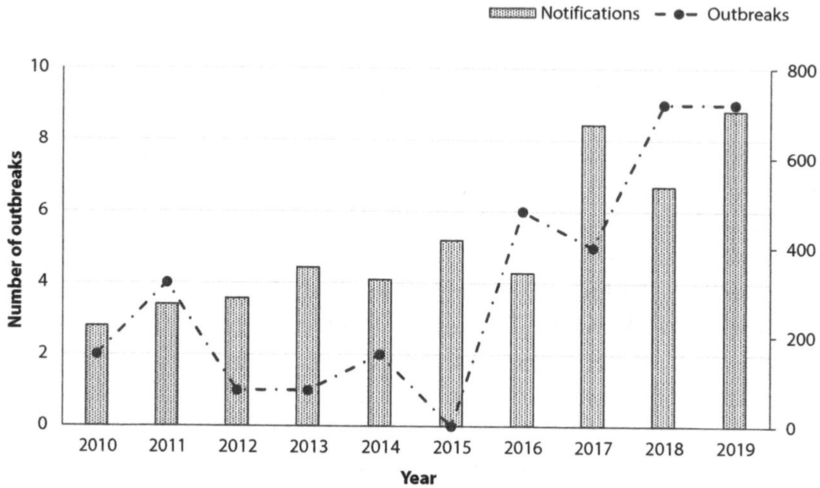

Figure 9.8 Notifications and outbreaks caused by *Salmonella* spp. and different serovars of *S. enterica* from the RASFF Portal in the years 2010-2019.

a total of 224 and 272 notifications were presented, respectively. Regarding the foodborne outbreak notifications, 29 (74.4%) were alert, and recall or withdraw of the contaminated foods were the most common action taken by the involved EU Member State. Among such outbreak episodes, 16 were caused by S. Enteritidis and 12 were due to S. Typhimurium, and the more involved food categories corresponded to meat and meat products (n = 15) and eggs and egg products (n = 14), as shown in Figure 9.9.

A total of 984 notifications for *L. monocytogenes* were reported in the same period, with the highest number (n = 139) again in 2019, while the lowest number (n = 81) was observed in 2016. Among 139 total notifications in 2019, 94 (67.6%) corresponded to alert (Figure 9.10) and 6 (4.3%) were linked to foodborne outbreaks due to meat and meat products (n = 4), and the remaining two were caused by milk and milk products, and prepared dishes and snacks (data not shown). The further outbreaks reported in the same period were due to milk and milk products and fruits and vegetables, in 2018 and 2016 respectively, while a serious episode of listeriosis occurred in 2014, causing the death of 12 out of 20 involved people, after the consumption of lamb-roll sausages from Denmark.

The third most notified microorganism in the RASFF Portal during the period 2010-19 is *E. coli*, with a total of 906 notifications, distinguished as "too high count" in live bivalve mollusks (n = 175) and other foods (n = 290), or Shiga toxin-producing *E. coli* (n = 441). In Table 9.7, the notifications for

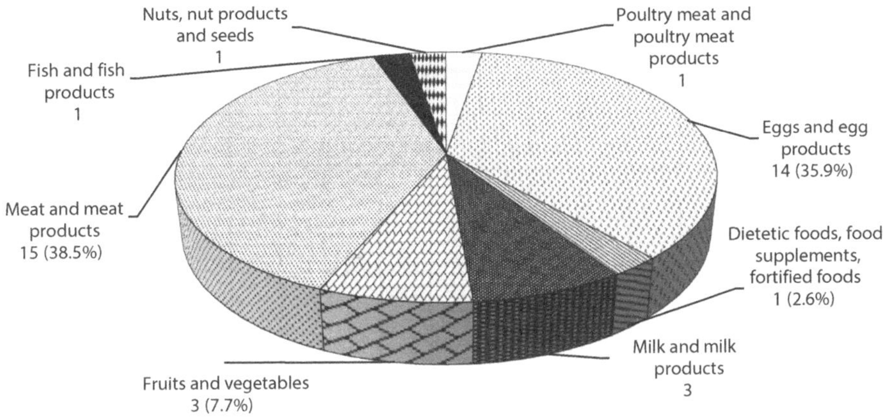

Figure 9.9 Classification of food categories causing foodborne outbreaks by *Salmonella* spp. or different serovars of *S. enterica* in the years 2010-2019.

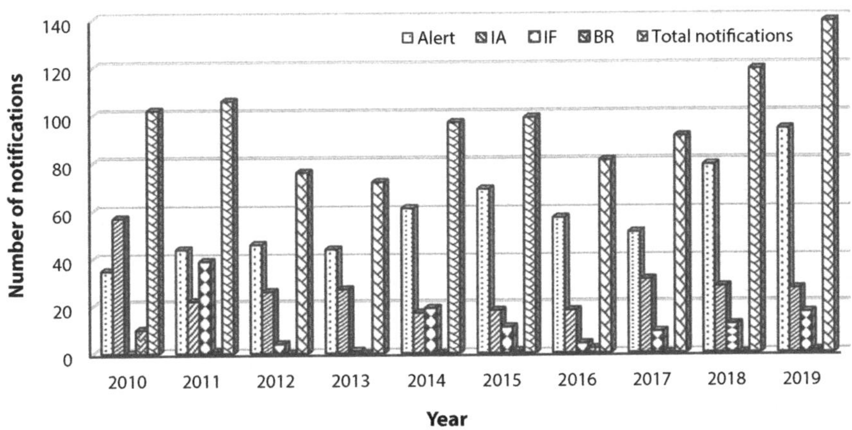

Figure 9.10 Number of notifications for *L. monocytogenes* from the RASFF Portal in the years 2010-2019.

the other pathogens reported in the same period are shown. The highest number (n = 107) is referred to *B. cereus*, followed by *C. jejuni* (n = 49) and *Campylobacter* spp. (n = 37). The latter microorganism is also the most cited in the EFSA-ECDC Reports since 2005, as previously described in paragraph 9.3 and associated with the highest number of zoonoses.

Table 9.7 Number of notifications for pathogens from the RASFF Portal in the years 2010-2019.

Year/ notification	Foodborne pathogens								
	B. cereus[a]	*Campylobacter coli*	*C. jejuni*	*Campylobacter* spp.	*Cl. botulinum*	*Cl. perfringens*	*Clostridium* spp[b]	*S. aureus*	*Y. enterocolitica*
2019									
Alert	6	1	-	-	-	-	-	-	-
IA[c]	1	-	-	-	-	-	1	-	1*
IF[d]	-	1	-	1	-	-	-	-	-
BR[e]	7	-	2	-	-	-	-	-	-
2018									
Alert	4*	1	1	-	-	-	1	1	-
IA	3	7	17	3	-	-	-	1	-
IF	3	1	1	-	-	-	-	-	-
BR	-	-	-	-	-	-	1	-	-
2017									
Alert	3	-	-	-	3	-	-	-	-
IA	2	4	-	3	1	-	1	-	-

(*Continued*)

Table 9.7 Number of notifications for pathogens from the RASFF Portal in the years 2010-2019. (*Continued*)

Year/ notification	Foodborne pathogens								
	B. cereus[a]	*Campylobacter coli*	*C. jejuni*	*Campylobacter* spp.	*Cl. botulinum*	*Cl. perfringens*	*Clostridium* spp[b]	*S. aureus*	*Y. enterocolitica*
IF	-	-	-	-	-	-	-	-	-
BR	3	-	-	-	-	-	1	-	-
2016									
Alert	-	-	-	2	2	-	1	-	-
IA	2	-	7	2	-	-	-	-	-
IF	2*	-	-	-	-	-	1	-	-
BR	1	-	-	-	-	-	-	-	-
2015									
Alert	4	-	-	2	2*	2	-	-	-
IA	-	-	5	3	-	-	-	-	-
IF	1	-	-	-	-	-	-	-	-
BR	2	-	-	-	-	-	-	-	-

(*Continued*)

Table 9.7 Number of notifications for pathogens from the RASFF Portal in the years 2010-2019. (*Continued*)

Year/ notification	Foodborne pathogens								
	B. cereus[a]	*Campylobacter coli*	*C. jejuni*	*Campylobacter* spp.	*Cl. botulinum*	*Cl. perfringens*	*Clostridium* spp[b]	*S. aureus*	*Y. enterocolitica*
2014									
Alert	6	-	-	1	2	1	-	-	-
IA	6	1	8	2	1	-	-	-	-
IF	-	-	-	-	-	-	-	-	-
BR	1	-	-	-	-	-	-	-	-
2013									
Alert	2	-	-	1	3	-	-	-	-
IA	5	2	4	2	-	-	-	-	-
IF	3	-	-	-	-	-	-	-	1
BR	5	-	-	-	-	-	-	-	-
2012									
Alert	6	-	-	-	1	-	-	-	-
IA	1	-	2	4	-	-	-	-	-

(*Continued*)

Table 9.7 Number of notifications for pathogens from the RASFF Portal in the years 2010-2019. (*Continued*)

Year/ notification	Foodborne pathogens								
	B. cereus[a]	*Campylobacter coli*	*C. jejuni*	*Campylobacter* spp.	*Cl. botulinum*	*Cl. perfringens*	*Clostridium* spp[b]	*S. aureus*	*Y. enterocolitica*
IF	-	-	-	1	-	-	-	-	-
BR	2	-	-	-	-	-	-	-	-
2011									
Alert	6	-	-	1	2	1	-	-	1*
IA	5	-	1	5	1	-	-	1	1
IF	-	-	-	3	-	-	-	-	-
BR	1	-	-	-	-	2	-	-	-
2010									
Alert	2	-	1	-	2	-	-	-	-
IA	5	-	-	2	-	-	1	1	-
IF	0	-	-	-	-	-	-	-	-
BR	7	-	-	-	-	-	-	1	-
TOTAL	107	18	49	37	20	6	8	5	4

With regards to parasites, two only notifications regarded *Trichinella* spp. in the period 2010-2019, specifically in chilled wild boar carcasses from Hungary and pork carcasses from Poland in 2017 and 2012, respectively. However, the distribution was restricted to the notifying countries and action taken was the destruction of the product. Some trichinellosis episodes have been recently described in Europe. In 2017, the infestation of pork meat with *Trichinella spiralis* (51 and 62 larvae per gram in sausages and dried meat, respectively) gave origin to two human cases in Serbia and eight in France, while the consumption of hunted wild boar meat not inspected by the official veterinarians caused a large outbreak due to *Trichinella britovi* affecting 114 individuals, with 19 of them hospitalized (Cacciò *et al.*, 2018).

The notifications for the parasitic infestation with *Anisakis* spp. from 2010 to 2020 (n = 392) are shown in Figure 9.11. It can be highlighted that no alert notification is reported, while border rejection is the highest represented category during the investigated period.

Fish and fish products are also involved in other common notifications regarding the presence of histamine. A total of 407 notifications with 18 foodborne outbreaks were obtained from the RASFF Portal during the period 2010-20 (data not shown). The most representative were the tuna species, as a consequence of many favorable factors, such as the high levels of histidine in muscle tissue for a speed and long duration swimming as predator, and the delay of bringing the caught fish onboard the vessel (Visciano *et al.*, 2020).

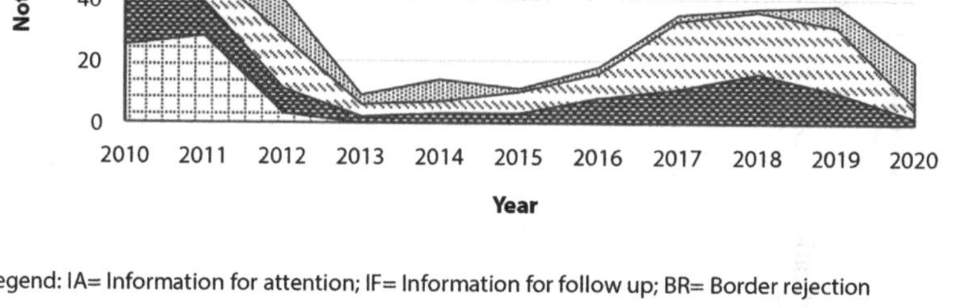

Figure 9.11 Number of notifications for the parasitic infestation with *Anisakis* spp. from the RASFF Portal in the years 2010-2020.

9.5 Conclusions

This chapter focused on foodborne outbreaks as an important component of the global food safety implications affecting public health worldwide. The risk for human illnesses occurrence can be favored by many situations, such as poor or insufficient education of farmers, food business operators, as well as consumers with regard to the basic hygiene practices in food production and preparation, lack of official controls and/or improper programs of their activities, but also illegal actions of irresponsible or even fraudulent producers. The scenario was finalized with the description of the most representative etiological agents of foodborne diseases, their symptoms, and the frequency of detection in foods as well as in humans related to zoonoses.

References

A.T. Åberg, I. Karisson, and M. Hedeland, *Analytical and Bioanalytical Chemistry*, Vol. 413, pp. 345-354, 2021.

M. Berti, L. Teodori, O. Portanti, A. Leone, I. Carmine, N. Ferri, P. Visciano, M. Schirone, and G. Savini, *Italian Journal of Food Science*, Vol. 32(1), pp. 125-132, 2020.

S.M. Cacciò, R.M. Chalmers, P. Dorny, and L.J. Robertson, *Food and Waterborne Parasitology*, Vol. 10, pp. 1-5, 2018.

S. Chen, S. He, X. Xu, and H. Wang, *Current Opinion in Food Science*, Vol. 42, pp. 23-30, 2021.

G. Di Cola, A.C. Fantilli, M.B. Pisano, and V.E. Ré, *International Journal of Food Microbiology*, Vol. 338, 108986, 2021.

European Food Safety Authority, 2014. Update of the technical specifications for harmonised reporting of food-borne outbreaks through the European Union reporting system in accordance with Directive 2003/99/EC. *EFSA Journal* 2014;12(3):3598, 25 pp.

EFSA and ECDC (European Food Safety Authority and European Centre for Disease Prevention and Control) 2021. The European Union One Health 2019 Zoonoses Report. *EFSA Journal* 2021;19(2):6406, 286 pp.

T. Di Febo, M. Schirone, P. Visciano, O. Portanti, G. Armillotta, T. Persiani, E. Di Giannatale, M. Tittarelli, and M. Luciani, *Food Analytical Methods*, Vol. 12, pp. 322-330, 2019.

E.M. Glowacki, J.B. Glowacki, A.D. Chung, and G.B. Wilcox, *American Journal of Infection Control*, Vol. 47, pp. 1280-1282, 2019.

F. Hajikarimi, M.M. Solan Dallal, M.R. Pourmand, and M. Abdi, *Gene Reports*, Vol. 19, 100632, 2020.

C. Hennechart-Collette, S. Martin-Latil, A. Fraisse, F. Niveau, and S. Perelle, *Food Microbiology*, Vol. 91, 103546, 2020.

P. Hu, C. Liu, J. Ruan, M. Yuan, C. Ju, Y. Ma, Y. Yuan, H. Chen, M. Yu, and Y. Duan Y, *Infection, Genetics and Evolution*, Vol. 86, 104607, 2020.

L. Iannetti, P. Visciano, C. Marfoglia, G. Iannitto, G. Parisciani, M. Sericola, D. Petrone, M.S. Mangieri, F. Pomilio, and M. Schirone, *Italian Journal of Food Science*, Vol. 31, pp. 618-625, 2019.

L. Iannetti, M. Schirone, D. Neri, P. Visciano, V. Acciari, G. Centorotola, M.S. Mangieri, M. Torresi, G.A. Santarelli, V. Di Marzio, C. Marfoglia, G. Migliorati, and F. Pomilio, *Food Microbiology*, Vol. 91, 103533, 2020.

A. Igwaran and I. Okoh, *Heliyon*, Vol. 5, e02814, 2019.

Y. Li, S. Zhang, J. Li, M. Chen, M. He, Y. Wang, Y. Zhang, H. Jing, H. Ma, Y. Li, L. Zhao, H. Zhao, B. Kan, and B. Pang, *Food Microbiology*, Vol. 4, 103233, 2019.

M. Luciani, M. Schirone, O. Portanti, P. Visciano, G. Armillotta, R. Tofalo, G. Suzzi, and T. Di Febo, *Food Microbiology*, Vol. 73, pp. 85-92, 2018.

Z.R. Nadi, T.Z. Salehi, I.A. Tamai, A.R. Foroushani, M. Sillanpaa, and M.M. Soltan Dallal, *Microchemical Journal*, Vol. 155, p. 104660, 2020.

A. Osimani, L. Aquilanti, and F. Clementi, *International Journal of Hospitality Management*, Vol. 72, p. 145-153, 2018.

Z. Qiao, Y. Fu, C. Lei, and Y. Li, *Food Control*, Vol. 112, 107116, 2020.

B.Q. Saeed, T. M. Osaili, and S. Taha, *Food Control*, Vol. 125, 108024, 2021.

I. Sánchez-Alonso, N. Carballeda-Sangiao, M. González-Muñoz, S.C. Arcos, A. Navas, and M. Careche, *Food Control*, Vol. 124, 107837, 2021.

M. Schirone, P. Visciano, A.M.A. Olivastri, M.P. Sgalippa, and A. Paparella, *Foods*, Vol. 8(195), pp. 1-9, 2019.

A.A. Sher, M.A. Ashraf, B.E. Mustafa, and M.M. Raza, *Food Microbiology*, Vol. 97, 103751, 2021.

K. Umeda, H. Nakamura, K. Yamamoto, N. Nishina, K. Yasufuku, Y. Hirai, T. Hirayama, K. Goto, A. Hase, J. Ogasawara, International Journal of Food Microbiology, Vol. 256, pp. 30-35, 2017.

B. van der Berg, C. Walgaard, J. Drenthen, C. Fokke, B.C. Jacobs, P.A. van Doom, *Nature Reviews Neurology*, Vol. 10(8), pp. 469-482, 2014.

P. Visciano, M. Schirone, and A. Paparella, *Foods*, Vol. 9 (1975), pp. 1-15, 2020.

Y.-n. Wu, X.-m. Liu, Q. Chen, H. Liu, Y. Dai, Y.-j. Zhou, J. Wen, Z.-z. Tang, and Y. Chen, *Food Control*, Vol. 84, pp. 382-388, 2018.

A.J. Rodrigues Caldeira, C.P. Pereira Alves, M. Santos, Food Control, Vol. 124, 107913, 2021.

J.A. Kase, G. Zhang, Y. Chen, Current Opinion in Food Science, Vol. 18, pp. 56-63, 2017.

10

Emerging Techniques in Food Preservation

Sunita Aggarwal[1]* and Shalini Sehgal[2]

[1]*Department of Microbiology, Institute of Home Economics, University of Delhi, New Delhi, India*
[2]*Department of Food Technology, Bhaskaracharya College of Applied Sciences, University of Delhi, New Delhi, India*

Abstract
Today, the major challenge faced by the food industry is to meet the consumers' needs. The growing awareness among consumers of the importance of safe and high-quality food has pressured the food industry to adopt novel food processing techniques that maintain food quality without losing its freshness. Different conventional methods like heating, freezing, drying, canning, etc., which are being used for preservation, degrade thermolabile nutrients, besides changing foods' texture, size, and color. This has led to different innovative food preservation technologies that provide biologically safe food without compromising its quality. There are various new thermal and non-thermal techniques based on mechanical, electromagnetic, ultrasonic, plasma, or biological processes which are currently being adapted for food preservation. These techniques inactivate or kill spoiling microorganisms, interfering with food enzymes and pathogens, with minimal deterioration of the food's nutritional, sensory, functional and organoleptic attributes. Besides, most of the techniques are fast and environmentally friendly.

Keywords: Food preservation, advanced food preservation technologies, high-pressure processing, cold plasma, irradiation

10.1 Introduction

Food preservation is the process of preventing or slowing down food spoilage and extending its shelf life by killing or inactivating the microorganisms and endogenous enzymes. Evidence suggests that food preservation has

**Corresponding author*: sunita.aggarwal@ihe.du.ac.in

Tejpal Dhewa, Anil Kumar Puniya and Anil Panghal (eds.) Nutritional Science and Technology: Concept to Application, (223–244) © 2023 Scrivener Publishing LLC

been practiced for delayed consumption, distribution, and storage since ancient times. Salting and fermentation were the commonly employed methods to preserve food in the ancient past. The breakthrough in food preservation was the development of a canning process in the early 19th century (1809) by a French chemist, Nicolon Appert, known as the father of canning. Another turning point in preserving food came with pasteurization, the technique developed by Louis Pasteur in 1863. After that, other physical and chemical methods like drying, heating, freezing, smoking, the addition of preservatives, etc., were developed and used to improve the quality of preserved food and make it safe by reducing the number of contaminating microorganisms [1].

The first and foremost requirement to sustain healthy human life is safe and good-quality food. Food, a perishable commodity, can be contaminated from spoilage and disease-causing microorganisms at any stage of processing and preparation from harvesting until consumption. Food contamination can occur from diverse sources, including soil, water, air, contact surfaces, equipment, and food personnel [2]. The presence of microbes in food is one of the major factors responsible for chemical or biological changes in food to make it unfit for consumption. Moreover, these microbes can be vehicles for various foodborne diseases, namely botulism, salmonellosis, bacillary dysentery, etc. The activity of these microbes in food has to be curtailed to make it safe for consumption.

There are a number of methods commonly used to prolong shelf life and improve food quality. These include pasteurization, heating at a temperature of more than 100° C, canning, freezing, chilling, drying, smoking, asepsis, and chemical preservatives. These methods prevent self or microbial decomposition of food by:

(i) inactivating or killing the microorganisms
(ii) denaturing food and microbial enzymes
(iii) delaying the growth and increasing the lag phase of growth
(iv) preventing chemical reactions

However, the use of these conventional methods has certain limitations since they [3]:

(1) bring changes in nutrition, sensory and functional attributes of the food
(2) denature proteins, essential enzymes, and other heat-sensitive compounds
(3) lead to loss of volatile compounds

(4) alter color, texture, and flavor of food
(5) induce the formation of other new compounds

The supply of high-quality and safe food is a big concern of the food sector. The limitations of conventional methods emphasize developing new technologies for food preservation, which can increase food quality and safety without compromising their structure, taste, color, and other attributes. Over the years, the demand for these advanced techniques has increased significantly to control microbial growth and spoilage with minimum alteration in food quality. The research has resulted in several such advanced preservation techniques based on mechanical processes, electromagnetic and acoustic technologies, ohmic heating, or chemical processing in the last few decades.

The various novel, innovative techniques that can be used for food preservation include:

1. High-Pressure Processing (HPP)
2. Pulsed Electric Field (PEF) Processing
3. Pulsed Light Technology (PLT)
4. Ultrasonic Food Processing
5. Ohmic Heating
6. Cold Plasma or Plasma light
7. Oscillating Magnetic Field (OMF)
8. High-Pressure Thermal (HPT) Processing
9. Bacteriocins
10. Dielectric Heating
11. Microwave
12. Irradiation

These techniques are based on mechanical processes, electromagnetic and acoustic technologies, irradiation, and chemical processing, each having its merits and limitations (Table 10.1). In the upcoming sections, we are going to discuss these techniques in detail.

10.2 High-Pressure Processing (HPP)

High-pressure or Ultra-high-pressure processing is a non-thermal preservation technology used for the first time in Japan in 1992 to produce acid foods commercially. After that, the technology has been increasingly used worldwide by the food industry to produce healthy, fresh food with

Table 10.1 Merits and limitations of various advanced process techniques.

S. no.	Process technique	Merits	Limitations
1.	High-Pressure Processing (HPP)	• Enhances food safety by reducing pathogen number • Maintain freshness and nutritional quality of food as no effect on covalent bonds by high pressure • No development of unpleasant flavor • No preservatives are required	• May alter the structure of macromolecules like starch may gelatinize on pressurizing • Resistance may develop in bacteria to high pressure • Not suitable for dry foods like cereals, spices, and on leafy vegetables like spinach
2.	Pulsed Electric field (PEF)	• Increases the extraction and pressing yield of sugar, oils, and other metabolites from plant cells (electroporation) • There is less consumption of energy • Processing time is reduced • Freezing rates increases due to cell disintegration	• It is used for liquid food as electric pulse flow is easy • The liquid product should be free from air bubbles & have low electrical conductivity • Process is expensive • Not suitable for solid foods
3.	Pulsed Light Technology (PLT)	• Decontamination is very fast • Consumption of energy is much less • Maintain physical, sensorial, and nutrition attributes of the food	• Suitable for smooth, non-reflecting surfaces or clear liquids without suspended particles • May lead to lipid oxidation in meat • Heating of samples may occur

(Continued)

Table 10.1 Merits and limitations of various advanced process techniques (*Continued*)

S. no.	Process technique	Merits	Limitations
4.	Ultrasound	• Ultrasound treated food is generally safe as free radicals short-lived • Efficient extraction of fruit juices • Short processing time • Minimal alteration in nutrients, color and flavor	• High cost • Sometimes, off-flavor and discoloration due to physicochemical changes
5.	Ohmic heating	• Processing time is less because of faster heating • Treatment is homogeneous and effective • Efficient alternative to conventional thermal method with less energy consumption	• Monitoring and controlling the process is difficult • Frequency band is narrow • Control on heating is difficult
6.	Cold plasma	• Minimal heat production so food is nutritionally healthy and functionally intact • Suitable for thermally sensitive materials • Pre-packaged food in flexible film can be treated • Large-scale processing is possible at atmospheric pressure	• Most suitable for food surfaces & contact surfaces because of low penetrating power • Decrease in bactericidal effect by quencher effect of proteins and lipids in food • The cavities on the matrix surface protect microbial cell

(*Continued*)

Table 10.1 Merits and limitations of various advanced process techniques (*Continued*)

S. no.	Process technique	Merits	Limitations
7.	Oscillating Magnetic Field (OMF)	• Less requirement of energy input • Does not require any special preparation of food for treatment. • Very minimal thermal denaturation of food characteristics	• Intensity of the magnetic fields to be used depends on the electrical resistivity and thickness of food. • Cell survival is more under heterogeneous magnetic fields.
8.	High Pressure Thermal (HPT) Processing	• Less food damage due to accelerated heating and reduced processing time. • Adverse effect activities of endogenous enzymes like polyphenol oxidases peroxidases.	• Capital cost is high • Machinery required is complex
9.	Bacteriocins	• Application in food is presumably harmless as degraded by proteolytic enzymes while passing through the digestive tract • Being colorless, odorless, and tasteless does not cause off flavor and off-color in food • Effective over a wide range of temperature and pH	• Use of a single bacteriocin may not be enough to provide safe food. Several bacteriocins should be used in combination to increase the activity range.

(*Continued*)

Table 10.1 Merits and limitations of various advanced process techniques (*Continued*)

S. no.	Process technique	Merits	Limitations
10.	Dielectric Heating using radiowaves	• Processing time is less because the entire material is heated fast • Can be used for non-conductive substances • Very fast heating rates or super heating occur	• The equipment is costly
11.	Microwaves	• Large volume can be heated in a short duration • Energy consumption is less • Improvement in product quality	• Initial installation cost is high at the industrial level • May change texture and aroma
12.	Ionizing radiations	• Decontamination of packed frozen food is possible to buy irradiation • Only a little change may occur in proteins, carbohydrates, lipid and other nutrients of food.	• May result in rancidity in high-fat foods because of the breaking of inter-atomic bonds • Some nutrients maybe lost as vitamins A, B, and E are irradiation sensitive. • Special processing room lined with thick concrete walls is required to avoid escape of radiations

(*Continued*)

Table 10.1 Merits and limitations of various advanced process techniques (*Continued*)

S. no.	Process technique	Merits	Limitations
13.	Non-ionizing radiations	• Very minimal change in taste, texture, or appearance of food, if occur.	• Can cause skin burns & eye disorders in process workers if exposed to UV • May increase rancidity in food with high unsaturated fatty acids • The treatment cannot be used in the dairy industry because of the adverse effect of UV on various light-sensitive nutrients, vitamins, fatty acid and phospholipids

improved quality. Hundreds of food products processed commercially through HPP are available globally in the markets. The FDA has approved HPP as an alternative to pasteurization. Also, pressure-assisted thermal sterilization (PATS – a combination of heat and pressure) is reported to be a better substitute over sterilization.

The HPP allows the extension of food products' shelf life without heating and compromising their nutritional and sensory qualities. The treatment effectively inactivates the endogenous enzymes, spoilage organisms, and pathogens, thus reducing the chances of oxidative browning, food spoilage and foodborne illnesses. Several log reductions have been noted in counts of *E. coli*, *Salmonella* sp., and *Listeria monocytogenes* in foods processed through HPP [4, 5]. Also, HPP does not affect the low molecular weight components, vitamins, peptides, color, and flavoring components in food. Moreover, the ripening of cheese becomes faster by high-pressure treatment.

HPP technology involves applying very high pressure, usually 300-700 MPa, to the food for a short duration, usually 1-20 minutes. The process is also called cold pasteurization or pressurization. Its work is based on two fundamental principles [6]:

1. Pascal's isostatic principle – This states that food samples are acted upon by the high pressure equally from all sides. The action is instant and does not depend on the product's shape and size. Also, no damage occurs to the food product because of uniform pressure acting throughout the product.
2. Le Chatelier's principle – According to this principle, any system at equilibrium always resists the disturbance by minimizing it. While the reactions, conformational changes, or transition in phase involving a decrease in the volume are favored under high pressure, those accompanying increase in volume are inhibited.

The process includes the following steps (Figure 10.1):

(i) High pressure bearing steel vessels for HPP treatment are filled with pressure transmitting fluid, usually water. Other fluids like ethanol, glycol, castor oil, etc., can also be used alone or in combination with water as specified by the manufacturer.

(ii) The food to be processed is submerged in the fluid. Both the packaged/unpacked liquid and solid food samples can be treated suitably with high pressure. The flexible or semi-flexible packaging material that can sustain high pressure should be used for packaged food.

(iii) The pressure is built up in the liquid vessel by pressuring the hydraulic fluid with a pump. The buildup pressure acts instantaneously from all the sides of food, independent of its shape and size.

As pressure application is uniform throughout the food product, the latter can return to its original shape and structure without any distortion on decompression. The temperature may increase due to adiabatic heating during the pressure build-up. The rise depends on the pressure rate of pressurization and fluid type.

(iv) The food product is kept at the desired pressure level for the stipulated time. The pressure is released (Decompression) after a few minutes. Simultaneously product temperature also falls below its initial temperature with a decrease in pressure.

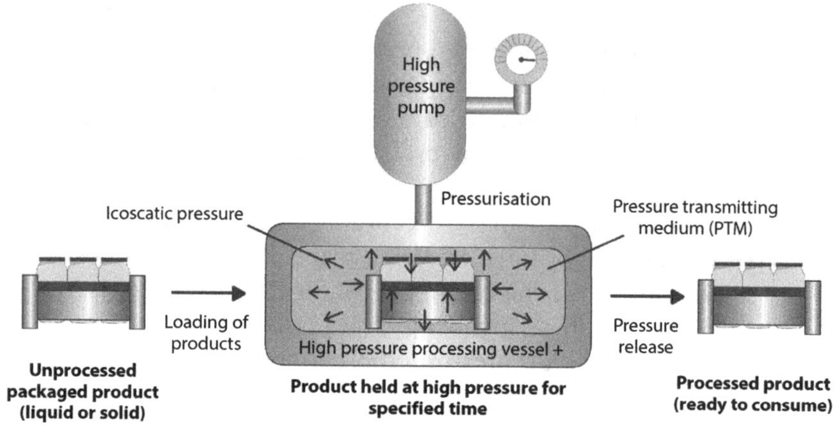

Figure 10.1 High-pressure processing technology for food preservation.

The pressure of approximately 300MPa without heating damages viruses, yeast, and vegetative bacterial cells and endogenous enzymes, while the pressure of > 600 MPa at 60°C to 70°C is required for spore formers destruction. The mechanism of action probably involves

(1) Breakdown of non-covalent bonds
(2) Inactivation of metabolic enzymes
(3) Changes in structure and permeability of cell membrane.

The process is carried out at an ambient temperature, and it does not have any impact on lipid, peptides, and saccharides because of the low compressibility of covalent bonds. The processing time is less because of equal and instant pressure distribution throughout the food product. The use of the technique helps in almost 100% meat removal from the shellfish.

Application: The technique is widely used to preserve beverages, apple cider, purees, jam, jellies, meat, chicken, chopped onions, sausages, juices, fish, salad, dressing, rice cakes, yogurt, vegetable smoothies, raw pet food, etc.

10.3 Pulsed Electric Field (PEF) Processing

Pulsed electric field or high electric field pulses processing is another efficient non-thermal technique for food preservation. Here the food placed between the two electrodes is exposed to short high voltage pulses at various

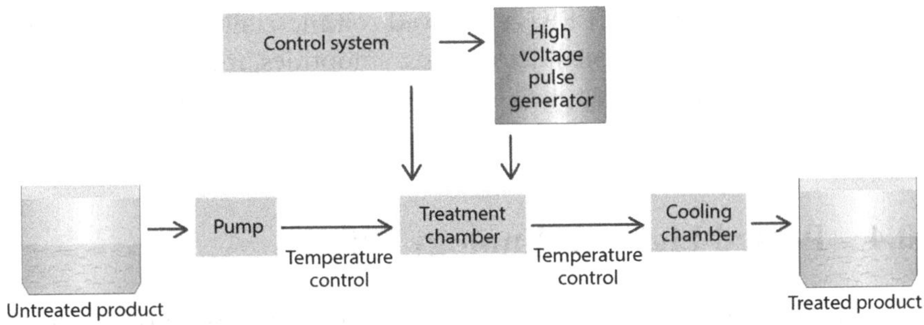

Figure 10.2 Pulsed electric field processing.

temperatures (usually 60°C). The electric current flows in food because of charged ions. The treatment inactivates microorganisms present in food without altering its color, taste, freshness, integrity, and nutrient value. Typically, the PEF processing equipment consists of (Figure 10.2):

(1) A high-intensity electric pulse generator generates a high voltage that conducts a high-intensity electric field (force per unit charge) between the electrodes.
(2) Treatment chamber – Food is placed in a static or continuous state in the treatment chamber
(3) Fluid handling system and
(4) Control and monitoring devices.

The process is usually suitable for liquid foods and involves exposure of food to pulsed voltage of several kilovolts per cm (5-50 KV/cm) for a few micro- to macro-seconds (20 – 1000 ms) with a pulse frequency of 200 to 400 Hz [7]. However, the threshold value of these parameters depends on other factors, including water activity, pH, temperature, nutrient content, soluble solids, and conductivity of food components, besides the shape and size of microbial cells. The underlying mechanism of PEF processing is the irreversible destruction of PM's integrity and functionality, probably by depolarization and pore formation (perforation). The treatment also changes the structure of enzymes and conformation of cell membrane (CM) phospholipids. Yeasts are more sensitive in companion to bacteria. Bacteria and mold spores are very resistant to PFE treatment. After PEF treatment food is packed and stored at low temperature. The shelf life of the products increases with minimal or no detrimental effect on food quality.

Use of PEF and conventional pasteurization in combination decreases the required pasteurization temperature and improves food quality.

Application: Currently, PEF is employed commercially to preserve food items like fruit and vegetable juices, soups, smoothies, milk, egg, milk and egg products, etc. It is also applied for increasing the extraction of bioactive components from plants.

10.4 Pulsed Light Technology (PLT)

Pulsed light technology, also called high light intensity, is a non-thermal, photonics-based food preservation technique performed by applying short pulses of light (1-20 flashes per second) to food. The pulses are of short duration (1 to 0.15) and high frequency and energy (.01 50 J/cm^2). These cover a broad light spectrum of wavelengths ranging from 180 to 1100 nm and consist of ultraviolet rays (180 – 400 nm), visible light (400-700 nm), and infrared rays (700-1100 nm). The technique works on the germicidal effect of UV rays. Because of the low penetration power of UV rays, the process is more effective for sterilizing food surfaces, equipment, and packaged food [8].

The pulse light treatment unit includes pulse generator (inert gas-filled lamps) and a sterilization unit. On applying the high voltage, high and short electric current, collision happens between the electron and inert gas molecules, resulting in excitation of gas molecules. The light is released either continuously or in pulses to return excited atoms to lower energy states. Xenon discharge lamps can be used for generating pulsed light. With this technique, electric energy is transferred to light pulses and released for a fraction of a second as a short pulse of light. The ultraviolet (UV) rays in light pulses are the key component of decontamination. Exposure to UV rays results in intra-strand pyrimidine dimers, primarily thymine-thymine (T-T) dimers. The presence of dimers distorts the DNA structure and interferes with the formation of daughter DNA molecules. This ultimately leads to the death of a cell. UV light also results in single and double-stranded DNA breakage in spores. PEF polychromatic light with a wavelength range from 200 – 300 nm is used for microbial inactivation. This is in contrast to classic UV treatment, where low-pressure mercury lamps produce monochromatic light of wavelength 254 nm and is used for microbial death. In PEF, medium pressure lamps are used. The technique is eco-friendly as no toxin or residual chemical is left on treated material. The technique, however, is not suited to rough surfaces as cells may be hidden in rough surfaces.

Applications – It is used for sterilization/pasteurization of foods such as fruit juices, sugar syrup, meat products, milk, vegetables, apple cider and fruits, and food contact surfaces and packaging material decontamination.

10.5 Ultrasound Food Processing

Another non-thermal approach that can be exploited for its preservative effect on foods is the application of ultrasonic waves. Based on their frequency, ultrasound waves belong to two categories [9]:

1. Low Power Ultrasounds – These have low power, so they do not cause any chemical or physical change in the material through which they pass. Because of their non-disruptive effects, low-power ultrasounds are used for monitoring and ensuring the quality and safety of food during processing and storage.
2. High Power Ultrasounds – These have high power or energy that alter food's physical and biochemical properties through their disruptive effect. Frequency of these waves is low (20-100 kHz). The high power ultrasounds found their food processing, preservation, and extraction applications.

As part of the process, high-intensity sound waves propagate through the processing material using ultrasound equipment. This results in gas bubbles in the material, which on collapse causes mechanical disruption and release of cellular content from the cells. Ultrasonic cavitation results in the destruction of various spoilage causing microorganisms, enzymes, and pathogens like *E. coli*, *Salmonella*, and viruses, thereby increasing their shelf life. The treated samples can be preserved for longer due to increased temperature and pressure surrounding the food particles.

Application: The method is applied to preserve fruit juices, jam, sauces, meat, dairy products, etc. Besides, the technology extracts proteins and lipids and their homogenization and emulsification in liquid foods. The shear forces created by ultrasonic cavitation cause a reduction in particle size and increase in surface area, thus making the extraction process more efficient than conventional ones.

10.6 Ohmic Heating of Food

It is an advanced thermal process where the alternating electric current (AC) passes through the food material. The food acts as electrical resistance between the electrodes. The flow of electric current results in generation of internal energy within the food material which in turn causes their quick and uniform heating. The other terminologies used for the process are electro heating, Joule heating, or electrical resistant heating. The process has an upper edge over the classical thermal process. The problem of protein denaturation and coagulation associated with conventional heating can be avoided by ohmic heating because of electricity.

The process is based on Ohm's law. The ionic salts and water in food conduct electricity. Electrical conductivity in food is linearly related to temperature and increases with the water content and temperature. Inactivation and killing of microbes occur by heating and pore formation in cell envelope (electroporation), leading to leaching of cell components. The color, texture, and nutrients of foods are maintained.

Application – The process can be used for a variety of purposes in the food industry, e.g., heating, thawing, cooking, sterilization, blanching, and extraction, and used for soups, patties, juices, purees, stews type food, and so on. It is useful for heat-sensitive liquids and proteinaceous food.

10.7 Cold Plasma

Plasma light or cold plasma is novel energy-efficient non-thermal technology that has gained importance in food processing wherein electricity (energy) is used to produce the plasma – the fourth state of matter [10].

Unlike natural thermal plasma, where constituent particles (ions, electrons, and neutral atoms, ozone, UV rays, reactive oxygen and nitrogen species) are in thermodynamic equilibrium, in cold plasma these plasma species are in a non-equilibrium state because of the temperature difference between various gas particles [11]. Under an applied electric field, out of a billion neutral atoms or molecules, one undergoes ionization to produce an electron. The generated electrons are at a very high temperature and transmit a small part of this energy to neighboring neutral atoms and molecules, causing slight heating of neutral particles.

Plasma is formed by passing electric energy into air, nitrogen, or oxygen for food processing. The reactive plasma species being antimicrobial in nature, kill and reduce the microbial growth via different mechanisms.

As discussed earlier, UV photons interact with DNA and interfere with DNA replication by forming pyrimidine dimers. Various plasma species are absorbed on the cell membrane and cause cell membrane distortion and damage by electroporation. Oxidation reactions occurring within the microbial cell cause its inactivation. The endogenous enzymes, especially peroxidases and polyphenoloxidases responsible for browning reactions, also become inactive. The Gram-positive bacteria are more resistant to cold plasma in comparison to Gram-negatives. The process is an energy saver as well as environment friendly. The inexpensive ambient air is used to produce plasma. The energy requirement is much less, and also no chemical waste or effluent water is generated.

Application – The process is effective for sterilization of surfaces, disinfection of equipment, floor, and air in the processing plant, pasteurization of liquid food, and disinfection of packaging material.

10.8 Oscillating Magnetic Field (OMF)

Oscillating Magnetic Fields generated with electromagnets of alternative current has a potential application in food preservation and quality improvement. OMF is applied in pulses (1 to 100) of intensity 5-50 T (Telsa) and frequency 5 to 500 kHz to food for a short duration, i.e., 25 to 100 microseconds. The treatment is given at a normal atmospheric pressure and at a temperature of 0-50°C. The food to be processed is packed in a plastic bag and is subjected to the magnetic field of intensity 5-50 T, which generates electric energy of 10^{-2} to 10^{-3} EV per oscillation and increases food temperature by 2 to 5°C. The energy so released inactivates the microbes by breaking the covalent bonds in DNA, weakening bonds between proteins, and affecting membrane fluidity. This inhibits microbial growth and reduces microbial number at least by 2 logs. Before treatment, the food can be packed in flexible plastic film to avoid post-process contamination.

Application – The OMF process can be used to preserve both solid and liquid foods. Examples of OMF treated foods include juice, milk, yogurt, etc.

10.9 Higher Pressure Thermal (HPT) Processing

HPT processing technique combines high pressure and high temperatures to inactivate microorganisms, particularly spore formers like *C. botulinum*.

The process is a better alternative for microbes' removal than thermal process alone, which takes a long time for decontamination of food. HPT process causes less damage to food as it receives less heat. Also, the quality attributes of food like nutrients, texture, taste, etc., remain unaltered or improve. In the HPT process, high pressure (600 MPa) is applied along with an elevated temperature of more than 60°C, which results in quick and uniform heating of food, thereby reducing the exposure time to heat. The process involves preheating food products to a set temperature by placing them in the holding unit, pressuring the holding unit, and increasing the temperature. The food is kept at a required temperature for a specific time. Decompression results in quick cooling of the product, thus avoiding degradation of food quality attributes.

HPT treatment is particularly used to increase the stability and safety of low acid foods (LAF), where the main concern is the presence of *C. botulinum* spores. HPT at a temperature of more than 90°C is vital for low acid food. Less stringent conditions are generally required for the inactivation of non-proteolytic *C. botulinum* spores (e.g., temp. <90°C) as compared to proteolytic strains (>90°C). The underlying mechanism of spore inactivation involves spore activation and germination by maximum compression heating.

Depending on the pressure applied, spores can germinate via two routes:

(1) Germination via nutrient trigger pathway where spore nutrient receptors get activated. It occurs at a moderate pressure range of 50 – 300MPa.
(2) Germination via release of calcium dipicolonic acid. It occurs at a high-pressure range of 400 – 800 MPa, which probably triggers the opening of DPA channels in the inner membrane.

The kind and number of microorganisms and the intrinsic properties of food affect treatment efficiency. HPP, which is carried at low or ambient temperature, is unable to prevent the growth of spore farming bacteria like *Bacillus* and *Clostridium* in low acid food. HPT, however, can inactivate these bacterial endospores.

Application – HPT is used for ready-to-eat meat and less acid food.

10.10 Bacteriocins

Certain microorganisms secrete antimicrobial heat-stable bioactive peptides and protein substances that suppress the growth of other similar or closely related bacteria. These bacteriocins are highly active against foodborne pathogens and apply them in food sectors as bio preservatives [12]. Though both Gram-positive and Gram-negative bacteria have released a number of bacteriocins, many, e.g., nisin, bulgarian, and lactain have been investigated for their role as food preservatives. These bacteriocins are mainly secreted by lactic acid bacteria like *Lactococcus lactis*, *Lactobacillus casei*, *Lactobacillus plantarum*, etc. Nisin is the first bacteriocin approved by FDA and is used globally as a food additive. It is a 3500 Dalton polypeptide of 34 amino acids produced by *Lactococcus lactis* sp. and contains many rare amino acids. Nisin, by attacking the cell membrane, inactivate or kill the bacteria. Gram-positives are more sensitive as compared to Gram-negatives because of the presence of the outer membrane. Many other bacteriocins having potential as bio preservatives are yet to be commercialized.

On the basis of structure and physicochemical properties, bacteriocins have been categorized into three classes: Lantibiotics, Non-lantibiotics, and small heat-stable proteins and may be narrow-spectrum or broad-spectrum depending on their anti-bacterial range. Some bacteriocins inhibit peptidoglycan synthesis for cell wall, while others make pores in the cell membrane. Many also inhibit gene expression and protein synthesis.

Application - Bacteriocins can be used to preserve juices, beverages, meat, dairy, and seafood, along with bioactive food packaging.

10.11 Dielectric Heating Using Radio Waves

Dielectric heating is a novel approach for food preservation that depends on radio frequency heating. It is a process where the alternating electric field is used to heat the material through dielectric rotation of the polar molecules. The polar molecules align themselves in the electric field direction and change their alignment direction with the electric field's charging direction, resulting in dielectric rotation. This dielectric rotation, in turn, increases kinetic energy and thus the temperature of the molecules. All parts of the food heat up thoroughly on the flow of this kinetic energy to the neighboring molecules. For Dielectric heating, radio waves or electromagnetic radiations can be used. Radio waves can travel in all directions (omnidirectional) and penetrate more deeply in the material than electromagnetic waves.

The process is also called radiofrequency, electronic, or high-frequency. The radio frequencies commonly used range between 10-100 MHz. The technique has a better edge in pasteurizing low moisture foods as compared to conventional thermal treatment. This is because of the fast and more uniform distribution of heat. The treatment finds its application for treating solids and semisolid foods with low thermal conductivity.

Applications: Radio frequencies are being used to control pests after harvest. It is also used for quick processing of food, e.g., used in biscuit production post-baking and for baby cereal products and packet/packed milk.

10.12 Microwaves

Unlike radio waves, microwaves are unidirectional electromagnetic radiations used in the food processing industry for drying and microbial decontamination. The most commonly used microwave frequencies in food processing are 2450 MHz (2.45 GHz) and 915 MHz. Inactivation of Microorganisms occurs due to rapid heating by microwaves. The principle of heating is similar to radio frequency. Under alternating microwave fields, water and other polar molecules present inside the food align and start rotating continuously. The rotation causes fast and thorough heating of food. However, a temperature gradient does develop in the food with the edge hotter than the center—microbial decontamination results from a change in cell membrane permeability due to lipid degradation and blocking of DNA replication [13].

Application: The process finds application in the blanching of fruits and vegetables and pasteurization of high acid food and baking goods. It is also applied for the heating of liquid, suspension, and solids.

10.13 Irradiation

The low level of ionizing (gamma rays, X rays, electrons) and non-ionizing (UV rays) radiations can also be applied for the preservation of food and making it safe for consumption (Figure 10.3). The process is called irradiation and is found to be safe by FDA. No or minimal unnoticeable changes may occur in food attributes. The process leaves no radioactivity after the treatment.

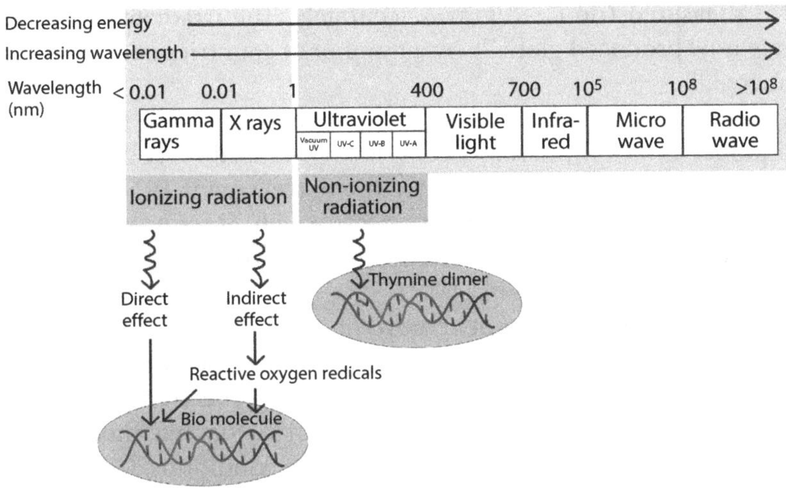

Figure 10.3 Effect of radiations on microbial cells.

10.13.1 Ionizing Radiations

Ionizing radiation includes high-energy gamma rays, electron beams, and X rays. These have high penetration power and are capable of breaking chemical bonds and converting molecules into ions. Among α, β and γ rays, γ rays have the greatest penetration power with the least ionization capacity. In contrast, α rays have very low penetration power and very high ionizing power. β rays have an intermediate level of penetrations and ionizing ability.

The gamma rays are electromagnetic radiations with no mass or charge and have sufficient energy to convert food molecules into ions by dislodging electrons. The sources of gamma rays used in the food industry are radioisotopes cobalt–60 (^{60}Co) and cesium- 137 (^{137}Cs) [9]. Gray (GY) denotes the amount of rays absorbed by a material; one gray is equal to 1 joule of energy absorbed by one kilogram of food.

The electron accelerators generate electron beams, whereas bombardment of heavy metals, i.e., tantalum and tungsten, create X-rays. As X-rays and electron beams generation do not use radioisotopes, these can be turned 'on and off' as per requirement, unlike gamma rays. Switching on and off is not possible using radioisotope; it has to be lowered into a water reservoir after use. The unit electron volt (eV) denotes the amount of energy possessed by an electron. One eV means the energy gained by an electron when it passes through a potential difference of 1 volt. The process of irradiation is highly regulated. The amount of radiation absorbed by the food is measured (unit rad/grays) accurately. The irradiated food products are now produced globally by a number of countries and are available commercially.

The microbicidal effects of ionizing radiations are due to their direct and indirect effect on microbial cells. Ions produced on collision of these rays with atoms inside the cells alter the nucleic acids (RNA and DNA) irreversibly and inhibit cell growth. The free radicals (e.g., hydroxyl radical) created by water radiolysis indirectly bring structural changes in the cell membrane and damage it.

However, these rays cannot remove neutrons from the atom nucleus because of insufficient energy and therefore do not impart any radioactivity in the food after treatment. The dose and effectiveness of rays are decided by the food composition and external factors like moisture, temperature, and availability of oxygen and storage conditions. Generally, frozen foods need more radiation dose as compared to chilled food or food at room temperature. Foods treated with ionizing radiations are labeled with the international Radura logo.

Applications: The treatment can be used for sterilizing/pasteurizing fruits, milk, vegetables, food and for controlling insects and delaying the ripening process.

10.13.2 Non-Ionizing Radiations

Ultraviolet Rays (UV) are the non-ionizing radiation used effectively in the decontamination of food, particularly food surfaces and packaging material, due to the low penetration capacity of these rays. UV rays cover a wavelength range of 180-400 nm in the electromagnetic spectrum and can be divided into three parts: UV–A (315-400nm), UV-B (280-315nm), and UV-C (180-280nm). UV-C is most effective in the inactivation of microbes, particularly wavelength around 260 nm that is strongly absorbed by the nucleic acids and shows the greatest lethality. The action of UV rays on cells occurs through the formation of intra-strand pyrimidine dimers that damage the DNA and block the replication and transcription during the cell division. Sensitivity to UV rays varies with the type of microbes, their ability to repair the damage, food composition, and the surrounding environmental conditions. The efficacy of UV rays is reduced by parameters like UV absorbing proteins, color compounds, organic solutes, and suspended matter in food. Generally, bacteria are more sensitive to UV rays as compared to protozoa and algae. This may be probably because of more efficient DNA repair mechanisms in the latter.

Moreover, Gram-negatives are more susceptible than Gram positives. Viruses are very resistant to UV rays. Generally, the resistant pattern shown by microbes to UV rays is gram-negatives < gram positives ~ yeasts < bacterial spores < mold spores << viruses.

Applications: UV rays are used in surface decontamination and water disinfection is used in shellfish deplete and meat chills storeroom.

10.14 Conclusion

Lifestyle change has changed the perception of consumers on food quality. Health-conscious consumers are now interested in nutritionally healthy and microbiological-safe food. Several traditional methods, viz., high heat treatment, pasteurization, drying, freezing, and canning, are used to increase the shelf life of food products. These food preservation methods have certain disadvantages, including loss of vitamins, denaturation of proteins, and alteration in food's sensory attributes (color, flavor, and texture). To overcome these disadvantages and meet the consumer's demand for minimally processed and nutritionally less altered food with long shelf life, a number of novel food preservation techniques have evolved. These advanced thermal and non-thermal processing techniques include high-pressure processing, high-pressure thermal processing, cold plasma, ohmic healing, ultrasound processing, irradiation, and many others. These techniques allow fast food processing with minimal or no alteration of food characteristics. Also, many of these techniques are eco-friendly. These techniques do offer great potential in the future for the food industry.

References

1. Adams, M.R. and Moss, M.O. (2000). *Food Microbiology.* 2nd edition. New Age International Publishers, India.
2. Frazier, W.C., Westhoff, D.C. and Vanitha, N.M. (2013). *Food Microbiology.* 5th edition. Tata McGraw-Hill Publishing Company Ltd, India.
3. Ray, B. (2004). *Fundamental Food Microbiology*. 3rd edition. CRC Press.
4. Gouvea, F.S., Padilla-Zakour, O.I., Worobo, R.W., Xavier, B.M., Walter E.H.M., and Rosenthal, A. Effect of high-pressure processing on bacterial inactivation in açaí juices with varying pH and soluble solids content. *Innovative Food Science and Emerging Technologies*. Volume 66, 2020, 102490, https://doi.org/10.1016/j.ifset.2020.102490.

5. Daher, D., LeGourrierec, S. and Pérez-Lamela C. Effect of High Pressure Processing on the Microbial Inactivation in Fruit Preparations and Other Vegetable Based Beverages. *Agriculture.* 2017; 7(9):72. https://doi.org/10.3390/agriculture7090072
6. Ambrose D.C.P. Why High Pressure Processing (HPP) Considered a Better Alternative to Thermal Processing? *Curr Agri Res* 2018;6(3). Doi: http://dx.doi.org/10.12944/CARJ.6.3.01.
7. Roman, B. and Michelle, B. (2013). Advanced Food Preservation Technologies. *Microbiology Australia.* 34. 108-111. 10.1071/MA13037.
8. Aguilar, C.N., Ruiz, H.A., Rubio Rios, A., Chávez-González, M., Sepúlveda, L., Rodríguez-Jasso, R.M., Loredo-Treviño, A., Flores-Gallegos, A.C., Govea-Salas, M. Ascacio-Valdes, J.A. (2019) Emerging strategies for the development of food industries, *Bioengineered,* 10:1, 522-537, DOI: 10.1080/21655979.2019.1682109
9. Sarika, K. and Bindu, J. (2018) An overview of non-thermal preservation techniques in food. In: *Protocols for the production of high value secondary products from industrial fish and shell fish processing* (Eds. Bindu, J., Sreejith, S. and Sarika, K.), ICAR-Central Institute of Fisheries Technology, Cochin, pp. 99-108.
10. Thirumdas, R., Sarangapani, C. & Annapure, U.S. Cold Plasma: A Novel Non-Thermal Technology for Food Processing. *Food Biophysics* **10,** 1–11 (2015). https://doi.org/10.1007/s11483-014-9382-z
11. Cold Plasma in Food Technology. Wiley analytical science. May 2019. https://analyticalscience.wiley.com/do/10.1002/gitlab.17224
12. Abbasiliasi, S., Shun Tan, J., Tengku Ibrahim, T.A., Bashokouh, F., Ramakrishnan, N.R., Mustafa, S and Ariff, A.B. Fermentation factors influencing the production of bacteriocins by lactic acid bacteria: a review. *RSC advances,* 2017. 7, 29395-29420. DOI: 10.1039/C6RA24579J
13. Jay, J.M., Loessner, M.J. and Golden, D.A. (2006). *Modern Food Microbiology. 7th* edition. CBS Publishers and Distributors, India.

11

Food Omics and Its Implications in Nutritional Sciences

Somnath Mandal[1], Nandita Sahana[1], Ramesh S.V.[2] and Veda Krishnan[3]*

[1]*Department of Biochemistry, Uttar Banga Krishi Viswavidyalaya, Pundibari, Cooch Behar, West Bengal, India*
[2]*Division of Physiology, Biochemistry and Post-Harvest Technology, ICAR-Central Plantation Crops Research Institute (ICAR-CPCRI), Kasaragod, Kerala, India*
[3]*Division of Biochemistry, ICAR-Indian Agricultural Research Institute (ICAR-IARI), New Delhi, India*

Abstract
Food omics is the science dealing with the specialized domain of food and nutrition integrated with advanced omics technologies to improve stakeholder's well-being, health, and confidence. The discipline is addressing the stakeholder's needs in terms of food quality, traceability, sustainability, legal issues, authenticity, traceability, safety, integrity in a broad perspective. This chapter deals with tremendous progress made in analytical chemistry techniques used in food matrix characterization, over the past few years, unresolved issues and challenges remain to be addressed in this rapidly progressing area of research.

Keywords: Food omics, nutrition, health, analytical techniques

11.1 Food Omics: An Overview

Advancement in different omics, from genomics to metabolomics, has revolutionized biology and it has paved new insights through the possibility of connecting food components, and diet of an individual with health and diseases. This emerging field of omics is known as "Food omics". Development of new holistic analytical approaches suitable for the study

**Corresponding author*: vedabiochem@gmail.com

Tejpal Dhewa, Anil Kumar Puniya and Anil Panghal (eds.) Nutritional Science and Technology: Concept to Application, (245–272) © 2023 Scrivener Publishing LLC

of food and nutritional sciences led to the development of this field. Food omics is a new science wherein researchers working in food chemistry, analytical chemistry, biochemistry, microbiology, molecular biology, food technology, clinical sciences, and human nutrition work collaboratively with an ultimate objective of improving human health and well-being.

11.2 Techniques in Food Omics

Food is a very complex combination of different biomolecules such as carbohydrate, protein, fat, vitamins, and minerals that has entirely different molecular structure and physiochemical characteristics. In such a complex matrix identification of compounds are not easy, especially when the compounds are present in minute quantity. Various techniques are used in quality analysis of food which is essential for nutritional value determination, medicinal value, quality assurance, and identification of adulteration or contaminants of the food. The steps involved in different analysis include several techniques right from the preparation of the samples to final interpretation of the data generated. Principally, spectroscopic, chromatographic, biological, immunological, electrophoretic and electrochemical techniques are available for analysis of various food components. Apart from the above-mentioned techniques there are several sample preparation and extraction methods which are crucial for detection and quantification of analytes that are present in very minuscule quantity in food samples. Extraction techniques such as, Solid Phase Extraction (SPE), Super Critical Fluid Extraction (SFE), Headspace Extraction, Flow Injection Analysis (FIA), Microwave Assisted Extraction (MAE), Automatic Thermal Desorption (ATD), Hyphenated extraction etc., are applied in food industries for quality checking. Each of these extraction methods has its advantage over the other based on the analyte to be isolated from the food sample.

11.2.1 Chromatographic Techniques in Food Analysis

Chromatographic technique was first invented by Tswett (1906) when he used liquid chromatography for isolation of plant pigments. The principle of all the chromatographic techniques involves partitioning of the analytes between mobile and stationary phases based on their physiochemical properties like solubility, affinity, charge, shape and size and adsorption to other materials. The stationary phase is generally either solid or liquid compound packed in columns and the mobile phases are generally either liquid or gas. Earlier Paper chromatography, Thin layer chromatography etc. were used for detection of different compounds. Presently Column chromatography is the most frequently used technique for food analysis where partitioning of the analytes in stationary and mobile phase is the basis of separation. In column chromatography, High Performance Liquid Chromatography (HPLC) and Gas Chromatography (GC) are mostly applied in analysis of food matrix (Fanali *et al.*, 2016).

Liquid chromatography has numerous applications in analysis of food components like proteins, carbohydrates, amino acids, phenolic compounds, triglycerides, phospholipids additives, chiral compound and pigments (Ramirez, 2004). The HPLC technique is widely used technique where liquid mobile phase is pumped in high pressure through the solid stationary phase packed in a column. The instrumentation is simple, composed of a reservoir for the mobile phase solvent, a sample injector, a column and a detector coupled with a data acquisition system. The length and diameter of the column varies based on the application. Silica, modified silica, carbon sorbents, metal oxides etc. are used as column material. The Normal phase HPLC (NP-HPLC) is used to separate polar and moderately polar compounds whereas Reverse phase-HPLC (RP-HPLC) is used for separation of for a wide range of polar and non-polar compounds. When the particle size of the column material is less than 2μm then it is categorised as Ultra-High-Pressure Liquid Chromatography which has significantly improved the detection of very small analytes. UHPLC has been used for estimation of plant pigments like carotenoid (Bohoyo-Gill *et al.*, 2012). Other liquid chromatographic techniques where the separation is based on charge of the ions (Ion exchange chromatography) or on the molecular weight of the analytes (Size exclusion chromatography) are also regularly used in food analysis. The recent development of chiral chromatography has enabled the separation of enantiomers and separation of isomer mixtures (Zhao *et al.*, 2021).

Unlike liquid chromatography the GC uses gas in mobile phase and are mostly used for analysis of volatile/gaseous compounds. The samples in GC are vaporised using high temperatures and then mixed with carrier gas which is applied as mobile phase. GC has wide range of applications in several food composition analyses which include profiling of fatty acids and off-flavors, sterols and alcohols, volatiles and aroma compounds. Other polar compounds such as amino acids, hydroxy (poly)carboxylic acids, fatty acids, phenolic compounds, sugars, vitamins, may also be analysed by GC but they need appropriate chemical derivation before analysis (Lehotay, 2002). In GC, if solid stationary phase is used then the technique is termed as Gas solid Chromatography (GSC) and if the stationary phase is liquid then it is termed as Gas Liquid Chromatography (GLC)

(Cordero *et al.*, 2012). The GC instruments are consist of a carrier gas system, injector port, column, detector and data acquisition and processing unit (Forgacs and Cserhati, 2003). The column in GC system can be of packed type or open tubular type. Open tubular type columns are used more often as it provides lower flow resistance and efficient detection than packed columns.

Detectors are integrated part of the chromatographic techniques; the GC systems are generally coupled with Flame Ionisation Detectors (FID) or Mass Spectroscopy (MS) detectors. In many cases non-destructive Electron Capture Detectors (ECDs) are also used. In HPLC usually, UV-vis detectors, Photodiode Array (PDA) detectors, fluorescent detectors, Refractive Index (RI) detectors etc. are used.

11.2.2 Spectroscopic Techniques in Food Analysis

The major advantage of the spectroscopic method over the chromatographic method for analysis of food components is that in spectroscopic method often no or minimum sample preparation steps are involved. Also, such spectroscopic equipments are capable of running multiple tests on a given sample at the same time. Near Magnetic Resonance (NMR), Near Infrared Spectroscopy (NIR), Mid Infrared Spectroscopy (MIR), Infrared Spectroscopy (IR), UV-Vis Spectroscopy, Fluorescence Spectroscopy and Raman spectroscopy etc. have wide application in the food quality assessment and in monitoring of food authenticity.

UV-Vis spectroscopy is based on the Lambert-beers law where absorption of the ultra violet and visible light by a particular compound at a given wavelength is measured (McGorrin, 2006). UV-Vis spectroscopy combined with flow injection analysis is widely used for oil quality assessment and different parameters like ansidine value (AV), peroxide value (PV), saturation level colour etc. can be determined (Labrinea *et al.*, 2001). On the other hand, Fluorescence spectrometer measures the emission of light by different fluorophore like, fluorescein, rhodamine B, acridine arrange, quinine, pyridine-based dyes etc. (Lakowicz, 2006). The combination of HPLC and the fluorimeter detector has enabled the detection and estimation of very low concentration contaminants such as presence of antibiotics, microbial contaminants specially *Salmonella* and *E. coli*, food additives and mycotoxins in foods (Nakai and Horimoto, 2006). This technique is often used for the determination of causative agents of food poisoning. Fluorescence detector is also used in determination of adulteration of the virgin oils of olive, conr, soyabean, sunflower etc. (Engelsen, 1997).

The IR spectroscopy uses the electromagnetic infrared radiation that ranges from 780 nm to 1000 μm. The IR range has been further divided in three categories such as Near Infrared where upto 780 nm of the wavelength is considered), mid infrared where bigger wavelength ranging from 5-30 μm is considered, and far infrared where wavelengths ranging from 30-1000 μm is considered. When different radiation of varying wavelength is passed through a given sample part of the radiation is absorbed at a particular energy based on the functional groups present in the sample. The detected peak in the absorption spectrum generally corresponds to the vibrations in atomic level of the molecule. Thus, the generated IR spectrum is actually the fingerprint of its molecular composition (Shurvell, 2006).

Near Infrared spectroscopy is a technique for quantitative analysis of foods such as fat, protein, carbohydrate etc. All these parameters can be estimated simultaneously in NIR from a wide variety of samples but prior calibration with known composition is required in each case, which is both expensive and time taking (Li-Chan *et al.*, 2006). In past, NIR and MIR have been used for batch checking of wheat and other food grains before their despatch (Norris, 1989). MIR has been employed in research of fat and oil chemistry. Beside NIR and MIR, Fourier Transform Infrared Spectroscopy (FTIR) is another infrared spectroscopic technique which has been employed for analysis of processed food emulsion like butter, mayonnaise etc. (Van de Voort, 1992; Van de Voort, 1993). FTIR also has wide application in determination of protein secondary structures and protein quality (Mejri *et al.*, 2005). Raman spectroscopy is an associated technique of IR spectroscopy which is based on Raman Effect. Raman Effect is a result of the interaction between the incident photons and electrons of the molecules in the sample. Raman spectroscopy is employed to study the changes in the structure of starch after processing or after long time storage. Like IR, Raman spectroscopy has also been used to study carbohydrate, proteins, fat and oils but the disadvantage of the method is that the signals are often weak in many cases (Niaura, 2008).

The principle of NMR spectroscopy is based on the inherent magnetic properties of the nucleus on any atom that arises due to the spinning of the positively charges nucleus around its axis. NMR basically consist of a created magnetic field which is homogenous through pole ends, a very steady radio frequency transmitter and receiver along with appropriate signal recorder system (Keeler, 2010). Nuclear Magnetic Resonance spectroscopy is the most advance technique in food analysis till date. NMR has immense potential in nutritional research in investigating the structure of organic compounds. Other than that NMR is applied to identify and estimate different metabolites present in food and vegetables. NMR has been used to analyse fish oil and vegetable oils, milk and milk products, fish and meat, beverages like green tea, fruit juice, coffee, wine and beer. This technique is the most suited technique for the study of food omics (Parlak and Guzeler, 2016).

11.2.3 Electrochemical Biosensors in Food Analysis

Biosensors are more recently introduced technique to ensure the food safety using electrochemical devices coupled with biological reactions (Bunney *et al.*, 2017). This advanced technique can detect any biologically sensitive recognition element like enzymes, antibodies, and nucleic acid or whole cells accurately and promptly. The electrochemical biosensors generally use an electrode transducer which can detect electron released from a biological reaction after the sample analytes are applied to bioreceptors. It is an efficient and high-speed technology to detect pathogenic substances in food like bacteria or fungus and pesticides or other toxic elements. Further improvement of this technique has been done by introduction of novel bio-recognition molecules like synthetic aptamers, DNA, protein, viruses etc. (Ali *et al.*, 2017; Chandra *et al.*, 2017).

11.2.4 Biological Methods in Food Analysis

Application of DNA-based detection methodology, Polymerised Chain Reaction (PCR) in food analysis and food safety protocols is increasing (Klancnik *et al.*, 2012). The detection of bacterial contaminants like *Salmonella* and *E. coli* is done by PCR now a day since it offers many advantages against the conventional culturing methods (Perry *et al.*, 2007). PCR technology is prompt, accurate, more specific and sensitive to detect spoilage microorganism from food samples. PCR is also used to detect allergens, to assess food authenticity and also to investigate presence of GMOs in food (Lantz, *et al.*, 2000). The introduction of different advancement of the simple technique of PCR, to multiplex PCR (mPCR) for multiple pathogen detection from one sample orreal time PCR, for quantifying the level of contaminants has been proved very useful. Parallelly Nucleic Acid Sequence based Amplification (NASbA) or easy on field detection by Loop Mediated Isothermal Amplification (LAMP) and microarray technology for thorough screening for food contaminants has modernised food contaminant detection techniques. The only disadvantage of these techniques is that PCR has a long and tedious sample preparation step and requires positive controls and reference samples for all experiments.

The immunological methods like Enzyme Linked Immuno Sorbent Assay (ELISA) and Lateral Flow Immune Assay uses the natural antigen antibody interaction process for detection of food contaminants such as pathogens and allergens by monoclonal or polyclonal antibodies (Law *et al.*, 2015). In ELISA the target antigen i.e part of pathogen or toxins interacts with the surface bound specific primary antibody and after that the enzyme conjugated secondary antibody is then applied. The resultant primary antibody-antigen-secondary antibody sandwich is finally detected with substrate which is converted to coloured substance by the activity of the conjugated enzyme on secondary antibody. Generally, Horseradish peroxidase and Alkine phosphatase enzymes are used in this technique (Yeni *et al.*, 2014). ELISA has been used for detection of many food spoilage bacteria like *Salmonella*, *Clostrodium*, Staphylococcal enterotoxins, botulinum toxin, *Vibrio* toxins etc. To simplify the tedious ELISA assay modern Lateral Immune Assays have been developed and strips based on-site detection strategies has been developed. The strips are usually made of sample pad for receiving the sample input, conjugate pad for reaction to occur, nitrocellulose membrane and absorbent pad. The sample is applied in sample pad which moves to conjugate pad by capillary action where the antigen-antibodies binding takes place and produces coloured substance for detection. This is a very rapid method where detection can be done in few minutes by visualisation of coloured band on the strip (Ngom *et al.*, 2010; Zhao *et al.*, 2014).

11.2.5 Electrophoretic Methods for Food Analysis

Electrophoresis is defined as the movement of charges particle in an electric field. There are two types of electrophoresis technique that are used in food analysis. Gel electrophoresis, like Poly Acrylamide Gel Electrophoresis (PAGE), non-denaturing PAGE, Isoelectric focussing (IEF), Two-Dimensional Gel Electrophoresis (2-D PAGE), Isotachophoresis (ITP), and Agarose gel electrophoresis are research techniques and has limited application in food analysis. These techniques are majorly used to separate and analyse protein components (Autran, 1996). Some reports of wine assessment with electrophoretic techniques have been reported in the past (Carnevillier *et al.*, 1997). Capillary electrophoresis is the advanced electrophoretic technique which offers wide application in food analysis due to the availability of the high throughput commercial scale instrument. The old electrophoresis method has been modified in capillary electrophoresis method and can be used to analyse carbohydrates, amino acids and proteins, lipids, inorganic and organic ions, vitamins, secondary metabolites and additives, toxin drugs and pesticides (Cancalon, 2006)

11.2.6 Sample Extraction Techniques in Food Omics

11.2.6.1 Solid Phase Extraction (SPE)

Solid Phase extraction (SPE) method is an improvised technique from the classical liquid-liquid extraction process. Unlike the liquid extraction method, it utilizes the adsorption of the analytes to solid phase rather than separation on the basis of only solubility (Thurman and Mills, 1998). The type of sorbent applied in the solid phase is selected based of the analytes to be separated and the food matrix or interferent substances present in the sample. Generally, the sorbent is packed in a column which is considered as stationary phase and the analytes or the samples are introduce in aqueous mobile phase. The partitioning between these two phases is based on binding of the analytes to stationary phase by interactions like vander-walls interaction, hydrogen bonding, dipole-dipole or electrostatic interactions etc. Based on these principles SPE can be of many types. Normal phase SPE generally utilises stationary phase which is polar in nature and a mobile phase which is non-polar whereas reverse phase SPE involve non-polar stationary and a moderately polar mobile phase (Otles and Kartal, 2016). In both cases the vander-walls interaction between the sorbent and analytes plays crucial role in separation. For charged compounds ion exchange SPE may be employed where the electrostatic forces are the basis of separation. A combination of these interaction can also be utilized, referred as mixed-mode SPE where properties of different functional groups of the sorbent are used (Majewska *et al.*, 2008; Zwir-ferenc and Biziuk, 2006). Most common sorbent used in SPE is silica bonded with non-polar C18, C8 chains or cyclohexyl groups or polar chains such as hydroxyl, cyano or diol groups. Among the other sorbents, polymeric raisins, florisil, polar carboxylic or amino group, alumina, magnesium, graphitized carbon are commonly used. For some specific compound Restricted Access Media (RAM) or Molecularly Imprinted Polymers (MIPs) are used (Ridgway *et al.*, 2008). SPE is one of the most used sample preparation technique to identify pesticidal residues in vegetable sample. Carcinogens like acrylamides and amines in protein and other processed food samples can also isolated in this method. Protocols for isolation of sterols from fat food and purity of anthocyanins has been established using SPE.

11.2.6.2 Super Critical Fluid Extraction (SFE)

A super critical fluid can be defined as dense gas which has both the properties of a liquid and a gas (Hawthrone and King 1999, Luquede Castro *et al.*, 1994). Although a super critical fluid has the density of a liquid but it does not settle down in the bottom of a container like liquid, instead it expands and takes the shape of the container, for any gaseous compound a supercritical liquid phase is attainable at a critical temperature and critical pressure. The critical temperature is a given temperature above which a liquid cannot be in its condensed liquid phase (King, 1989). The parameter of critical pressure is defined as the vapor pressure of the substance created at its critical temperature. CO_2 is the most preferable gas in SFE process because of its nontoxic, non-flammable environment compatible nature. The critical temperature of CO2 is very low i.e 31.1°C and the critical pressure is 72.84°atm (Femenia, 2001) which is easily available. The instrumentation in SFE is simple and commercially available. The main components are a source of the fluid which is generally a high-pressure CO_2 cylinder, a pump to regulate the pressure of the fluid, extraction chamber which is placed in an oven so that the sample can be heated to the required temperature for extraction, a restrictor which controls the flow rate and pressure of the fluid in the chamber and a collector where analytes are collected after extraction. Usually, the mixing of the supercritical fluid and solvent is done before its entry to the extraction chamber by using two parallel pumps and a mixing chamber. The extract collection can either be offline or in online mode. The offline extraction is performed by evaporating the mobile phase from the extracted analytes and collecting them in collection chamber for processing them later. In this case loss of analytes and procedural error may occur while collecting the analytes. In online mode the collected analytes are directed towards a chromatographic or spectrometric instrument, generally a GC, HPLC or FTIR coupled with appropriate detector system where further separation and simultaneous detection of the analytes are performed without removing them from the mobile phase. SFE is commonly used for analysis of fats, oils, fat soluble vitamins and essential oils. SFE based protocols has been developed for pesticidal residue analysis in food matrix (deValle and Uquiche, 2002; Lehotay, 2002; Turner, 2001).

11.2.6.3 Headspace Extraction

Head space extraction is applicable for the analysis of Volatile Organic Compounds (VOCs) (Falaki, 2019). VOCs are the major contributor to the aroma and flavor of the food. The term Headspace refers to the volume of vapor or gas present above the sample. Headspace extractor is generally combined with GC or GC-MS where the volatile component can be directly injected into a column for detection and quantification. This technique is divided in two categories, the simpler Static Headspace Extraction (SHE) and improved Dyamic Headspace Extraction (DHE) or purge and trap method (Mitra, 2003). In Static Headspace Extraction (SHE) the sample is contained in the sample vial or autosampler and subjected to constant temperature and pressure which release and diffuse the VOCs in the headspace. When an equilibrium is reached between the VOC concentration in the sample and in the headspace vessel then the vial or autosampler is connected to GC column for detection and quantification. This process is repeated by programming. The static process is suitable for both liquid and solid samples matrixes and does not need any initial preparatory steps (Salemi *et al.*, 2006).

In purge and trap method VOCs are swept from the sample using a gas flow before any equilibrium is reached making this process an exhaustive extraction process. Dyanamic headspace method is suited for analytes present in minute quantity. The purge vessel, sorbent trap, six-port valve and transfer lines are main components of the system where sample is given in purge vessel and mixed with purge gas (generally helium). The helium gas sweeps the volatiles into the trap where the sorbent retains the VOCs and the gas is released from the system. The VOCs are released from the sorbent by heating and directed to GC column for further analysis.

11.2.6.4 Microwave Assisted Extraction (MAE)

Microwaves are electromagnetic waves consisting of electrical and magnetic oscillating perpendicular fields. Microwave energy was used for digestion of different environmental, biological, geological samples. In plant sample preparation the microwave energy is used for heating of the sample in a solvent with high dielectric constant. Usually, the heating rapture the cell wall and release the analytes in the solvent. But now this technology is used for rapid extraction of natural products from plants samples like secondary metabolites and valuable medicinal compounds from vegetables and medicinal plants. The easy upscaling of this technology enables large volume sample preparation and makes this extraction process feasible to industrial scale. The Microwave assisted extraction process is a useful technique for industrial scale secondary metabolite and aroma compounds extraction (Destandase *et al.*, 2013).

11.2.6.5 Analytical Thermal Desorption (ATD)

Thermal desorption is another technique to concentrate the Very Volatile Organic Compounds (VVOCs) before subjecting the sample to GC. Here a packed column or a tube with selected sorbents are used which can adsorb volatile compounds. Later small volume of heated gas flow is passed through the column where the desorption will take place. Thus, the volatiles get more concentrated in a smaller volume of sample which is injected in GC (Woolfenden, 2012). This method is also termed as single stage operation. Now more advanced two stage operation are used where the gas flow from sorbent tube is collected in a cold trap or focusing trap which is nothing but a smaller narrower tube. This approach reduces the volume of the sample significantly and improves the peak width of the chromatographic technique. Thermal desorption is used in profiling aroma compound in food and drinks. It is also used to detect volatiles from milk products like cheese (Valero *et al.*, 2001).

11.2.6.6 Flow Injection Analysis

Flow injection analysis in a modern fully automated wet-chemical method for analysis of food, oil and beverages. In this system samples are applied in a continuous, non-segmented flow in the same length tubing system where the reagents are added at appropriate timing and position while the samples are moving towards flow through detector or auto analyser system. The system is error free, uses less reagent. It is a high throughput sample extraction and analysis technique and applicable in industrial scale (Andersen *et al.*, 2014).

11.2.6.7 Hyphenated Extraction

The term 'hyphenated' was introduced for coupled extraction, separation and detection method (Azqhandi *et al.*, 2020). The first of its type was GC-MS where GC was coupled with mass-spectroscopy which is detection technique. Hyphenated methods reduce the cost of analysis, handling errors and time required for the analysis. It also protects the user from the toxicity of the analytes (Wilson & Brinkman, 2003).

11.3 Food Omics Studies & Their Challenges

Although tremendous progress has been made in analytical chemistry over the past few years, quite a few unresolved issues and challenges remain to be addressed in this rapidly progressing area of research.

11.3.1 Detection of Food-Borne Infections

The burden of food borne diseases are quite substantial. According to World Health Organization (WHO) estimates; consumption of spoiled food causes 600 million cases worldwide (almost one in ten people living in the world) of foodborne diseases. One third of deaths due to food spoilage occur below five years age children group (WHO, 2020). Food safety, nutrition and food security are interlinked, compromising socioeconomic development with underdeveloped health care systems, by creating roadblock to progress of national economies, tourism and trade (Weiler & Fernendez, 2019). The microbiological techniques of pathogen detections are cumbersome and time-consuming and are now replaced with more advanced and sensitive polymerase chain reaction (PCR), sequencing, and Microarray-based technologies. Among the frequently used molecular biology techniques, conventional and real-time qPCR, are the most commonly used for detection of foodborne pathogen due to its high sensitivity and specificity. Although this technology is having the advantage of rapid detection methods but the accuracy depends on purity of the culture to be tested. Enrichment of pathogen concentration from the sample matrix is done by pre-analytical sample processing method which requires additional time. PCR technique suffers from limitations like inability to distinguish the dead and viable cell's DNA, a crucial factor for, regulatory authorities and consumers associated with the food industry. One of the solution can be processing of the samples before DNA extraction with biological dyes like ethidiummonoazide (EMA) and propidiummonoazide (PMA) process. They intercalates in dead DNA which facilitates in identifying dead and live samples (Zeng *et al.*, 2016). Modification of old techniques or discovery of rapid tests with more accurate results in compliance with official guidelines maintaining recovery, accuracy, sensitivity, selectivity, and specificity still remains elusive.

11.3.2 Detection of Food Allergens

According to the latest reports, food allergy is the most talked about problem arising from food components. Hypersensitive reactions can be prevented by susceptible consumer by following the practice of food avoidance from consuming suspected allergenic substances (Andreozi *et al.*, 2019). However, search for specific, rapid and cost-effective methods for the high-throughput screening of specific allergen content in food products is rising globally. Immunoanalytical methods like ELISA (both sandwich and competitive), Western Blot, lateral flow immunoassay, nanobody facilitated detection methods, High-Resolution Melting analysis (HRM), Mass spectrometry based methods, Nucleic acid-based methods, Droplet Digital-PCR, LAMP, Biosensor based detection methods have progressed rapidly over the years for detection of food allergens present even in very minute level (García-Cañas *et al.*, 2012). More advanced allergen extraction and solubilisation methods, simultaneous detection of several allergens using universal buffer system, removal of the undesired competition between several bio reagents which can interfere with signal development and resolutions of biosensors is the need of the hour in this aspect. Combinations with different food matrix is another challengein the area of food processing and allergen detection from ultra-processed food samples with robust applicability and immunoanalytical methods are still in developing stage (Amadeo Sena-Torralba *et al.*, 2020). There are several reports that food processing methods can alter allergen structure and interfere with allergenic potential of the suspected protein, thereby reducing its impact. In such scenarios uncertain allergen detection may cause unnecessary hue and cry. Another concern for detecting allergens in food products is the robustness of the chosen analytical methods. Unavailability of high standard certified reference material (CRM) limits development of analytical methods for several allergens (de la Cruz S *et al.*, 2017).

11.3.3 Detection of Food Associated Viruses

Detection of food spoilage associated viruses is of paramount importance in any food analysis method as they are the causal organisms of gastrointestinal diseases and hepatitis including many other deadly diseases. With the emergence of several fast, accurate molecular tools and methods, the detection of noroviruses, rotaviruses, hepatitis A viruses, and hepatitis B viruses and other relevant viruses in different food matrixes is relatively easy now (Foddai *et al.*, 2020, Bosch *et al.*, 2018). However further improvement is necessary for the designated methods since their efficiency varies considerably depending on the type of method used, the nature of food matrix, and the nature and family of virus. Standardized and improved test procedures are need of the hour. With the emergence of viral pandemic due to SARS-CoV-2, avian influenza virus-commonly known as bird flu, their impact analysis on food industry are yet to be known. The rapid and easy detection technique of such type of unknown, emerging virus type influencing social norms will be major challenge in future.

11.3.4 Detection of Food Adulterations

Food adulteration is defined as the process in which the value of the food is decreased either by the intentional or unintentional addition of inferior quality material or by removal of valuable ingredient which was supposed to be present in the concerned food matrix. According to the Food Safety and Standard Authority of India (FSSAI, 2012), adulteration of food commonly deals with the changing the natural composition by adding or removing any ingredient and thereby compromising the quality of food. Globally, approximately two-third people are suffering from health problems due to consumption of adulterated or contaminated foods. Approximately around one-fifth of foods are adulterated per year by some means (Pal & Meena, 2020). With the help of chromatography based analytical detection techniques like HPLC, GCMS the nature and concentrations of food adulterants from different food matrices can be precisely quantified. In the contrary, high cost of the equipment's, low efficiency, complex sample preparation, requirement of technical supremacy is limiting the mass use of these techniques. More rapid, non-destructive and accurate detection techniques of food adulteration is coming up fast. Optical techniques (computer vision using colour characteristics), Infrared spectroscopy (FTIR), Raman spectroscopy and Raman imaging including Hyperspectral imaging, Laser induced breakdown spectroscopy, Terahertz spectroscopy and terahertz imaging, Fluorescence spectroscopy, Photoacoustic spectroscopy, Electrical techniques including Electronic nose and electronic tongue, Electrical impedance spectroscopy, Nuclear magnetic resonance spectroscopy and magnetic resonance imaging were used extensively to detect adulterations in food matrix like milk and milk products, meat, fish, honey, edible oil, beverages etc (Bansal *et al.*, 2017). Unfortunately, none of these techniques are ready yet for large scale applications. Low limit of detection or (LOD), absence of proper calibration models, research gap between laboratory discoveries and real-world applications are major bottlenecks for wider and panworld applications (Mcgrath *et al.*, 2018). Rapid advancement of sensor technology can benefit the detection procedure and overcome the challenges (Campuzano *et al.*, 2020).

11.3.5 Automation and Miniaturization

The success of food omics will primarily depend on availability of easy, fast, robust, reproducible, accurate detection methods from different food matrix. Most of the high throughput analytical methods depend on costly and big sized equipment having large space requirement (Ayda Ghahary &

Rambod Abiri, 2021). Miniaturization of analytical methods is progressing rapidly under every emerging technology. These new generation bio analytical techniques are expected to be more robust with improved and efficient monitoring of food and environmental adulterants. The emergence of planar-array and suspension-array technologies can be game changer in detection offood borne pathogens, food associated allergens and adulterants, food toxins, indiscriminate use of antibiotics, and environmental contaminants (Virginia *et al.*, 2012, Alejandro Cifuentes, 2012). Microfluidics technology is also growing in stature for food analysis, with emergence of multipurpose microfluidic platforms to control multiple unit operations simulteniously.

11.3.6 Detection of Food Toxins

The major food toxins can be divided into (a) naturally occurring toxins which include mycotoxins, marine and feed product derived toxins, cyanogenic glycosides, glucosinolates and toxins arising from poisonous mushrooms. Cereals like maize, pulses, and oilseeds like groundnut possessed high levels of mycotoxins, namely aflatoxin and ochratoxin, produced by fungal pathogens on food spoilage (WHO, 2018). (b) Persistent organic pollutants (POPs) where compounds that accumulate in the environment and human body by the bio magnification process. Known examples can be synthetic pesticides applied in the farming system, unwanted by-products of industrial processes and waste incineration. This is global phenomenon where chemical compounds are accumulating in environment and ecosystem, thereby negatively influencing food chains. (c) Heavy metals such as lead, cadmium, mercury, arsenic Contamination in food occurring mainly through pollution of air, water and soil. The type and number of toxic residues in food commodities is increasing alarmingly as a consequence of rapid industrialization, new agricultural practices, environmental pollution, and climate change. Analytical techniques which arerapid, robust, sensitive, and accurate to detect food toxins are emerging globally (Tahir *et al.*, 2019).

11.3.7 Detection of Genetically Modified Foods and Feeds

In many developed countries of the world, to safeguard the consumer's right of choice, genetically modified organisms (GMO) related legislations have been *in vogue*. Predominantly DNA based technologies are in practice to analyse presence or absence of transgene, to implement these legislations. Quantitative PCR (qPCR) is the preffered method for the laboratories involved in detection of GMO for its day to day routine analysis due to its many advantages. But many challenges remained to be solved due to development of diversified GMOs. To overcome these challenges, alternative GMO detection techniques are emerging, for single transgene target (e.g., loop-mediated isothermal amplification), of multiple transgene targets (e.g., PCR capillary gel electrophoresis, microarray, and Luminex), quantification of transgene (e.g., digital PCR), or valorization of partially known (e.g., DNA walking and Next Generation Sequencing (NGS) or unknown (e.g., NGS) GMO (Marie-Alice Fraiture *et al.*, 2015, Debode *et al.*, 2019). However, for those GMO with partial known sequences, these strategies develop false signals interfering with the accurate interpretation of the results. In general, promoter elements like p35S and NOS terminator (tNOS) are frequently used key target elements in developing transgenic plants, flanking sequences or genetic elements like transposons also have important roles to play in DNA walking or transgene enrichment strategies. Another major area needs to be addressed are GM feedstuffs, despite that fact that between 70 to 90% of all GM crops and their biomass are used as animal feed worldwide. The same techniques used for molecular characterization, traceability, environmental safety studies and horizontal gene transfer (HGT) of GM food matrices can be operational for detection of both GM food and feed (Giraldo *et al.*, 2019). Toxicological, and allergenicity studies for GM feed designated for animal consumption are scarce.

11.3.8 Detection of Nanomaterials

Nanotechnology applications have major implications in the food industry with method development for cutting edge processing, packaging, and long-term storage, with improvement in its flavour and texture. The nanosensors offer the stakeholders by providing accurate information on nutritional status and alarming through pathogen detection. In general, food bioactive compounds are hydrophobic in nature having poor bioavailability and stability. Nanotechnology-based delivery systems provide an enhanced bioavailability and targeted delivery of food bioactive compounds. Utilization of nano-colloidal particles in food quality, safety, processing, and packaging, has been widely reported recently (Alejandro Cifuentes, 2012). In spite of global progress in production technology of nanoparticles the standard regulatory rules for the utilization of nanotechnology in food products are yet to be standardized. The applications of nanoparticles as food packaging material are better suited than the utilization of nanoparticles as a food ingredient. Nanomaterials has to be dealt with utmost safety due to the threat that it may enter the food chain through the air, water, and soil during their manufacture leading to serious consequences. Scientific *in vivo* studies on the effects of nanofoods in human and animal health have been limited (Naseer *et al.*, 2018). Proper labelling and regulations for marketing of nanofoods need to be formulated to increase consumer acceptability and confidence.

11.3.9 Emergence of Green Analytical Methods

Green analytical chemistry implementation in food analysis laboratories is the need of the hour to conserve environment. Newer techniques to reduce or eliminate use of solvents, reagents, preservatives, and other chemicals which are hazardous to human health or the catastrophic to environment. Enforcement of faster and more energy-efficient analysis including sample preparation techniques with the use of new green solvents, miniaturization, or employment of solvent-less techniques, emergence of non-destructive method of food analysis and the combination with new and cleaner separation techniques and chemo metrics will greatly contribute to reaching the goals of this emerging green era (Virginia *et al.*, 2012, Alejandro Cifuentes, 2012).

11.4 Food Omics: A Platform to Investigate Health Benefits

The concept of development of personalized nutrition, adopting a particular dietary practice to tackle a physiological state or disease and due to various other reasons, led to the exploration of high-throughput omics-based approaches to understand the intricacies of diet and human health (Rubio-Aliaga *et al.*, 2012). Comprehending the effect of food or diet on the health or wellbeing of an individual is a demanding task since the food as such or its individual bio-constituents exert complementary or antagonistic effects (Putignani and Dallapiccola, 2016). Furthermore, it is imperative to understand the interactions of genetic factors of individuals and gut microbiome with the diet to assess the effects of nutritional intervention through metabolic profiling (Claus, 2017). Moreover, the role of food omics in dietary intervention or human health would have multipronged objectives. Majority of the applications of omics techniques in food science involve deciphering the molecular effect(s) of food constituents in the diet (Fito *et al.*, 2016). Also, the idea that identification of novel biomarkers linked to the dietary exposure or health, or nutritional status of individuals offers enormous application potential for the science of food omics. It is also pertinent to assess the bioavailability, bioactivity and bioefficacy of food components to effectively formulate a healthy diet based on food omics approaches.

Metabolomic approaches have been effectively utilized to uncover the potential biomarkers linked to specific dietary intake and has greatly aided in the process of dietary standardization (Edmands *et al.*, 2011; Pujos-Guillot *et al.*, 2013). Untargeted metabolomic studies of urinary samples of subjects who were under different diets [namely Average Danish Diet or New Nordic Diet] using UPLC-qTOF-MS supported the applicability of food omics in assessing the compliance with definite food regimens (Andersen *et al.*, 2014). Though this study has not undertaken the targeted quantitation of metabolites, it has clearly accentuated the potential of metabolites as biomarkers for compliance to specific or complex dietary regimens (Andersen *et al.*, 2014). In this context, application of nutrimetabonomics approach is indispensable to quantitatively measure the metabolic biomarkers following a dietary plan or to identify food induced metabolic changes in the endogenous pathways (Stamler *et al.*, 2013; Claus, 2017).

Food intervention studies have identified many food-based biomarkers. Instances for these approaches include identification of plasma biomarkers for dietary intake of cranberry juice (Liu *et al.*, 2017) and urinary biomarkers for consumption of red wine (Edmands *et al.*, 2015). Nevertheless, considering the complex mixture of food, and food matrices identification of specific biomarker for the diet is a challenging task. Hence, the concept of nutrimetabolomics has emerged to measure the metabolic effects of a food (van Duynhoven and Jacob, 2016). Preliminary nutrimetabonomic investigations on diet-metabolite profile-disease link utilize NMR-based profiling of urine and serum metabolites to correlate the food intake and cardiovascular health, type 2 diabetes, diabetic nephropathy (Stamler *et al.*, 2013; Soininen *et al.*, 2015). Developments in the field of multi-omics have proven the suitability of Mediterranean diet in preventing the cardiovascular diseases (Fito *et al.*, 2016). Similarly, comparative urinary NMR-metabolomic analysis of subjects under Mediterranean and low-fat diet revealed that the former is associated with changes in primary metabolism and microbial co-metabolites whereas the latter diet shows characteristic urinary excretion of histidine, xanthosine and hippurate metabolites (Vazquez-Fresno *et al.*, 2015). Also, Mediterranean diet is associated with suppressed levels of biomarkers linked to the oxidative stress (Whalen *et al.*, 2016).

NMR-based urinary metabolite profiling of volunteers who were given four different diets, designed based on WHO guidelines to prevent the non-communicable diseases, led to the development of mathematical models for each diet and were further statistically validated in independent cohort studies (Garcia-Perez *et al.*, 2017). This study thus provided an association between urinary metabolite profile and healthy diet (Garcia-Perez *et al.*, 2017). A system-level interactome analysis utilizing computational frameworks and network-based inference has provided an excellent basis to establish diet-disease correlation and the feasibility of designing suitable dietary regimen for countering certain diseases (Zheng *et al.*, 2017). Despite the limitations such as the expensive nature of multi-omics studies and restrictions of the computational capabilities, it is possible to arrive at the diet-disease relationship and to effectively design a diet to overcome the disease (Braconi *et al.*, 2017).

Food omics studies deciphering the relationship between food and human health have largely been confined to investigating the effect of diet on non-communicable diseases such as CVD. However, (O'Keefe *et al.*, 2015) explored the role of dietary differences in Americans of African descendents and the inhabitants of rural Africa in the incidence of colon cancer. Microbiota and metabolomic profiles of the subjects suggest that African diet has lower propensity for the development of colon cancer due to its higher colonic saccharolytic fermentation, butyrogenesis and reduced levels of secondary bile acids (O'Keefe *et al.*, 2015).

Besides genetic differences, gut microbiome is an important factor that contributes to the response of an individual to the dietary intervention. A large number of metabolites profiled in the food omics analysis are microbial co-metabolites (eg. hippurate, *p*-cresol etc.) and thus signifying the profound interaction of gut microbiota and dietary nutrients (Mancano *et al.*, 2018). Modulations in the gut microbiota have been correlated with the development of diseases such as obesity, inflammatory bowel syndrome, neurodevelopmental impairments among others (Goto *et al.*, 2015; Kelly Jr. *et al.*, 2017). Dietary interventions cause modulations on the richness and diversity of the gut microbiota and their metabolites thereby profoundly influences the human health (Portune *et al.*, 2017). Food omics is suggested as a most comprehensive approach to investigate the effect of diet-microbiota on induced molecular modulations and in the manipulation of diet for improved health (Putignani and Dallapiccola, 2016). The influence of gut microbiome metabolism on human obesity, food-derived polyphenols and microbiome-mediated metabolome modifications, characteristic metabolic fingerprints associated with specific dietary preferences have been unravalled through the application of food omics science (Moco *et al.*, 2012; Beger *et al.*, 2016; Portune *et al.*, 2017). A combination of omics approaches including 16S rRNA genomics, metaproteomics and metatranscritpomics was utilized in capturing the changes in intestinal microbiota due to the intake of dietary resistant starch (Maier *et al.*, 2017). However, it must be acknowledged that the host genetic factors exert little influence on the composition of gut microbiome as proven by the correlation analysis of host genome SNP, genetic ancestry, microbiome profile (Rothschild *et al.*, 2018).

Food omics play a significant role in proving (or disproving) the potential health claims of a particular dietary regimen. This facet of food omics is important since there has been large scale upsurge or explosion of diverse products in the nutraceutical industry along with their unproven health claims. For instances, omega-3 fatty acids, Coenzyme Q, alpha-lipoic acid, curcumin, medium chain fatty acids of edible oils (coconut oil) etc. The claims are diverse ranging from disease reduction, therapeutic, natural foods, etc (Srinivasan, 2020). Not only clinical trials but also regulatory clearance is mandatory to bring the products with such novel claims in to the market (Srinivasan, 2020). In this context application of food omics tools could greatly aid in navigating the complex regulatory regimen (Ferrocino and Cocolin, 2017). Changing life style and food habits led to the greater reliance on processed, pre-cooked or packaged foods which pose serious health implications. Further transport, packaging and storage components of food processing add one more dimension to the nutrition from food. Application of multi-omics in food science could decipher the effect of these processes in food so that it is possible to design dietary compositions to prevent the disease and promote healthy living (Cozzolino, 2015; Cifuentes, 2017).

Thus, it is certain that besides analysing the food matrix, food omics has been helping in deciphering the health impacts of dietary manipulations (as done in personalized nutrition) and to study the impact of gut microbiome in dietary health. The high inter-individual variability in terms of metabolism and interactions due to the gene-diet-microbiome pose a serious challenge to the currently followed model of universal public health nutritional guidelines. Thus, application of high-throughput techniques and multi-omics approaches in food science is critical to develop the concept of precise individual-centric nutrition.

Conclusions

Food being an important component of human health and well-being, application of multiple omics-based approaches in characterizing the food per se and its metabolic fate in human body warrants utmost importance.

In this context, application and scope of food omics concept is enormous as many queries pertaining to food, food composition, and its metabolic fate in human body in the backdrop of emerging links between gut microbiota, food metabolites etc, inter-individual variability and its manifestation in the development of diseases/disorders required to be uncovered. The search for novel biomolecules and key food constituents from plant and animal dietary sources with serious implications for promoting human health and wellbeing is on the rise (Álvarez-Rivera *et al.*, 2021). Therefore, application of food omics concepts is indispensable in combining the findings of nutraceuticals research with that of finalizing a tailor-made personalized nutrition and therapeutic interventions. Also, the convergence of multiple tools employed in food omics would enable the development of appropriate strategies to define precision nutrition.

References

Albert Bosch, Elissavet Gkogka, Françoise S. Le Guyader, FabienneLoisy-Hamon, Alvin Lee, Lilou van Lieshout, BalkumarMarthi, MetteMyrmel, Annette Sansom, Anna Charlotte Schultz, Anett Winkler, Sophie Zuber, Trevor Phister, (2018) Foodborne viruses: Detection, risk assessment, and control options in food processing, *International Journal of Food Microbiology*, 285, 110-128.

Alejandro Cifuentes (2012) Food Analysis: Present, Future, and Food omics International Scholarly Research Network *Analytical Chemistry*, Article ID 801607, 16, doi:10.5402/2012/801607

Ali J, Najeeb J, Ali MA *et al.* (2017) Biosensors: Their fundamentals, designs, types and most recent impactful applications: A review. *Journal of Biosensors & Bioelectronics.* 8(1): 235. doi: 10.4172/2155-6210.1000235

Amadeo Sena-Torralba, Yeray Pallás-Tamarit, Sergi Morais, Ángel Maquieira, (2020) Recent advances and challenges in food-borne allergen detection, *Trends in Analytical Chemistry*, Volume 132, 2020, 116050,

Andersen Mb, Rinnan A, Manach C *et al.* (2014). Untargeted metabolomics as a screening tool for estimating compliance to a dietary pattern. *J. Proteome Research* 13(3):1405-1418.

Andreozzi, L., Giannetti, A., Cipriani, F., Caffarelli, C., Mastrorilli, C., & Ricci, G. (2019). Hypersensitivity reactions to food and drug additives: problem or myth? *Acta Biomedical Atenei Parmensis*, 90(3-S), 80–90. https://doi.org/10.23750/abm.v90i3-S.8168

Autran JC (1996) Electrophoresis, in *Analytical Techniques for Foods and Agricultural Products*, ed. G. Linden, VCH Publishers, New York, 121–148.

Ayda Ghahary and RambodAbiri (2021). Food omics; Principles, Challenges, and Applications - A Promising Tool for Food Analysis. *Acta Scientific Nutritional Health* 5.3 (2021): 01-05

Azqhandi MHA, Khezeli T, Ghaedi M (2020) New methodologies and equipment used in new-generation separation and preconcentration methods. New Generation Green Solvents for Separation and Preconcentration of Organic and Inorganic Species, 149–206. doi:10.1016/b978-0-12-818569-8.00004-8

Bansal S, Singh A, Mangal M, Mangal AK, Kumar S. (2017) Food adulteration: Sources, health risks, and detection methods. *Crit Rev Food Sci Nutr.* Apr 13;57(6):1174-1189. doi: 10.1080/10408398.2014.967834. PMID: 26054861.

Bazila Naseer, Gaurav Srivastava, Ovais Shafiq Qadri, Soban Ahmad Faridi, Rayees Ul Islam and Kaiser Younis (2018) Importance and health hazards of nanoparticles used in the food industry *Nanotechnol Rev* 2018; 7(6): 623–641.

Beger Rd, Dunn W, Schmidt Ma *et al.* (2016) Metabolomics enables precision medicine: A White Paper, Community Perspective. *Metabolomics* 12(10), 149.

Bohoyo GD, Dominguez VD, Parra JP, Gonzalez GD (2012). UHPLC as a suitable methodology for the analysis of carotenoids in food matrix. *European Food Research and Technology* (235):1055. 10.1007/s00217-012-1838-0.

Braconi D, Bernardini G., Millucci L., & Santucci A. (2017): Food omics for human health: current status and perspectives, *Expert Review of Proteomics*, DOI: 10.1080/14789450.2018.1421072

Bunney J, Williamson S, Atkin D, Jeanneret M, Cozzolino D, Chapman J, Power A, Chandra S (2017). The Use of Electrochemical Biosensors in Food Analysis. *Curr Res Nutr Food Sci* 5(3): doi : http://dx.doi.org/10.12944/CRNFSJ.5.3.02

Campuzano, Susana, Víctor Ruiz-Valdepeñas Montiel, Verónica Serafín, Paloma Yáñez-Sedeño, and José M. Pingarrón (2020). Cutting-Edge Advances in Electrochemical Affinity Biosensing at Different Molecular Level of Emerging Food Allergens and Adulterants, *Biosensors* 10, no. 2: 10. https://doi.org/10.3390/bios10020010

Cancalon PF (2006) Electrophoresis and Isoelectric Focusing in Food Analysis. *Encyclopedia of Analytical Chemistry*. doi:10.1002/9780470027318.a1007.

Carnevillier V, Feuillat M, Charpentier C (1997) 'L' Electrophorese, Principle et Applications en Oenologie', *Rev. Oenolo.*, 81, 35–38.

Chandra S, Siraj S, Wong DKY (2017) Recent advances in biosensing for neurotransmitters and disease biomarkers using microelectrodes. *Chem Electro Chem*. 4(4):822-833.

Cifuentes A (2017) Food omics, foodome and modern food analysis. *Trends Anal Chem* 96:1. https://doi.org/10.1016/j.trac.2017.09.001

Claus (2017) Sp. Development of personalized functional foods needs metabolic profiling. *Curr Opin Clin Nutr Metab Care* 17(6), 567-573 (2014).

Cordero, C., Liberto, E., Sgorbini, B., Rubiolo, P., Bicchi, C., (2012). Gas chromatography. In: Pic o, Y. (Ed.), Chemical Analysis of Food: Techniques and Applications. Academic Press, San Diego, CA, pp. 311e374.

Cozzolino D (2015) Food omics and infrared spectroscopy: from compounds to functionality. *Curr Opin Food Sci* 4:39–43. https://doi.org/10.1016/j.cofs.2015.05.003

de la Cruz S., López-Calleja I., Martín R., González I., Alcocer M., García T. (2017) Recent Advances in the Detection of Allergens in Foods. In: Lin J., Alcocer M. (eds) *Food Allergens. Methods in Molecular Biology*, vol 1592. Humana Press, New York, NY. https://doi.org/10.1007/978-1-4939-6925-8_20

Debode, F., Hulin, J., Charloteaux, B. *et al.* (2019) Detection and identification of transgenic events by next generation sequencing combined with enrichment technologies. *Sci Rep* 9, 15595. https://doi.org/10.1038/s41598-019-51668-x

del Valle JM, Uquiche EL (2002) Particle size effects on supercritical CO2 extraction of oil-containing seeds. *J Amer Oil ChemSoc* 79, 1261–1266. https://doi.org/10.1007/s11746-002-0637-9.

Destandau E, Michel, Elfakir C (2013) Microwave-assisted Extraction. RSC Green Chemistry No. 21 Natural Product Extraction: Principles and Applications. *Royal Society of Chemistry*, www.rsc.org

Edmands WM, Ferrari P, Rothwell JA, Rinaldi S, Slimani N, Barupal DK, Biessy C, Jenab M, Clavel-Chapelon F, Fagherazzi G *et al.* (2015): Polyphenol metabolome in human urine and it's association with intake of polyphenol-rich foods across European countries. *Am J Clin Nutr* 2015, 102:905-913.

EdmandsWm, Beckonert Op, Stella C *et al.* (2011) Identification of human urinary biomarkers of cruciferous vegetable consumption by metabonomic profiling. *J Proteome Res* 10(10), 4513-4521 (2011).

Engelsen SB (1997). Explorative spectrometric evaluation of frying oil deterioration. *Journal of the American Oil Chemists' Society* (74): 1495-1508.

Falaki F (2019) Sample Preparation Techniques for Gas Chromatography DOI: http://dx.doi.org/10.5772/intechopen.84259

Fanali C, Dugo L, Mondello L (2016). Advances in chromatographic techniques for food authenticity testing. Advances in Food Authenticity Testing, 253-284. doi:10.1016/b978-0-08-100220-9.00010-2

Femenia A, Garcia-Marin M, Simal S *et al.* (2001) Effects of Supercritical Carbon Dioxide (SC-CO2) Oil Extraction on the Cell Wall Composition of Almond Fruits, *J. Agric. Food Chem.* (49): 5828–5834.

Ferrocino I, Cocolin L (2017) Current perspectives in food-based studies exploiting multi-omics approaches. *CurrOpin Food Sci* 13:10–15. https://doi.org/10.1016/j.cofs.2017.01.002

Fito M, Melander O, Martinez Ja, Toledo E, Carpene C, Corella D (2016). Advances in integrating traditional and omic biomarkers when analyzing the effects of the Mediterranean diet intervention in cardiovascular prevention. *Int J Mol Sci* 17(9).

Foddai, A.C.G., Grant, I.R. Methods for detection of viable foodborne pathogens: current state-of-art and future prospects. *Appl Microbiol Biotechnol* 104, 4281–4288 (2020). https://doi.org/10.1007/s00253-020-10542-x

Forgacs E, Cserhati T (2003) Gas chromatography. In: Lees, M. (Ed.), Food Authenticity and Traceability. CRC Press, Boca Raton, FL, pp. 197e217.

FSSAI (2012). Quick tests for some adulterants in food. Food Safety and Standards Authority of India, New Delhi, India.

G. Álvarez-Rivera, A. Valdés, C. León and A. Cifuentes, (2021) Chapter 1: Food omics – Fundamentals, State of the Art and Future Trends , in Food omics: *Omic Strategies and Applications in Food Science*, pp. 1-53. DOI: 10.1039/9781839163005-00001

G. H Weiler & J.L Fernendez, (2019) Food safety is everybody's responsibility, Commentry published in Phillipine Daily Inquirer.

Garcia-Perez I, Posma JM, Gibson R, Chambers ES, Hansen TH, Vestergaard H, Hansen T, Beckmann M, Pedersen O, Elliott P *et al.* (2017): Objective assessment of dietary patterns by use of metabolic phenotyping: a randomised, controlled, crossover trial. *Lancet Diabetes Endocrinol,* 5:184-195.

Giraldo PA, Shinozuka H, Spangenberg GC, Cogan NOI and Smith KF (2019) Safety Assessment of Genetically Modified Feed: Is There Any Difference From Food? *Front. Plant Sci.* 10:1592. doi: 10.3389/fpls.2019.01592

Goto Y, Kurashima Y, Kiyono H (2015). The gut microbiota and inflammatory bowel disease. *Curr Opin Rheumatol* 27(4), 388-396.

Hawthorne SB, King JW (1999). Principles and Practice of Analytical SFE in Practical Supercritical Fluid Chromatography and Extraction. M. Caude and D. Thiebaut (Eds.), Harwood Academic,Chur, Switzerland, pp. 219-282.

Keeler J (2010) Understanding NMR Spectroscopy. A John Wiley & Sons, Ltd. Publication, Cambridge, Second Edition:

Kelly Jr, Minuto C, Cryan Jf, Clarke G, Dinan Tg (2017). Cross Talk: The Microbiota and Neurodevelopmental Disorders. *Front Neurosci*11, 490.

King, JW (1989) Fundamentals and applications of supercritical fluid extraction in chromatographic science. *J. Chromatogr. Sci.* 27: 355-364.

Klancnik A, & Minka K, Nataša T *et al.* (2012) PCR in food analysis. 10.5772/38551.

Labrinea EP, Thomaidis NS, Georgiou CA (2001). Direct olive oil anisidine value determination by flow injection. *Analytica Chimica Acta* (448): 201-206.

Lakowicz JR, (2006). Principles of Fluorescence Spectroscopy. New York; Springer Science+ Business Media, LLC;. Determination of Food Quality by Using Spectroscopic Methods. http://dx.doi.org/10.5772/52722363

Lantz PG, Abu Al-Soud W, Knutsson R, *et al.* (2000). Biotechnical use of the polymerase chain reaction for microbiological analysis of biological samples. *Biotechnology Annual Review* (5):87–130.

Law JWF, AbMutalib NS, Chan KG *et al.* (2015) Rapid methods for the detection of foodborne bacterial pathogens: principles, applications, advantages and limitations. *Front. Microbiol* (5)770: doi: 10.3389/fmicb.2014.00770.

Lehotay SJ (2002) Determination of pesticide residues in nonfatty foods by supercritical fluid extraction and gas chromatography/mass spectrometry: collaborative study. *J AOAC Int.* 85(5):1148-66. PMID: 12374416.

Lehotay SJ (2002). Determination of pesticide residues in nonfatty foods by supercritical fluid extraction and gas chromatography/mass spectrometry: collaborative study. *J AOAC Int.* 85(5):1148-66. PMID: 12374416.

Li-Chan ECY, Ismail AA, Sedman J, Van de Voort FR (2006). Vibrational Spectroscopy of Food and Food Products. Handbook of Vibrational Spectroscopy.

Liu H, Garrett TJ, Su Z, Khoo C, Gu L (2017) UHPLC-Q-Orbitrap-HRMS-based global metabolomics reveal metabolome modifications in plasma of young women after cranberry juice consumption. *J Nutr Biochem* 2017, 45:67-76.

Luque de Castro MD, Valcarel M, Tena MT (1994) Analytical Supercritical Fluid Extraction, Springer-Verlag, Berlin.

Maier TV., Lucio M, Lee Lh *et al.* (2017) Impact of Dietary Resistant Starch on the Human Gut Microbiome, Metaproteome, and Metabolome. *MBio* 8(5).

Majewska M, Krosowıak K, Raj A *et al.* (2008). Solid phase extraction in food analysis. *Food Chem. Biotechn.* (72):1–13.

Mancano, G., Mora-Ortiz, M. and Claus, S.P., (2018). Recent developments in nutrimetabolomics: from food characterisation to disease prevention. *Current Opinion in Food Science, 22*, pp.145-152.

Marie-Alice Fraiture, Philippe Herman, Isabel Taverniers, Marc De Loose, Dieter Deforce, and Nancy H. Roosens (2015) Current and New Approaches in GMO Detection: Challenges and Solutions BioMed Research International Volume, Article ID 392872, 22 Pages. http://dx.doi.org/10.1155/2015/392872

McGorrin RJ (2006) Food analysis techniques: Introduction. Encyclopedia of Analytical Chemistry.

Mejri M, Roge B, Ben Souissi A, Michels F, Mathhlouti M (2005) Effects of some additives on wheat gluten solubility: A structural approach. *Food Chemistry* 92,7-15.

Mitra S (2003) Sample Preparation Techniques in Analytical Chemistry. Vol. Hoboken, New Jersey: John Wiley & Sons, Inc. (162):1-439. ISBN: 0-471-32845-6.

Moco S, Martin Fp, Rezzi S (2012). Metabolomics view on gut microbiome modulation by polyphenol-rich foods. *J Proteome Res* 11(10), 4781-4790

Nakai S, Horimoto Y (2006). Fluorescence spectroscopy in food analysis. Encyclopedia of Analytical Chemistry.

Ngom B, Guo Y, Wang X *et al.* (2010) Development and application of lateral flow test strip technology for the detection of infectious agents and chemical contaminants: a review. *Anal. Bioanal. Chem.* (397): 1113–1135. doi: 10.1007/s00216-010-3661-4.

Niaura G (2008) Raman spectroscopy in analysis of biomolecules. Encyclopedia of Analytical Chemistry.

Norris KH, Hruschka WR, Bean MM, Slaughter DC (1989). A definition of wheat hardness using near infrared reflectance spectroscopy. *Cereal Foods World* (34): 696-705.

O'Keefe, S.J., Li, J.V., Lahti, L., Ou, J., Carbonero, F., Mohammed, K., Posma, J.M., Kinross, J., Wahl, E., Ruder, E. and Vipperla, K., (2015). Fat, fibre and cancer risk in African Americans and rural Africans. Nature Communications, 6, 6342. https://doi.org/10.1038/ncomms7342.

Ötles S, Kartal C (2016) Solid-Phase Extraction (SPE): Principles and applications in food samples. *Acta Sci. Pol. Technol. Aliment.*, 15(1): 5–15. DOI: 10.17306/J. AFS.2016.1.1

Pal, M. & Meena M (2020) Food adulteration: A global public health concern: Food and Drink Industry, Issue 3, vol 1, 38-40

Parlak Y, Güzeler (2016) N Nuclear Magnetic Resonance Spectroscopy Applications in Foods. *Curr Res Nutr Food Sci*; (4): doi: http://dx.doi.org/10.12944/ CRNFSJ.4.Special-Issue-October.22

Perry L, Heard P, Kane M, *et al.* (2007). Application of multiplex polymerase chain reaction to the detection of pathogens in food. *Journal of Rapid Methods and Automation in Microbiology* (15)2:176–198.

Portune Kj, Benitez-Paez A, Del Pulgar Em, Cerrudo V, Sanz Y (2017). Gut microbiota, diet, and obesity-related disorders-The good, the bad, and the future challenges. *Mol Nutr Food Res* 61(1).

Pujos-Guillot E, Hubert J, Martin Jf *et al.* (2013). Mass spectrometry-based metabolomics for the discovery of biomarkers of fruit and vegetable intake: citrus fruit as a case study. *J Proteome Res* 12(4), 1645-1659.

Putignani L, Dallapiccola B (2016). Food omics as part of the host-microbiota-exposome Interplay. *Proteomics* 147, 3-20.

Ramirez P, Senorans FJ, Ibanez E, *et al.* (2004), Separation of rosemary antioxidant compounds by supercritical fluid chromatography on coated packed capillary columns, *J. Chromatogr. A*, (1057): 241-245

Ridgway K, Lalljie SPD, Smith RM (2008). Microextraction methods in food analysis. In S. Otles (Ed.), Handbook of food analysis instruments. Boca Raton: Taylor and Francis, NW.

Rothschild D, Weissbrod O, Barkan E, *et al* (2018). Environment dominates over host genetics in shaping human gut microbiota. *Nature*; 555:210-5.

Rubio-Aliaga I, Kochhar S, Silva-Zolezzi I. (2012) Biomarkers of nutrient bioactivity and efficacy: a route toward personalized nutrition. *J Clin Gastroenterol* 46(7), 545-554.

Salemi A, Lacorte S, Bagheri H *et al* (2006) Automated trace determination of earthy musty odorous compounds in water samples by on-line purge-and-trap-gas chromatography-mass spectrometry. *Journal of Chromatography A.* (1136):170-175. DOI: 10.1016/j.chroma.2006.09.087

Shurvell HF (2006). Spectra-structure correlations in the mid- and far-infrared. Handbook of Vibrational Spectroscopy.

Soininen, P., Kangas, A.J., Würtz, P., Suna, T. and Ala-Korpela, M., (2015). Quantitative serum nuclear magnetic resonance metabolomics in cardiovascular epidemiology and genetics. *Circulation: Cardiovascular Genetics*, 8(1), pp.192-206.

Srinivasan M., (2020). Food omics: The What, Why and How of It. In. S. Singh (ed.), Metagenomic Systems Biology, https://doi.org/10.1007/978-981-15-8562-3_9

Stamler J, Brown Ij, Yap Ik *et al.* (2013) Dietary and urinary metabonomic factors possibly accounting for higher blood pressure of black compared with white Americans: results of International Collaborative Study on macro-/micronutrients and blood pressure. *Hypertension* 62(6), 1074-1080.

Tahir Rasheed, Muhammad Bilal, Faran Nabeel, Muhammad Adeel, Hafiz, M.N. Iqbal (2019), Environmentally-related contaminants of high concern: Potential sources and analytical modalities for detection, quantification, and treatment, *Environment International*, Volume 122, 2019, Pages 52-66, ISSN 0160-4120

Terry F. McGrath, Simon A. Haughey, Jenny Patterson, Carsten Fauhl Hassek, James Donarski, Martin Alewijn, Saskia van Ruth, Christopher T. Elliott, (2018) What are the scientific challenges in moving from targeted to non-targeted methods for food fraud testing and how can they be addressed? – Spectroscopy case study,*Trends in Food Science & Technology*, Volume 76, 2018, Pages 38-55, ISSN 0924-2244, https://doi.org/10.1016/j.tifs.2018.04.001.

Thurman EM, Mills MS (1998) Solid-phase extraction principles and practice. New York: John Wiley.

Tswett, M, (1906). Physikalisch-Chemische Studien €uber das Chlorophyll. Die Adsorption (Physical-chemical studies chlorophyll. Adsorption*). Berichte der Deutschenbotanischen Gesellschaft* 24, 316e326.

Turner C, King JW, Mathiasson L (2001) Supercritical Fluid Extraction and Chromatography for Fat-Soluble Vitamin Analysis, *J. Chromatogr. A* (936): 215–237.

Valero E, Sanz J, Martínez-Castro I (2001) Direct thermal desorption in the analysis of cheese volatiles by gas chromatography and gas chromatography-mass spectrometry: comparison with simultaneous distillation-extraction and dynamic headspace. *J Chromatogr Sci.* 39(6):222-8. doi: 10.1093/chromsci/39.6.222. PMID: 11396685.

Van de Voort FR, Sedman J, Emo G, Ismail AA (1992). A rapid FTIR quality control method for fat and moisture determination in butter. *Food Research International* (25): 193-198.

Van de Voort FR, Sedman J, Ismail AA (1993) A rapid FTIR quality-control method for determining fat and moisture in high-fat products. *Food Chemistry* (48): 213-221.

vanDuynhoven JP, Jacobs DM: Assessment of dietary exposure and effect in humans: the role of NMR. *Prog Nucl Magn Reson Spectrosc* 2016, 96:58-72.

Vazquez-Fresno R, Llorach R, Urpi-Sarda M, Lupianez-Barbero A, Estruch R, Corella D, Fito M, Aros F, Ruiz-Canela M, Salas-Salvado *J et al.* (2015): Metabolomic pattern analysis after mediterranean diet intervention in a nondiabetic population: a 1- and 3-year follow-up in the PREDIMED study. *J Proteome Res*, 14:531-540.

Virginia García-Cañas Carolina Simó Miguel Herrero Elena Ibáñez Alejandro Cifuentes (2012) Present and Future Challenges in Food Analysis: *Food omics Anal. Chem.*, 84, 23, 10150–10159.

Whalen KA, McCullough ML, Flanders WD, Hartman TJ, Judd S, Bostick RM: Paleolithic and Mediterranean diet pattern scores are inversely associated with biomarkers of inflammation and oxidative balance in adults. *J Nutr* 2016, 146:1217-1226.

WHO (2018) Natural toxins in food -https://www.who.int/news-room/factsheets/detail/natural-toxins-in-food

WHO (2020) Food safety -www.who.int › Newsroom › Fact sheets.

Wilson I, Brinkman UAT (2003) Hyphenation and hypernation. *Journal of Chromatography A*, 1000(1-2), 325–356. doi:10.1016/s0021-9673(03)00504-1

Woolfenden E (2012) Thermal desorption for gas chromatography, in: Gas Chromatography, ed. C.F. Poole, Elsevier Analytical thermal desorption: History, technical aspects and application range, *Thermal Desorption Technical Support* (10):235–289; http://www.markes.com/Downloads/Application-notes.aspx.

Yeni F, Acar S, Polat OG *et al.* (2014). Rapid and standardized methods for detection of foodborne pathogens and mycotoxins on fresh produce. *Food Cont* (40):359–367. doi: 10.1016/j.foodcont.2013.12.020,

Zeng D, Chen Z, Jiang Y, Xue F and Li B (2016) Advances and Challenges in Viability Detection of Foodborne Pathogens. *Front. Microbiol.* 7:1833. doi: 10.3389/fmicb.2016.01833

Zhao X, Lin CW, Wang J (2014) Advances in rapid detection methods for foodborne pathogens. *J. Microbiol. Biotechn.* (24): 297–312. doi: 10.4014/jmb.1310.10013.

Zhao Y, Zhu X, Jiang W, *et al.* (2021) Chiral Recognition for Chromatography and Membrane-Based Separations: Recent Developments and Future Prospects. *Molecules* (26): 1145. https:// doi.org/10.3390/molecules26041145

Zheng T, Ni Y, Li J, Chow Bkc, Panagiotou G (2017). Designing Dietary Recommendations Using System Level Interactomics Analysis and Network-Based Inference. *Front Physiol* 8, 753.

Żwir-Ferenc A, Biziuk M. (2006). Solid phase extraction technique – trends, opportunities and applications. *Polish J. Environ. Studies*, 15(5): 677–690.

12

Consumer Viewpoints Regarding Food and Risk Assessment

Ananya Rana[1], Shivangi Mishra[2], Komal Soni[3], Mrinal Samtiya[3], Neetu Kumra Taneja[1] and Tejpal Dhewa[3]*

[1]*Department of Basic and Applied Sciences, National Institute of Food Technology Entrepreneurship and Management (NIFTEM), Sonepat, Haryana, India*
[2]*Department of Food Science and Technology, National Institute of Food Technology Entrepreneurship and Management (NIFTEM), Sonepat, Haryana, India*
[3]*Department of Nutrition Biology, Central University of Haryana, Mahendergarh, India*

Abstract
Controversy, conflict, and paradigm changes have frequently occurred in the investigation of risk perception about food safety. A few years back, food was considered safe and associated with low risk, and if the risk was considered, it was related to hygienic conditions. Consequently, studies have been done, and it was found that food risk is equally important as the other risk factors. It may be caused due to various factors such as environmental risk, technological hazards, chemical hazards etc. and attention move towards the safety and risk associated with food. Various private and public organizations came up with different laws and conducted various awareness programs about food safety, consumer rights, risk assessment and food contamination. Consumer attitude towards food safety and risk assessment should be directed towards explicitly establishing a framework for assessing consumer behaviour in these terms. Food risk assessment has mostly focused on attempting to explain the disparity in viewpoints between specialists and the general population while disregarding it. The chapter provides a brief summary of consumer viewpoints about food safety, as well as food safety laws and risk assessment.

***Keywords*:** Food safety, risk management, food safety regulations, risk assessment, consumer, risk analysis, food quality, quality control

**Corresponding author*: tejpaldhewa@gmail.com

Tejpal Dhewa, Anil Kumar Puniya and Anil Panghal (eds.) Nutritional Science and Technology: Concept to Application, (273–294) © 2023 Scrivener Publishing LLC

12.1 Introduction

Understanding consumers' viewpoints of risk and how it affects behavior is an important issue for both the food industry and consumers. The major requirement for any consumer is to get a constant supply of healthy, safe and good-quality food. Food is essential for survival and has a direct impact on human health and well-being. Therefore, food affects our lives and perhaps makes food safety and security a universal concern [1]. The relationship between consumers and the food industry depends on physical, psychological, and performance risks associated with food [2]. Risk is something which we face in every stage of our life so it can be said it is an inescapable part of life. Foodborne disease is becoming increasingly common, and recent high-profile illness/disease outbreaks connected to food have prompted calls for more effective political and economic regulations. As a result, throughout the recent decade, government regulation of food safety has expanded. Food safety is now being threatened by global features of food supply systems [3]. In addition to official activities, food safety management systems have improved and now play a vital role in producing safe, high-quality food [4].

Food safety and global aspects of the food system are becoming major concerns [5, 6]. Food in the international market is subjected to fraud because various parts of the distribution chain, including producers, packers, and distributors, are involved in national or international trade [7–9]. It affects people of all ages, races, genders, and income levels worldwide, so it is one of the key areas in the food industry. National and international food marketing continue to significantly impact food safety and public health [3].

Consumer decision-making is heavily influenced by concerns about food safety and the related dangers connected with food intake [10]. Consumers' faith in the food supply has been shaken by reports of food fraud. For example, the Bovine Spongiform Encephalopathy (BSE) issue in the late 1990s influenced consumer behaviour, resulting in lower beef consumption in the UK and throughout Europe. BSE is a disease that affects cattle and underlies a potential food safety threat of being accidentally transmitted to humans upon consumption of contaminated beef [11].

Consumers' perceptions are affected mainly by the psychological reviews about the product instead of the product's physical characteristics, mainly in the case of food safety [12].

12.1.1 How Consumers Form Different Points of View

Perception is an important factor affecting the consumer's evaluation of risks, making decisions, and acting [13]. According to consumers' psychology, there are two paths through which attitude are formed: central and peripheral. The central pathway helps form attitude when people have the ability, opportunity, and motivation to understand an issue. Peripheral pathways, on the other hand, assist those who lack the capacity, opportunity, or motivation to grasp the gist of the problem. In the latter case, all the messages they hear are processed peripherally [14, 15].

The formulation of food safety regulations relies heavily on the use of risk analysis approach, whereby information on physical, biological or chemical hazards with regard to food safety is systematically collected, analyzed and evaluated to determine the risk to ensure safety of food. The Codex Alimentarius framework for risk analysis focuses on a three-pronged approach to risk assessment, risk management, and risk communication. The Codex Alimentarius system's risk management methodology leads to the development of food safety standards, guidelines, and other recommendations. Many variables and a range of risk management methodologies are likely to influence national risk management decisions.

Codex's main objectives are consumer protection and international trade. When the risk assessment process is finished, risk managers can use the risk categorization to design risk management strategies. The necessity for functional separation between risk assessment and risk management was recognised in the 1995 consultation. Risk communication is the key interaction of information and risk views between risk assessors, risk managers, and other stakeholders. As per Codex, risk management is the process of analysing policy options in light of risk assessment findings and, if necessary, selecting and enacting suitable control measures like legislation.

The large rise in foodborne disease, particularly that caused by enteric microbes, that has been seen over the globe in recent years underscores the need of more effective control utilising globally approved risk management systems.

Consumers can divide food-related issues into two types: (i) negative issues (food risk), and (ii) positive issues (food benefit). These issues play a major role in consumers' decisions about what to eat more of, what to eat less of, or what not to eat [16]. High-benefit and low-risk perceptions of a food can affect consumer attitude towards purchasing and consuming [17].

Some scholars consider the advantages of biotechnology in food, while others consider the concerns [18]. Customers' awareness about genetically modified food is crucial to understand the advantages and hazards connected with genetically modified foods, as a result of which, decisions about whether or not it is acceptable are affected [19, 20]. In general, the public lacks appropriate information regarding genetically modified organisms (GMOs) and the foods they eat [21]. This is the biggest problem that leads to insecurity regarding genetically modified foods and the inability to decide about the benefits and risks of these foods [22, 23]. For this reason, consumer confidence is based on those who they believe are knowledgeable and trustworthy experts, also known as "social trust" [24].

12.1.2 What Do Consumers Want from Food Business Operators?

With the increase in food frauds and incidents, the concern of consumers and other stakeholders is increasing day by day. Food safety and the qualities of the food we purchase and eat are becoming increasingly important to consumers. Although food labeling is provided, it is insufficient to gain consumers' trust [25]. To continue their faith in good quality and safe food, consumers need assurance that a food business operator must [11]:

i. Be transparent: Transparency in the food chain includes open communication with consumers and all stakeholders and access to information they request without loss and delay [25, 26].
ii. Have protocols and procedures in place: Product quality needs to be strong and reliable while still affordable, so companies need to establish standard manufacturing and sales processes [27]. The company must have crisis plans, problem resolution standards, and surveillance [11].
iii. Be trustworthy: Trust is essential to the consumer in deciding food purchases [28]. By establishing continuous communication with consumers and providing information on social media, manufacturers strengthen consumers' trust [29].
iv. Put consumers first: Consumer needs, interests, and safety should be the priority for the food business operators. Also, if consumers seek low-cost, convenient food, the food industry must prioritise this and provide it to them without causing them to feel full [1].

v. Engage stakeholders in the process.
vi. Build your brand's reputation by educating and informing all of your stakeholders and customers, and
vii. Then be consistent in your approach.
viii. Don't break your promises.

12.2 Food Safety

Food safety, according to Henson and Traill, is the inverse of meal risk—the likelihood of not being harmed by taking a given food [30]. Biological (e.g., mycotoxins) and environmental toxins to agro-chemicals, veterinary medications, growth boosters, packing components, and many more are all potential unpleasant residues in meals. Because dangerous microbes have the potential to grow exponentially from really small numbers in foods or to spread in humans once consumed, microbiological issues pose an even larger challenge to food safety [31]. Millions of people are affected by foodborne illnesses. In England and Wales alone, 23,000 instances of salmonellosis were estimated to have cost around £40 and £50 million in 1991 (Center for International Trade Studies, 1997). According to the Health Department of Canada, there are more than 2 million incidents of foodborne disease in Canada each year, costing the country $1.3 billion yearly. Foodborne diseases harm between 6 and 33 million people in the United States each year, costing between $9.3 and $12.9 billion in medical costs and lost productivity.

Various countries have taken significant initiatives to strengthen the food safety delivered to consumers. The Food Safety Act of 1990 and the General Hygiene Act of 1995 in the United Kingdom have had a considerable impact on food. The food safety risk management practises in the food industry, its shifting focus beyond preventing deceit to a proactively methodology is based on science food safety strategy [32]. The UK commission adopted a communication on "Consumer Health and Food Safety" in April 1997. The rising role of impartial research scientists in the assessment of potential threats for the creation of community legislation was one of the highlights of this message. In April 2000, the United Kingdom formed a new Food Standards Agency. Food surveillance in retail outlets and food safety research coordination are among its responsibilities. President Bill Clinton of the United States announced the "National Food Safety Initiative" in May 1997, with the goal of improving monitoring, risk assessment, inspection, and compliance, as well as educating consumers and doing important new research [33]. In 2006, India passed the Food Safety and Standards Act (FSSA) to look after the issue related to food safety and regulation throughout the country. Under the FSSA, the Food Safety and Standard Authority of India (FSSAI) was established in 2008.

The majority of consumers are not able to understand the typical term food safety. In order to make them understand about food safety, we first need to know their views about food safety. Some consumers views and attitude towards food safety are discussed below.

12.2.1 Consumers View and Attitude Towards Food Safety

The basic diffence between view and attitude is that attitude is relatively rigid about any item but view is flexible to some extent and subject to change with awareness. Attitudes of consumers about food safety are classified on the basis of types of food safety concerns. Some literature reported some factors that affected the attitude of consumers about food safety, such as chemical issues, spoilage or contamination issues, health issues such as bad fat content and imbalance of nutrition and regulatory issues that include inspection and labelling. Issues about these problems were linked to rigidity in views about food safety in general. On the other hand, concerns about regulatory concerns, deceptive techniques, or ideal scenarios, the robustness of consumer attitude toward food safety did not differ significantly ($P > 0.05$). While consumers were worried about some aspects of food safety and quality, they were less interested in specific food safety practices.

People have a range of concerns about the food they eat, according to recent public opinion polls, and most customers believed the food purchased was safe to consume. Howsoever, the majority of the population (49%) ranked food rotting as the most serious hazard to food safety. Freshness and expiration dates were important to them (22%) while bacteria and contamination were becoming more of a concern (17%). Over time, consumers also consider many other aspects of the food like hygiene policy, labelling, food safety licenses, etc. [34].

According to previous studies, the majority of customers are worried about food safety, and some are willing to pay more for safer foods. McNutt discovered in 1986 that respondents' food safety concerns average 9.60 on a scale of one to ten, with ten being "extremely critical" [35]. According to a 1990 study, the most critical food safety concerns for Michigan consumers were food freshness and the absence of rotting. In comparison, a similar study conducted in 1984 found only 12% [35].

Some literature reported that consumers' views about the safety of the food varies with socioeconomic factors such as age, economic status and educational level. The high microbiological load of raw oysters in beverages have always been a source of worry in Trinidad and Tobago. According to Laloo and colleagues, 37.9% of Trinidad and Tobago residents avoided raw oyster cocktails because they were afraid of becoming sick, while others avoided them due to cultural differences, cost, and personal choices such as taste [36]. In a survey of 300 people in Trinidad and Tobago, Laloo and colleagues discovered that raw oyster cocktails were more popular among males (73.6%) than females (26.4%) (P 0.002), among East Indians (63.2%) rather than other respondents (36.8%) (P 0.001), and people under 40 years old (82.1%) rather than older people (17.9%) [36].

Johnson and colleagues conducted a study of food safety awareness among the elderly living at home (1998). A total of 800 senior citizens aged 65 and up were interviewed in Nottingham, UK. Approximately 70% of the refrigerators in their houses were not cold enough to store food safely. Food storage at unsuitable temperatures was not affected by socioeconomic or demographic status, but it was more common among the poor and those who did not live alone [37].

12.2.2 Consumers' View of Novel Techniques and Food Safety Regulations

Genetic alterations have produced a vast variety of novel foods or feedstuffs, potatoes with new leaves, soya, mazine, and oilseed rape [38, 39]. For food scientists and many others, consumer acceptability of biotechnology-produced foods remains a crucial but contentious topic. This is especially true in the case of overseas marketplaces [36].

Consumers' main concern regarding genetically altered foods is their safety, despite the fact that innovative foods undergo comprehensive safety testing before being approved. There are also concerns regarding the technology's long-term safety, for example the generation of super-weeds and the spread of dangerous illness. So far, no illnesses have been linked to the eating of genetically modified foods. Concerns about genetic food engineering vary widely among consumers around the world. In Japan and the United states, there appears to be far less worry about the eating of genetically modified foods than in Europe (2004).

In terms of specific biotechnology uses in food, Japanese consumers were largely positive in 1998, albeit there was a little decrease from the high levels recorded in 1995. Instead of traditional crop breeding, products with lower fat content or higher vitamin content were the most popular, as well as those that required less pesticide use. Consumers found disease-resistant farm animals, food enzymes, and higher-quality soy sauce or tofu to be generally acceptable. Biotechnology's application to culinary additives like flavourings was less well received. When respondents were given additional knowledge about the product, there was a considerable rise in acceptance of the production of biopharmaceuticals to soybeans [40].

Irradiation is recommended as an effective and safe food manufacturing process for minimising the risk of foodborne illness and preserving commodities without compromising health or nutritional quality when done according to GMPs. Irradiation of food looks to be gaining consumer acceptance in the United States, although it is taking longer in many European countries, notably the United Kingdom [41]. Consumer opinions of food irradiation have been assessed through numerous surveys and market studies. According to the findings of these investigations, many customers have misconceptions about the technique, believing that irradiation will render food radioactive [42].

Labelling of food products gives much information about the product so when consumers purchase the product they are able to receive all the information regarding the product, and it also shows how much consumers are aware of food safety and risk. According to the study conducted about labelling it was found that about two-thirds of customers always or usually read the labels but about 17% of people never do so. When asked if they preferred or supported identifying GM foods through labelling, 95% of those surveyed said they preferred such labels. Those who said yes to GM food labelling in the original question were then asked why they said yes. The primary argument for GM food labelling, according to nearly half of any and all respondents (48%), is "the right of consumers to know." Only a tiny proportion of people want such labelling so they can decide whether to buy (19%) or not buy (14%) certain things. In other words, the labelling issue is only related to specific issues about GM foods for a small percentage of responders (for instance, safety).

12.2.3 FBOs Took Crisis Steps with Any Food Safety Issue

FBOs assist in the development of a hierarchical view of the food production process, integrate multiple channels of communication, cater to the packaging and labelling demands of consumers, and appropriately address public concerns.

12.3 Food Standards and Regulations in India

In case of voluntary product certification, such as the ISI mark of BIS and Agmark, there exist grading and marking methods. Indian Standards (BIS) works with the standardisation of a wide range of consumer items, including food. They also administer an optional certification programme for processed foods with the "ISI" seal. Agmark is a voluntary certification programme for agricultural goods (raw and processed) designed to protect customers' health.

The food business had a difficult time complying with many government rules and legislation. Because of this, it was decided that all of these regulations controlling food quality should be brought together. In light of this, the Indian government approved the Food Safety and Standards Act (FSSA) 2006, which aims to consolidate all food safety regulations under one roof.

12.3.1 Food Safety and Standards Act, 2006

The purpose of the Food Safety and Standards Act, 2006 (FSSA), is to harmonise the laws governing food. According to a 2006 law known as the Food Safety and Standards Authority of India (FSSAI), food-related concerns have previously been addressed by several ministries and departments.

Scientifically based standards for food were created by FSSAI to guarantee safe and healthy food is available for human consumption. Premises and surrounding areas must be kept clean, as well as risk concerns to human health assessed and managed in a scientific way, which were not stated in the PFA. A trend away from compositional or vertical requirements toward safety or horizontal criteria is seen in the FSSA.

12.3.2 Codex Alimentarius Commission (CAC)

The major goal of the CAC's Combined FAO/WHO Food Standards Plan is to protect consumer health and ensure fair trade practises. CAC adopts Codex standards, guidelines, and other recommendations issued by its members. Subordinate CAC bodies also provide guidelines to various subsidiary bodies, such as that regarding risk management.

Expert advisory groups like the Joint FAO/WHO Expert Committee on Food Additives (JECFA) and the Joint FAO/WHO Meeting on Pesticide Residues (JMPR), as well as other expert bodies like the ICMSF (International Commission on Microbiological Specific requirements for Food), support the work of these committees. In particular, FAO and WHO frequently collaborate to organise discussions on specific Codex and member-country issues. JECFA and JMPR, however, are the most significant of these Codex advisory organisations. Other Codex committees, such as CCFAC, CCRVDF, and CCPR, utilise JECFA and JMPR recommendations in the establishment of Codex standards, guidelines, and other recommendations. Standards may be developed for global usage or for use by a single region or group of countries. All CAC members and other international bodies are requested to comment on the proposed standards, along with any economic implications. CAC "should have proper account for the Codex Alimentarius' aims" when considering such remarks. Members are urged to contact with relevant stakeholders and parties in their home countries [43, 44].

According to the CCGP, provisions in Codex standards that establish description, essential composition, and quality attributes should never be the subject of risk analysis. The CAC Procedural Manual, on the other hand, states that significant composition and quality criteria may overlap with food safety, adding that "quality features could include the raw material's quality, with the purpose of safeguarding the consumer's health" [43].

12.4 The Key Terms Involved in Risk Management

Managers of risk might utilise the categorization of risks produced by the assessment process to develop risk management strategies. Separating risk assessment from risk management as a distinct job was deemed essential in the 1995 consultation. The large rise in foodborne disease, particularly that caused by enteric microbes, that has been seen over the globe in recent years underscores the need of more effective control utilising globally approved risk management systems.

Risk assessment policy: Sometimes in the risk assessment process, there may be a requirement to utilise guidelines to make judgements about value and policy.

In order to maintain the objectivity of the risk assessment, policy makers have a duty to develop risk assessment guidelines in conjunction with risk assessors. To maintain uniformity and openness, the standards should be recorded. Establishing rules for rating dangers and recommendations for applying safety factors are some examples of risk assessment policy setting [34].

Risk profile: It is used to identify the features of hazards and risks that are most relevant to alternative risk management options when a food safety concern is discussed. Identification of hazards that are significant to prioritising and establishing the risk assessment policy, as well as characteristics of risk that are relevant to selecting safety standards and management alternatives, would be included in the risk profile.

It is critical to define the situation, product, or commodity in detail, as well as the values that may be at risk (e.g., human health or financial troubles), projected outcomes, customer perceptions of risks, and risk and benefit distribution.

ELEMENTS OF RISK MANAGEMENT

A. Analyzing the Risk

- An issue with food safety has been discovered.
- Identification of potential dangers.
- Risk assessment and risk management priority ranking of hazards.
- Policy for conducting risk assessment should be established.
- Ordering a risk assessment to be conducted.
- Taking into account the findings of the risk assessment.

B. Risk management option assessment

- Identification of management alternatives that are open to you.
- Choice of desired management solution, including evaluation of an adequate safety standard*
- The final decision of management.

C. Management decision implementation

D. Review and monitoring

- Risk management and/or risk assessment should be reviewed as appropriate to ensure its efficacy.

When it comes to risk management decisions, it is critical that all parties** who may be impacted have a chance to participate. Consumer organisations, representatives of the food business and commerce, educational and research institutions, and regulatory authorities may be included in this list, although they are not required to be. You may execute a consultative process in numerous ways. These include public meetings and chances for individuals or groups to provide feedback on publicly available papers. The risk management policy design process may include input from interested parties at any level, including assessment and review, and these inputs can be incorporated.

12.5 The Fundamental Principles of Food Safety Risk Management

Principle 1: An organised strategy to risk management is required.
Risk assessment, risk management option selection, management decision execution, and decision monitoring and methodical approach to risk management entails all of the above. The risk management operations may not include all of these aspects in certain circumstances (as when Codex establishes standards and national governments implement control measures).

Principle 2: The fundamental concern in risk management choices should be the protection of human health.
The fundamental consideration in calculating acceptable risk levels should be human health, and arbitrary or unjustifiable risk variations should be avoided at all costs.

In certain risk management scenarios, consideration of additional aspects (for example, economic costs, rewards, technological feasibility, and community preferences) might be relevant. Considerations such as these should not be left up to chance and should be clearly stated.

Principle 3: Decisions and procedures in risk management should be accessible.
Identifying and documenting in full any risk management process, including decision-making, is essential to ensure that the reasons behind the process are understood by all parties involved.

Principle 4: Risk management should have a particular component for determining risk assessment policies.
An alternative approach is to work with risk assessors to draft policies ahead of time. When making risk assessment decisions, it is important to know the boundaries of the value judgments and policy choices that may be made.

Principle 5: By preserving the functional separation of risk management and risk assessment, risk management may help to protect the scientific integrity of the risk assessment process.

Because risk management is responsible for putting policies into effect, it is critical that it be kept distinct from risk assessment in order to eliminate any conflicts of interest between the two departments. An iterative process in which risk managers and assessors collaborate and work in concert is required for successful risk analysis to take place.

Principle 6: Decisions on risk management should take into account the uncertainty in the outcome of the risk analysis.
In order to fully consider the potential consequences of the range of uncertainty in making decisions, risk managers should be provided with an intelligible numerical representation of uncertainty as part of the risk assessment.

Risk management decisions may be more cautious if the risk assessment is very unclear.

Principle 7: All components of the risk management process should involve clear, interactive communication with customers and other interested stakeholders at all stages of the process.

The practise of risk management requires constant communication between all parties involved. An important part of risk communication is making sure that all the information necessary for successful risk management process is included in the decision-making process.

Principle 8: Risk management should be an ongoing process that assesses and reviews risk management decisions based on any newly generated data.
A periodic review of a risk management decision should be conducted to see if it is successful in achieving food safety goals after it is implemented.

For the review to be successful, monitoring and additional actions are certainly going to be required [45].

12.6 Types of Food Risks

There are four types of food risks: food quality, food safety, food fraud, and food defence. Food defence refers to a deliberate or intentional act that is not intended to harm public health but is still harmful to the consumer. (Terrorism or malicious tampering are examples of purposeful acts on a food product that are intended to pose a public health danger.) [46].

Microbiological Risk: According to some surveys, it has been found that consumers were aware of various food microbial contaminants. *Salmonella* and *Campylobacter* are the most common microbiological contaminant [47, 48].

Food Infection/Food Poisoning: Food poisoning and food infection are both ingestion of harmful organisms that grow and spread within the body, resulting in disease. One of the most well-known examples is the bacteria *Salmonella*. Animals' intestines contain this microorganism. Also included in this list are raw milk and eggs. *Salmonella* is destroyed by heat, yet some germs may survive due to insufficient cooking. Cross-contamination is a common method of spreading *Salmonella*. In the case of raw meat or poultry, this might occur if the chopping board is not cleaned before being used for something that does not need cooking, such as salad. After using the restroom and before handling food, an infected food handler may spread *Salmonella* if they do not wash their hands with soap. *Salmonella* may multiply at a rate of one million to one million in only 20 minutes. Diarrhoea, fever, and stomach pains are some of the symptoms of *Salmonella* infection.

Food Intoxication: Toxins produced by bacteria may remain in food long after the pathogen has been eliminated, resulting in food poisoning. Toxins are produced by organisms when food is not heated or chilled to the proper temperature. You cannot smell, taste, or see the presence of food poisoning in your meal. Foods that smell and look beautiful are not always safe, so keep that in mind. *Staphylococcus aureus* is one such bacteria. Those creatures may be found in the air, the dust, and even the water. Fifty percent of healthy people have them in their nose, throat, and hair, in addition to their skin. When food is handled by people who are infected with this microorganism, certain areas of the body might get contaminated. Another indication of this contamination is diarrhoea.

Worm infestation by tape worms in pork is one example of a parasite-caused infestation. In addition, pests and insects may infest food. In terms of foodborne disease, biological dangers are among the most common. Microbial foodborne infections are still a major issue, and new pathogens continue to arise, despite all the efforts in food safety.

Let's speak about food quality now that you understand food safety.

Food Quality: Customers' perceptions of a product's value are influenced by its quality, which is measured by a set of qualitative criteria. As a result, this includes both negative traits like spoilage and contamination as well as positive ones like flavour, texture, and appearance. As a result, it is a holistic notion that incorporates nutritional characteristics, sensory aspects (colour, texture, form, appearance, taste, flavour, aroma), social concerns, and safety. Quality begins with safety, which is a must. Various countries and international agencies have established food standards that producers and suppliers must follow in order to guarantee that foods are safe and of excellent quality.

All food service providers (preparatory work, preparation, processing, packaging, and provision of service) need to adhere to good manufacturing practises to ensure the safety of their food products.

The following are important considerations:

1. Raw material and water quality
2. The cleanliness of the establishment, its personnel, its equipment, and the spaces where food is prepared, stored, and served
3. Food storage at the right temperature
4. Food safety
5. Excellent customer services.

Chemical Risk: Chemicals such as heavy metals, pesticides, food additives, food coloring used during food processing, are all considered as potential risks to human health [48, 49].

Technological Risk: Related to negative impacts of technological advancements in food products, i.e., genetically modified foods. Some potential toxic or allergic effects may also be associated with the food [47].

12.7 Factors that Modulate Consumers' Perception of Risk

As we enter the new millennium, the connection between the media, the food sector, and the customer is at an all-time low. Food scares appear to be more common, and news accounts can be exaggerated and polarising at times. Consumer confidence in the food business has been eroded by high-profile issues such as the development of bovine spongiform encephalopathy in the United Kingdom and dioxin contamination of poultry products in Belgium. The recent discussion over genetically modified foods has highlighted the growing divide between the food industry, food safety specialists, and the general public. Environmental pressure organisations have exploited this schism, which has been fueled by the media.

12.8 Conclusion

It is clear that consumer views/perception on food safety and the risk associated with food are not unrelated to one another. The consumers' demographic and socioeconomic background, culture, personal preferences, and experience are all factors which influence their perception with regard to food risk. Different attitude, according to the assessments, may not always lead to behaviour patterns that improve the safety of the food consumed. Customers, it can be stated, require expert advice when it comes to food safety issues. Consumers recognised a risk associated with food fraud, according to several researches, indicating that they conducted a risk assessment. However, various parameters skew and influence this risk assessment, and further research is needed to properly understand how these parameters influence risk.

It is most important to learn about consumer attitude and behaviours. To put it another way, before offering information to customers, one must first understand their attitude and behaviour. Risky food handling and food-consumption patterns can be used to detect large populations, such as the elderly, allowing experts to organise instructional programmes to reduce risk perception and analyse the results [50]. In this list one more point is important, and that is the need for awareness about handling of food. Consumers should be urged to embrace various food safety procedures in order to promote safe food handling behaviours. Quality control should be communicated in the most straightforward manner possible to aid comprehension.

Reduced risk exposure is an effective strategy to combat food fraud since customers are unaware of the consequences of food fraud and are unable to do a scientific risk assessment. It is possible that it is a means to feel in charge when information or expertise is not readily available. However, most customers do not consider food fraud while purchasing food, even if they are aware that they may be victims of food fraud. It is common knowledge that food is nutritious. The majority of consumers think it a good buy. Regular food fraud episodes, on the other hand, may result in behavioural changes, indicating that customers want to avoid risk and have conducted a risk analysis. Consumers' attitude about food safety are significantly linked to their level of trust in not only the food sector but also government institutions charged with maintaining food safety. Unfortunately, public trust is frequently destroyed by government agencies' unwillingness and indifference to implement or maintain consumer protection policies, owing to legislative changes and budget constraints [51]. Bill Clinton, the former president of the United States, was one of the first to raise this subject, emphasising the importance of pursuing the regulation of microbial risks in meals on a regular basis [52]. So it is very important that government authorities provide credible information about the product so that public trust increases. Customers would benefit from food safety education at home, such as education on temperature management, proper food preparation techniques, and cross-contamination. Product safety and efficacy data can also impact whether a food product is recommended or eaten by physicians and other health care professionals [53]. Knowing the repercussions of risky activities might help motivate people to follow safety rules. This is an area in which the medical community, the food business, government authorities, and the media all have a stake.

Consumer trust is a result of regulatory acceptability, according to Poortinga and Pidgeon, and the effects of new legislation intended to combat food fraud are still largely unclear. However, a transformation in customer views will take several years because the European and American frameworks are not yet well established in the business [54]. Food risk communication that is more personalised and takes into account diverse clusters of customers rather than considering them as a homogeneous group has been proposed as a technique to effectively reach customers [55]. As a result, instead of presenting a single risk communication strategy for all nations in Europe, it was suggested that several forms of risk communication based on people's preciseness be given. The communication platform is not transparent. Despite the fact that social media has been recognised as an excellent instrument for establishing a crisis plan [56], new research reveals that young customers do not want to be reached through such platforms [57].

References

1. Anderson WA. The future relationship between the media, the food industry and the consumer. *British Medical Bulletin*. 2000 Jan 1;56(1):254-68.
2. Mahon D, Cowan C. Irish consumers' perception of food safety risk in minced beef. *British Food Journal*. 2004 Apr 1.
3. Gizaw Z. Public health risks related to food safety issues in the food market: a systematic literature review. *Environmental Health and Preventive Medicine*. 2019 Dec;24(1):1-21.
4. Fearne A, Martinez MG. Opportunities for the coregulation of food safety: insights from the United Kingdom. *Choices*. 2005 Jun 1;20(2):109-16.

5. Hawkes C. Uneven dietary development: linking the policies and processes of globalization with the nutrition transition, obesity and diet-related chronic diseases. *Globalization and Health*. 2006 Dec;2(1):1-8.
6. Uyttendaele M, Franz E, Schlüter O. Food safety, a global challenge. International *Journal of Environmental Research and Public Health*. 2016 Jan;13(1):67.
7. Kruse H. Food safety in an international perspective. *Journal Für Verbraucherschutz Und Lebensmittelsicherheit*. 2015 Jun;10(2):105-7.
8. Negri S. Food safety and global health: an international law perspective. *Global Health Governance*. 2009;3(1).
9. Spink J, Moyer DC. Defining the public health threat of food fraud. *Journal of Food Science*. 2011 Nov;76(9):R157-63.
10. Sutherland C, Sim C, Gleim S, Smyth SJ. Consumer insights on Canada's food safety and food risk assessment system. *Journal of Agriculture and Food Research*. 2020 Dec 1;2:100038.
11. Wilson AM, Withall E, Coveney J, Meyer SB, Henderson J, McCullum D, Webb T, Ward PR. A model for (re) building consumer trust in the food system. *Health Promotion International*. 2017 Dec 1;32(6):988-1000.
12. Yeung RM, Yee WM. Consumer perception of food safety related risk: a multiple regression approach. *Journal of International Food & Agribusiness Marketing*. 2005 Dec 29;17(2):195-212.
13. Li Z, Sha Y, Song X, Yang K, ZHao K, Jiang Z, Zhang Q. Impact of risk perception on customer purchase behavior: a meta-analysis. *Journal of Business & Industrial Marketing*. 2019 Sep 25.
14. Petty RE, Fabrigar LR, Wegener DT. Emotional factors in attitudes and persuasion.
15. Wansink B. Consumer reactions to food safety crises. *Advances in Food and Nutrition Research*. 2004 Jan 1;48(24):455-79.
16. Ashwell M. Consumer perception of food-related issues. *Nutrition Bulletin*. 1991 Jan;16(1):25-35.
17. Gupta V, Sajnani M. Risk and benefit perceptions related to wine consumption and how it influences consumers' attitude and behavioural intentions in India. *British Food Journal*. 2019 Oct 21.
18. Frewer LJ, Howard C, Hedderley D, Shepherd R. Consumer attitudes towards different food-processing technologies used in cheese production—The influence of consumer benefit. *Food Quality and Preference*. 1997 Jul 1;8(4):271-80.
19. De Steur H, Gellynck X, Storozhenko S, Liqun G, Lambert W, Van Der Straeten D, Viaene J. Willingness-to-accept and purchase genetically modified rice with high folate content in Shanxi Province, China. *Appetite*. 2010 Feb 1;54(1):118-25.
20. Frewer LJ, van der Lans IA, Fischer AR, Reinders MJ, Menozzi D, Zhang X, van den Berg I, Zimmermann KL. Public perceptions of agri-food

applications of genetic modification–a systematic review and meta-analysis. *Trends in Food Science & Technology*. 2013 Apr 1;30(2):142-52.
21. Chen MF, Li HL. The consumer's attitude toward genetically modified foods in Taiwan. *Food Quality and Preference*. 2007 Jun 1;18(4):662-74.
22. Mingyang Z, Carine U. Trust, Perceived Risks and Consumers Attitude Towards Genetically Modified (GM) Foods: Evidence from Rwanda. *International Journal of Business Management & Economic Research*. 2021 May 1;12(3).
23. Lucht JM. Public acceptance of plant biotechnology and GM crops. *Viruses*. 2015 Aug;7(8):4254-81.
24. Trevathan-Tackett SM, Treby S, Gleason FH, Macreadie PI, Loke S. Cryopreservation methods are effective for long-term storage of Labyrinthula cultures. *Diseases of Aquatic Organisms*. 2018 Aug 28;130(1):65-70.
25. Beulens AJ, Broens DF, Folstar P, Hofstede GJ. Food safety and transparency in food chains and networks Relationships and challenges. *Food Control*. 2005 Jul 1;16(6):481-6.
26. Trienekens JH, Wognum PM, Beulens AJ, van der Vorst JG. Transparency in complex dynamic food supply chains. *Advanced Engineering Informatics*. 2012 Jan 1;26(1):55-65.
27. Savov AV, Kouzmanov GB. Food quality and safety standards at a glance. *Biotechnology & Biotechnological Equipment*. 2009 Jan 1;23(4):1462-8.
28. Coveney J. Food and trust in Australia: building a picture. *Public Health Nutrition*. 2008 Mar;11(3):237-45.
29. Li S, Wang Y, Tacken GM, Liu Y, Sijtsema SJ. Consumer trust in the dairy value chain in China: The role of trustworthiness, the melamine scandal, and the media. *Journal of Dairy Science*. 2021 Aug 1;104(8):8554-67.
30. Henson S, Traill B. Consumer perceptions of food safety and their impact on food choice. *Food Safety—The Challenge Ahead*. 1993:39-55.
31. Wilcock A, Pun M, Khanona J, Aung M. Consumer attitudes, knowledge and behaviour: a review of food safety issues. *Trends in Food Science & Technology*. 2004 Feb 1;15(2):56-66.
32. Sockett PN. Food poisoning outbreaks associated with manufactured foods in England and Wales: 1980-1989. *CDR* (London, England: Review). 1991 Sep 1;1(10):R105-9.
33. Tent H. Research on food safety in the 21st century. *Food Control*. 1999.
34. Covello VT, Merkhofer MW. Introduction to risk assessment. In *Risk Assessment Methods* 1993 (pp. 1-34). Springer, Boston, MA.
35. Brewer, M. Susan, Guy K. Sprouls, and Craig Russo. Consumer attitudes toward food safety issues. 1994 *Journal of Food Safety* 14.1 63-76.
36. Laloo S, Rampersad FS, La Borde A, Maharaj K, Sookhai L, Teelucksingh JD, Reid S, McDougall L, Adesiyun AA. Bacteriological quality of raw oysters in Trinidad and the attitudes, knowledge and perceptions of the public about its consumption. *International Journal of Food Microbiology*. 2000 Mar 10;54(1-2):99-107.

37. Johnson AE, Donkin AJ, Morgan K, Lilley JM, Neale RJ, Page RM, Silburn R. Food safety knowledge and practice among elderly people living at home. *Journal of Epidemiology & Community Health*. 1998 Nov 1;52(11):745-8.
38. Moseley, B. E. (1999). The safety and social acceptance of novel foods. *International Journal of Food Microbiology*, 50(1-2), 25-31.
39. Robinson C. Genetically modified foods and consumer choice. *Trends in Food Science & Technology*. 1997 Mar 1;8(3):84-8.
40. Hoban TJ. Consumer acceptance of biotechnology in the United States and Japan. *Institute of Food Technologists*. 1999;53(5):50-3.
41. Olsen SO. Strength and conflicting valence in the measurement of food attitudes and preferences. *Food Quality and Preference*. 1999 Nov 1;10(6):483-94.
42. Mossel DA, Drake DM. Processing food for safety and reassuring the consumer. *Food Technology*. 1990;44(12):63-7.
43. Di Caracalla VD. Codex Alimentarius Commission: Procedural Manual. 1997.
44. FAO/WHO, Report of the twelfth session of the Codex Committee on general principles. Paris, 25 - 28 November. ALINORM 97/33. Codex Alimentarius Commission. FAO, Rome, 1996.
45. World Health Organization. Risk Management and Food Safety: Report of a Joint FAO/WHO Consultation, Rome, Italy, 27 to 31 January 1997. Food & Agriculture Org.; 1997.
46. Spink J, Embarek PB, Savelli CJ, Bradshaw A. Global perspectives on food fraud: results from a WHO survey of members of the International Food Safety Authorities Network (INFOSAN). *npj Science of Food*. 2019 Jul 17;3(1):1-5.
47. Yeung RM, Morris J. Consumer perception of food risk in chicken meat. *Nutrition & Food Science*. 2001 Dec 1.
48. Kher SV, De Jonge J, Wentholt MT, Deliza R, de Andrade JC, Cnossen HJ, Luijckx NB, Frewer LJ. Consumer perceptions of risks of chemical and microbiological contaminants associated with food chains: a cross-national study. *International Journal of Consumer Studies*. 2013 Jan;37(1):73-83.
49. Huang X, Hites RA, Foran JA, Hamilton C, Knuth BA, Schwager SJ, Carpenter DO. Consumption advisories for salmon based on risk of cancer and noncancer health effects. *Environmental Research*. 2006 Jun 1;101(2):263-74.
50. Altekruse SF, Yang S, Timbo BB, Angulo FJ. A multi-state survey of consumer food-handling and food-consumption practices. *American Journal of Preventive Medicine*. 1999 Apr 1;16(3):216-21.
51. Day M. Food safety may have to wait. *New Scientist*. 1997 May 17;154(2082):6-.
52. Marwick C. Putting money where the US mouth is: Initiative on food safety gets under way. *JAMA*. 1997 May 7;277(17):1340-2.
53. Inniss S, Bahlo E, Kardinaal A. Beyond implementation: the role of physicians, health care professionals and consumers in the development and use of food products. *European Journal of Clinical Nutrition*. 1999 Dec 1;53:S25-9.

54. Poortinga W, Pidgeon NF. Exploring the dimensionality of trust in risk regulation. *Risk Analysis: An International Journal.* 2003 Oct;23(5):961-72.
55. Redmond, E. C., & Griffith, C. J. Consumer perceptions of food safety risk, control and responsibility. *Appetite.* 2005 43(3), 309–313. https://doi.org/10.1016/j. appet.2004.05.003.
56. Cope S, Frewer LJ, Houghton J, Rowe G, Fischer AR, de Jonge J. Consumer perceptions of best practice in food risk communication and management: Implications for risk analysis policy. *Food Policy.* 2010 Aug 1;35(4):349-57.
57. Burke T, Young I, Papadopoulos A. Assessing food safety knowledge and preferred information sources among 19–29 year olds. *Food Control.* 2016 Nov 1;69:83-9.

13

Application of Nanomaterials in Detection of Food Contaminants: A Food Safety Perspective

Sunny Dhiman*, **Prachi Singh, Anu Kumar and Gunjan Mukherjee**

University Institute of Biotechnology, Chandigarh University, Gharuan, Mohali, Punjab, India

Abstract
The most significant aspect of global food safety is the effective detection of food contaminants. Quick and on-site detection of food contaminants is of immense significance from a public health perspective. With the advancements in nanotechnology, nanomaterials have extended their utilization as novel and exciting tools in developing nano-sensing technologies. A vast variety of detection systems have been devised for ensuring rapid, efficient, and accurate detection of a wide variety of food contaminants. The nanomaterials possess characteristics that make them ideal for developing portable nano-sensors, have improved sensitivity, and better response time in detecting even very low levels of infection. The present review describes various nanomaterials used in the biosensing of food contaminants and advancements in nanomaterials-based biosensing methods for ensuring global food safety. Future prospects of nano-biosensing have been discussed along with challenges that need to be addressed to make nano-biosensing a more effective technology for rapid and accurate detection of major food contaminants.

Keywords: Food safety, contaminants, nanomaterials, nano biosensing

13.1 Introduction

The ever-increasing global population has put significant pressure on existing society food supplies. This situation is being worsened by continuously

**Corresponding author*: sdsunnydhiman473@gmail.com

Tejpal Dhewa, Anil Kumar Puniya and Anil Panghal (eds.) Nutritional Science and Technology: Concept to Application, (295–316) © 2023 Scrivener Publishing LLC

expanding populations, particularly in emerging countries. The widespread prevalence and recurrence of foodborne infections in developing and industrialized countries indicate that food safety concerns must be resolved [1]. The news portrayal of foodborne epidemics and contamination has undoubtedly raised awareness, leading to more disclosure and heuristic detection of foodborne diseases [2]. Thus, ensuring the availability of safe food to all consumers across the globe has become one of the most important and challenging tasks for the food industry and food authorities [3]. Several causes contribute to food deterioration, rendering it unhealthy and unsafe to consume. These factors include the use of fertilizers to increase crop yield, and the use of pesticides and antibiotics in agricultural practices. The accidental adulteration of foods with undesirable additives, the unintentional adulteration of foods with poor quality and hazardous ingredients, cross-contamination by food allergens, and unsanitary customer handling of foods also lead to spoilage. These variables have aggravated foodborne illnesses. The food and beverage industry is a multi-trillion-dollar global enterprise [4]. Before reaching the consumer's plate, the food sector requires innovative technologies to manufacture fresh, safe, and flavourful food products. As a result, all main food companies and food safety organizations are constantly looking for the best approaches to solve multiple unparalleled food safety concerns [4].

13.2 Global Scenario from Food Safety Perspective

The current global situation in the 21st century from a food safety perspective is dire. Unhealthy and unhygienic food can cause various dreadful diseases, eventually leading to deaths [5]. According to an estimate, approximately 2 million deaths occur every year due to around 200 diseases and disorders like diarrhoea, cancer, etc., caused by the intake of contaminated foods [6]. The current food production environment is facing several unprecedented challenges from a global food safety perspective. These challenges include variations in modern-day production approaches, change in climate, shortage of water, rise in demand for foods with organic labels, the trend for consuming food with lesser chemicals and undesirable additives from the consumer's end and global dimension of the food chain, etc. [7]. The expanding globalization of food commerce has raised the likelihood of accidents involving food poisoning. Now and then, fatal occurrences of foodborne infections have been observed. Food safety ensures that food will not harm the individual when consumed.

It is gradually becoming a vital consideration for sustainability since it is closely related to increased productivity and living.

According to the World Health Organization (WHO), about 600 million instances of foodborne infections are recorded worldwide, with 420,000 deaths, 30 percent of which are children under the age of five. This equates roughly 33 million years of productive and healthy life lost worldwide as a result of eating unhealthy food [8].

13.3 Current Food Safety Strategies and Food Safety

Pathogen-contaminated foods have become a serious problem across the world, as food safety has a direct interrelationship with the economy and society [9]. Thus, the foremost challenge for global food safety organizations is to ensure zero-degree tolerance to food contaminants. The assurance of food quality and food safety is accomplished via/through preventative measures adopted "from farm to plate" [3]. The best strategy is to follow preventative measures to minimize the risks at the consumer end without compromising the quality and nutritional values of the food [3]. The World Health Organization endorsed food safety as: "from farm to plate (and everywhere in between) make food safe" on World Health Day, 2015. A vast array of physical and chemical approaches exists for minimizing the microbial hazards which are likely to contaminate the food. These approaches include physical methods such as heating, freezing, refrigerated storage, radiation (UV, gamma), drying, filtration. In addition, various chemical methods like application of ozone, hydrogen peroxide and chlorine-based compounds are also employed [10]. However, all of them present serious drawbacks like reduction in the quality and texture of food, higher energy input requirements, serious occupational and health implications, significant environmental footprint and lesser compatibility with the "green" production philosophy [11–14]. The food industry demands advanced technologies for the production, processing and packaging of fresh, safe and flavourful food products. The increasing demand for "greener" and chemical-free strategies from the consumer end has put entire food industry under pressure to look for novel, sustainable and relatively cheaper methods with much better efficiencies to combat the emanating hazards of food contaminants without compromising the nutritional value of food products [15, 16]. An entirely new generation of methods are on board for ensuring food safety and quality so as to make the food industry capable of dealing with the inadequacies and inefficiencies of the conventional physical and chemical methods employed for the purpose [3]. A wide variety of innovative technologies have transformed the food industry in recent decades, and one of the most significant of these technologies is the development and implementation of nanotechnology in food [16, 17].

13.4 Application of Nanotechnology in Detection of Food Contaminants

Nanomaterials exhibit high potential to be utilized in various biosensing approaches for the quick, sensitive, and specific detection of a wide range of food contaminants. Biosensing based upon utilization of nanomaterials holds immense potential in fabricating extremely sensitive and selective food contaminant detection approaches indispensable for ensuring food safety [18]. Nano-biosensors are more highly reliable and sensitive in biosensing procedures than traditional sensors, which have a variety of detecting limitations [19]. Various nanomaterials, including gold NPs, nanorods, nanosheets, magnetic NPs, carbon nanotubes, graphene, etc., have been utilized in designing nano biosensing detection systems (Figure 13.1).

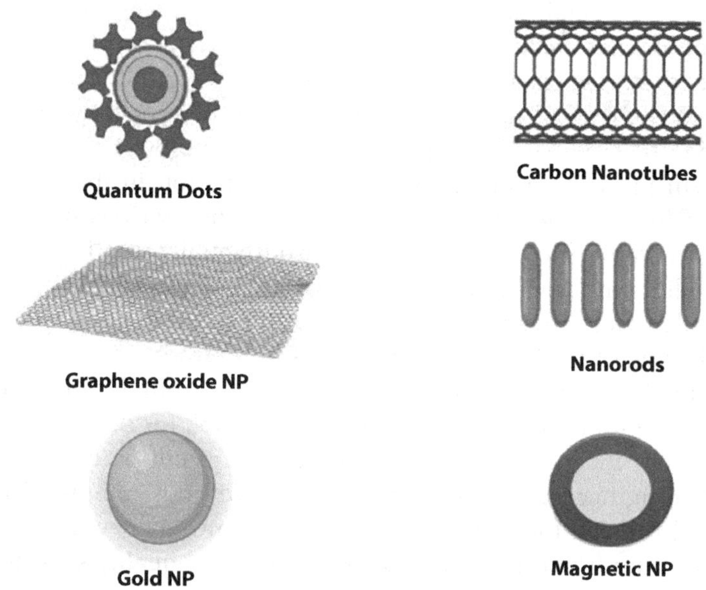

Figure 13.1 Nano materials used in designing nano biosensing detection systems.

Among these, graphene and gold NPs have been reported to have wider applicability.

Gold Nanoparticles

In the current scenario, gold nanoparticles are the material of choice for making nano-sensors as they can be synthesized with an ease. An active study has been done on the colorimetric properties of AuNP. AuNP's wavelength of absorption is based on the extent to which it is interspersed in the solution [20]. Maximum absorption wavelength is 530nm when AuNP's are finely interspersed in the solution; on the other hand, a blue shift is there when AuNP's are present in bulk. An operational example of a nano-sensor is melamine biosensor which is developed by using AuNP in which citric acid is used as surface stabilizer. There is a change in color from red to purple when surface stabilizer citric acid and melamine (present in solution) interact with each other [21]. The change in color from red to purple can be measured spectroscopically [22]. Change in color of the assay mixture is the indication of whether the particular protein is attaching or detaching from AuNP in the presence of target molecule. A simpler method was devised to detect *Bacillus subtilis* in which AuNP's were associated with lysozyme thus giving a purple color [23]. Lysozyme has great affinity for the peptidoglycan which is there in cell wall of gram-positive bacteria. There is a change in color from purple to red when *Bacillus subtilis* and lysozyme are there in solution because when cell wall of bacteria and lysozyme interact with each other there is separation of lysozyme from AuNP supramolecule [23]. Another example of a biosensor is when detection procedure is based on color change in AuNP supramolecule which is a complex of beta-galactosidase and AuNP. These supramolecules have great affinity for bacteria. Beta-galactosidase is released from AuNP based supramolecule and the separation of beta-galactosidase results in color transition from yellow to red in the presence of coloring substance, chlorophenol red beta-D galactopyranoside [24].

Gold NPs are oftenly utilized as noble metal in LSPR sensors since it can be efficiently functionalized with active biomolecules. Latest researches suggest that AuNPs-based quantitative platforms are smart and adaptable. In a recent research, two UiO-66/AuNP based substrates were synthesized. It was proven to detect a carcinogenic heterocyclic amine in barbeque meat [25].

Nanorods

Size ranges from 1-100nm in one dimension of nanoscale objects are called nanorods in nanotechnology. Metal or semiconducting materials are used to manufacture the nanorods by direct chemical synthesis. Nanorods can be used in displays, diagnosis, energy harvesting, light emitting devices, and biosensing because they have optical, electromagnetic, and photoelectric characteristics. The excellent properties of these nanorods make them suitable for biosensing which leads to more and more research in this area. Nowadays, surface-enhanced Raman scattering (SERS) biosensors are very popular because they help in the manufacturing of very efficient metal nanorods that increase Raman scattering signals [26, 27].

Anodized aluminium oxide (AAO) membranes, which are an example of metal rods, can be developed with the help of the seed mediated sequential growth method. There is relation between absorption spectra and aspect ratio and this is the basis of optical biosensors. The efficiency of the gold nanorods depends on the gap between nanorods, e.g., two-dimensional gold nanorods which have a gap of 0.8 nm are more efficient to apply for detecting chemical contaminants in food, e.g., melamine, BBP and DEHP which are present in very small amounts as low as 0.9fM by SERS. Secondly, the more the signal is scattered by the nanorod, the more is the detection power [27]. Harmful contamination can also be screened with the help of fluorescent metal nanorods, e.g., *Salmonella* contamination can be analysed fluorescently by using Au/Si hetero nanorods [28]. The functionalization of surface of hetero nanorod was accomplished with anti-*Salmonella* antibody and the functionalization of gold and silicon sides of the nanorod was done with fluorescent dye. Using this system, they documented that intensity of the fluorescence signal and *Salmonella* concentration are interlinked. The more the *Salmonella* concentration, the more will be the intensity of the fluorescence signal [28].

Nanorods are prospective materials for food safety. Chemicals like triethylamine and nitrite are widely used in industries in the synthesis of chemical drugs, pesticides, production of additives, emulsifiers, etc. These chemicals are as harmful when the body is exposed to them for longer durations. Triethylamine (TEA) is found to be mutagenic and possibly carcinogenic [29]. Needle-like tungsten oxide (WO_3) nanorods effectively detect triethylamine. This novel attempt could ease the traceability of triethylamine in marine food safety [30]. For nitrite detection, gold NRs have shown average sensitivity [31].

Quantum dots

Quantum dots (QD) are very minute semiconductor particles having the property of fluorescence and are made of two types of semiconducting material. Core of quantum dots are made of cadmium mixed with selenium or tellurium and coating is generally of zinc sulphide. Single cell DNA associated with QDs and magnetic particles are very much helpful for detection of pathogenic *E. coli* O157:H7 [32]. DNA probe holding the specific marker of *E. coli* O157:H7, i.e., eaeA gene were joined to QDs and magnetic microparticles with the help of sequence specific hybridization. The number of specific bacteria were measured with the help of QD fluorescence intensity after the separation of QD and magnetic microparticles [32]. In the same way, *Campylobacter* can be diagnosed in different foods by using aptamer-conjugated QD and magnetic microparticles with a detection limit of 10-250 CFU [33]. Recently, a prospective and versatile aldicarb sensor has been developed. This sensing platform is based on inner filter effect of AuNPs on fluorescence intensity of Carbon Quantum Dots [34].

Magnetic Nanoparticles

The advancements in developing magnetic nanoparticles with distinct sizes and functionalities has facilitated their proficient utilizations in a number of applications ranging from imaging and medicine to use in biosensors for pathogen [35–37]. (Fe_3O_4) nanoparticles have been reported to have the broader utilization. Removal of specific bacterial pathogens or cells from complex biological materials can be done with the help of these magnetic nanoparticles [38]. Specific food contamination from the food materials can be easily removed with the help of magnetic nanoparticles. Specific target bacteria can also be detected with the help of physical and chemical properties of the magnetic nanoparticles. Luo *et al.* devised a new approach for detecting *E. coli* O157:H7 by utilizing polyaniline-coated magnetic nanoparticles [39]. Target bacteria were mobilized onto a specific area of a lateral flow strip sensor. An exposure of the above system to acidic environment renders the polyaniline electrically conductive. A correlation among the concentration of bacteria under investigation and the electric current across the electrodes was utilized to detect *E. coli* O157:H7 at a concentration of 67 CFU/ml. Ravindranath *et al.* reported that bacterial contamination can also be detected with the help of magnetic nanoparticles infrared spectra [40].

Bisphenol A (BPA) can alter the functioning of the endocrine system. In the long term, its exposure can cause adverse health effects like cancer [41]. Recently, a Magnetic relaxation switch (MRS) was designed to detect BPA in packaging material and drinking water. This is based on self-assembling aptamer polystyrene and DNA functionalized magnetic NPs [42]. Magnetic NP's self-assembly property is not only beneficial in chemical detection but also in pathogen detection. Quantitative detection of lethal

H1N1 nucleoprotein can also be done by using MNPs in magnetic particle spectroscopy [43]. Typhoid causing *Salmonella typhimurium* was recently detected with the help of tryptamine-functionalized magnetic NP [44].

Nanomaterials based on carbon

Carbon nanotubes

Carbon has different forms, one of which is carbon nanotubes. A carbon nanotube is a small cylindrical structure and the length to diameter ratio is more than millions to one. They have some astonishing properties, e.g., high thermal conductivity, and electrical and chemical characteristics. So we can say that CNTs can be used in different types of functions, e.g., structural function as well as analytical function [45].

In order to make CNT-based biosensor, physical and chemical conjugation methods are used for attaching a probe to carbon-based nanotubes surfaces. When specific molecule binds to CNT, there is a remarkable change in the electrical characteristics of the CNT in the general environment. Any electrochemical change in CNT electrode can be easily analysed with the help of CNT sensing element. For example, an electrode combined with aptamer functionalized CNT can be used to analyze tetracycline in the milk [46]. Redox reactions of ferricyanide probe which was present on the shell of aptamer functionalized CNT declined drastically when aptamer underwent structural change in the existence of tetracycline. Variations in electrochemical reactions were investigated with the help of oxidation and reduction couples by differential pulse voltammetry. There was a direct relation between the amount of tetracycline and inhibition of electrochemical reaction. Maximum value of analysis of aptamer-functionalized CNT was 5nM for a specific cell in water and 50 nM for milk [46]. Sensitivity of the CNT-based sensors can be enhanced by administering enzymes which leads to the enhancement of signal intensity, e.g., sandwich ELISA utilizes Horseradish peroxidase (attached with the antibody) and results in production of color when it comes across a specific molecule. On the basis of this principle, a CNT-based sensor which is attached with HRP-tagged antibody was devised so as to analyse the presence of *Shigella flexneri* in food. The basic principle of this sensor is based upon the reduction in the sensitivity of the surface bound HRP when bound with specific bacteria. This sensor can analyse the amount 3.1×10^3 CFU/ml [47]. The outcomes obtained by means of this sensor were equivalent to that of traditional ELISA with an association of >95% regardless of the fact that it utilizes only one antibody in comparison with the two or more antibodies usually utilized for ELISA. One more example of such type of sensor is short-length CNTs which are combined with HRP-conjugated antibody for the detection of staphylococcal enterotoxin B (SEB) [48]. The use of HRP and an antibody considerably improved the sensor performance (detection limit 10 pg/ml). In spite of this, the sensitivity of the recent system for SEB in a food sample was equivalent to that of the ELISA method. The method is comparatively simpler and quick in comparison to ELISA and provides a potential substitute to traditional methodologies.

Inherent and native attraction of CNTs for definite impurity and dangerous particles can also be used so as to develop CNT-based electrochemical sensors in addition of using specific probes, such as antibody, aptamer, oligonucleotide, and specific receptors, which give high rate of sensitivity.

The constitutional and mechanical characteristics of CNTs are promising for biosensors. Highly sophisticated diagnostic instruments for the purpose of food safety can be developed by dynamically using characteristics such as, large surface area, high electrical conductivity and high sensitivity connected with the nanosize scale of CNTs [49].

Graphene

Graphene is also one of the different forms of carbon. Structure of graphene is two-dimensional honeycomb-like which has sp^2 hybridization [50]. Properties of graphene sheet includes 0.3 nm thick, almost translucent, highly conductive for electricity and the electrical characteristics are very much susceptible to the surroundings and to the analyte which is adsorbed on the surface [51, 52]. As a matter of fact, graphene nanoparticles have been most widely utilized for biosensing applications. Wan *et al.* utilized reduced graphene sheet nanofilm with chitosan film on its surface to develop a sulfate-reducing bacteria sensor [53]. In addition, the further functionalization of chitosan deposited surface of reduced graphene sheet nanofilm was accomplished using anti-SRB antibody. This detection system was utilized to detect bacteria under investigation using impedance spectroscopy with a detection limit upto 18 CFU/ml [53]. Graphene-based nanocomposites can also potentially serve in electrochemical detection of numerous pesticides [54]. The hybrid antibody biosensor enabled rapid visual screening and eradication of foodborne bacteria. To improve sensitivity, this device used conjugated application of AuNP layered with graphene oxide [55]. Apart from detecting pathogenic bacteria; graphene can also detect common oxidants like hydrogen peroxide. The combined effect of nanocomposite $MnFe_2O_4$/rGO (reduced grapheme oxide) improved H_2O_2 sensing [56].

13.5 Advancements in Nano-Biosensing of Food Contaminants

Nano-biosensing of pathogens

Various microorganisms like bacteria, viruses and protozoa, etc., cause one or the other types of food contaminations leading to ailments and severe illness of people who consume such contaminated foods. Conventional methods of foodborne pathogens detection by counting the colony forming units (CFU) are time consuming as well as quite expensive and laborious and take several days for culturing the microbes to get any measurable results [57]. Furthermore, it is not possible to culture each microbe under lab conditions. This necessitates the requirement and development of methodologies that do not rely on culturing of microbes. Nanobiosensing of pathogens based upon nanomaterials with high sensitivity might become an alternative to counting the colony forming units (CFU) using colony counters [58]. Three main strategies for pathogen detection through nano-sensing involve genomic DNA analysis, use of antibodies and use of aptamers (Figure 13.2).

The very first approach used for pathogen detection is analysis of genomic DNA which can be recognized utilizing its cDNA [59–63]. As a matter of fact, target DNA is present in very minute concentration in pathogens. So nanomaterials along with techniques such as polymerase chain reaction (PCR) could be specifically used for amplifying the target DNA for measurable results. In recent years, a nanowire-based electrical genomic DNA detection system has been devised utilizing rolling circle amplification for generating long ssDNA having plentiful repetitive sequences [60]. Genomic DNA analysis approach is promising; however, its usage is limited by several factors. Pathogens require one or the

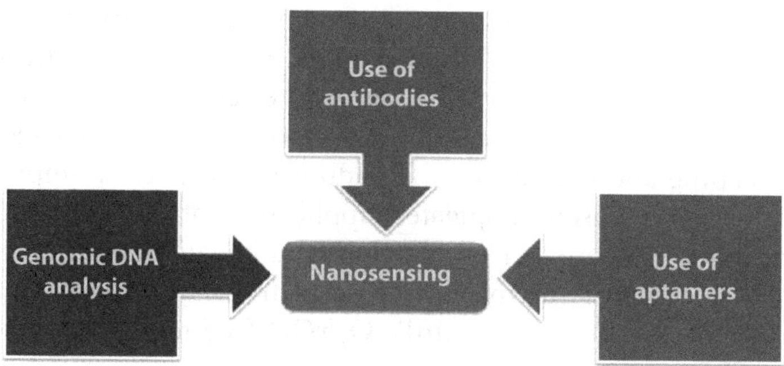

Figure 13.2 Strategies used for nano-sensing of pathogens.

other kinds of pre-treatments for extracting their genomic DNA which is a tedious and time-consuming process. The immunological approach for pathogen detection offers better convenience in comparison to genomic DNA analysis approach [64–67]. *E. coli* has been considered as one of the most infamous pathogens usually contaminating raw milk, fruits, vegetables and beef, etc. Jian *et al.* developed a rapid and highly sensitive method for detecting *E. coli* O157:H7 contamination in real time by utilizing Pt–Au bimetal NPs (possessing peroxidase activity) and immune-chromatographic assay [68]. Chen *et al.* developed an indirect immunofluorescence assay for quick detection of *E. coli* O157:H7 in beef by utilizing FITC (fluorescein isothiocyanate)-doped silica NPs [69]. De Oliveira *et al.* reported that polydiacetylene liposomes associated with antibody can be utilized for detecting *Salmonella* contamination [70]. In recent years novel, portable and automated detection systems based upon paper have been devised [71].

Aptamers, along with having a strong hold in molecular diagnostics and the biomedical field, are now also being considered in food safety. Aptamers are tiny, as RNA/DNA oligonucleotides which binds to target molecules. They are versatile; they have the ability to bind a wide range of substrates, from basic inorganic compounds to massive protein complexes. Aptamers can effectively detect pathogenic microbes in food. Recently, research proved that aptamer ID 12 can rapidly screen for pathogen *Vibrio parahaemolyticus* [72]. An innovative device, "aptasensor", is made from combining an aptamer with an ultrasensitive tranducer [73]. These apta-sensors can rapidly detect *E. coli* O157:H7, within 30 minutes [74]. On-the-spot detection of *Salmonella enterica* was also made possible with a lab-on-a-chip based nano-biosensor. This research carried out by Dehghani *et al.* involved a combination of Pt/Pd NP and an aptamer. Results can be procured fast, within the first three hours [75].

Introduction of an electrochemical antibody nanosensor has brough a significant development in nanobiosensing techniques. "Electrochemical antibody nanosensors are dependent on the conformational changes caused by biorecognition between the antibody and the antigen". When compared to other types of sensing devices, electrochemical sensing is a lot more favourable because direct data assessment is attained with the electrical signal output and no transduction is considered necessary. This significantly minimizes the amount of the signal accumulation and transduction interface, thus lowering the cost and enhancing the compact size of the sensor [76]. Quantum dot biosensors combined with antibody and an aptamer can be used to detect pathogens in food [77]. In essence, aptamers are nucleotide counterparts of antibodies, but producing aptamers is far simpler and less expensive than producing antibodies [78]. Utilization of antibodies as a component of a detection system has certain limitations such as tedious and time-consuming large-scale antibody production, costly purification procedures and inability to work in adverse conditions like high temperature [79]. These shortcomings can be overcome by utilizing an aptamer as a recognition element. A number of nanomaterials like magnetic NPs, nanorods, silver NPs, etc., have been used for developing apta sensing detection systems for quantifying the pathogens in different food samples [69, 80–84].

Nano-biosensing of toxins

Food poisoning can be caused by both pathogens and their secreted toxins. "Most commonly produced toxins by Gram-positive pathogens include emetic toxins and diarrheal enterotoxins secreted by *Bacillus cereus*, enterotoxins produced by *Clostridium perfringens*, neurotoxins produced by *Clostridium botulinum*, and an enterotoxins family produced by *Staphylococcus aureus* and other staphylococci" [85]. The majority of agricultural products and animal feeds get contaminated with the toxins generated by bacteria, filamentous fungi, etc., in the form of secondary metabolites. As an example, mycotoxins are responsible for contaminating approximately a quarter of globally available grains [86]. Toxins can create severe health issues like nervous disorder, hepatic disorder, nephritic disorders, cancer and can even cause deaths in their trace concentrations [87, 88]. Thus, the detection of toxins in the foods is of utmost importance from the public health perspective. Detection of toxins is advantageous over detecting whole cells since there is no need of culturing like whole cells.

A major fraction of biosensing techniques based on nanomaterials have been devised on immunoassays. An antibody-functionalized mesoporous carbon (MSC) NPs-based competitive-type biosensor has been devised for detecting AFB1 (aflatoxin B1) in peanuts [89]. AFB1 has been categorized as a first-class carcinogen by WHO. The second approach for detecting toxins consists of utilizing aptamers for measuring aptamer-toxin interaction. Ochratoxin A has been documented as the first mycotoxin which was detected by utilizing an aptamer-based assay in 2008. From then various methods have been devised which utilize nanomaterials and aptamers. A novel detection approach based upon fluorescent nitrogen-doped carbon dots on AuNPs has been devised for detecting AFB1 in peanut and corn [90]. Several metal-based nanomaterials, including silver nanoclusters, AuNPs doped Fe_3O_4 NPs, nanoceria tagged GOx, iridium oxide NPs [93] and CdTe QDs-GOx have also been utilized for assaying toxins

[81, 91–94]. Table 13.1 highlights the utilization of various nanomaterials in the detection of food contaminants.

Fatal saxitoxin, which is produced by algal blooms and shellfish, can also be detected with AuNP-aptamer biosensor [95]. Okadaic acid (OA) is mainly present in seafood. A novel napthalimide-Au-nanocomposite was newly designed capable of OA detection [96].

Table 13.1 Utilization of different nanomaterials in nano-sensing of food pathogens and toxins.

Contaminant	Nanomaterial utilized	Recognition molecule	Limit of detection	Reference
S. typhimurium	Upconversion NPs	Aptamer	15 CFU/mL	[97]
V. parahemolyticus	Upconversion NPs	Aptamer	10 CFU/mL	[97]
S. aureus	Upconversion NPs	Aptamer	25 CFU/mL	[97]
C. sakazakii	Fe_3O_4, liposomes	Antibody	10^3 CFU/mL	[67]
E. coli BL21	Ag-AuNRs	β-galactosidase	10^4 CFU/mL	[98]
Mycoplasma suis	AuNPs	Antibody	100 ng/mL	[64]
E. coli	AuNPs, Fe3O4	cDNA	1.8 aM	[61]
E. coli O157:H7	GOx, Au@ SiO_2	cDNA	0.01 nM	[63]
Shiga-like toxin 1	Al_2O_3-Fe_3O_4	Antibody	44 pM	[99]
Ochratoxin A	Nanoceria, GOx	Aptamer	0.1 nM	[92]
Ochratoxin A	Au doped Fe_3O_4	Aptamer	30 pg/mL	[91]
Aflatoxin B1	N-doped C dots, AuNPs	Aptamer	16 pM	[90]
Aflatoxin B1	AuNPs, SiO_2	Antibody	5 ppt	[100]

13.6 Potential Risks and Future Prospects of Nanomaterials in Food Safety

Nanomaterials have made noteworthy advancements. Ranging from pathogen detection to sensing lethal pesticides, they can be considered crucial. These days, involvement of nanoparticle-based technologies in food safety is inevitable. However, it is also important to understand its potential risks. Smart nanobiosensors still lack quantitative methodology to interpret complex data. This could be a big hurdle in accomplishing research requiring statistics. Electrochemical nanosensors are heat and pH hypersensitive, as well as intrusive. They likewise necessitate an on-site power supply. Thus, the main concerns stem from a lack of understanding concerning aspects like toxicology, biomagnification, and exposure risk related with the usage of nano-biosensors. Furthermore, the current uncertain situation is the result of insufficient budget for risk evaluations in research initiatives. The erratic behavior of nano-particulates in biological contexts, as well as the complexities of obtaining on-site data of variables, like size, shape, and surface properties, makes usage of nanoparticles problematic. Irrespective of all the discrepancies, technology can always find its way. Nanotech-based food safety methods can flourish well in future if risk assessments are analyzed properly. Expertise must be available for the evaluation of more advanced nano biosensor models to assure efficacy and safety. The nanoparticles utilized in the development of nano biosensors must be carefully evaluated and their results in biological settings, and their potential hazards must be investigated. By enhancing screening procedures, nanotechnology has the opportunity to make a positive influence on energy, the economy, and ecology. We anticipate that incorporating nanotech will be revolutionary in the realm of food safety by focusing on R&D activities, with coordinated efforts by governments and researchers in developing next-generation nano-products.

13.7 Conclusions

In conclusion, we report that AuNPs and carbon-based nanomaterials, especially graphene, have broader use and applicability for nanobiosensing of food contaminants. Several types of nano-biosensing methods have been devised based upon one or other of the types of NPs. However, from the sensitivity perspective nano-biosensing based upon fluorescent nanomaterials may be considered more appropriate for detecting food contaminants. A thorough literature survey clearly reveals the utilization of nano materials for designing a significant number of nano-biosensing detection systems for effective, unerring and real-time detection of food contaminants in various food samples. In spite of wider utilization and applicability of nanomaterials in nano-biosensing, this technology faces several challenges. These challenges include developing the nano-biosensing detection systems that should have ease in use, be able to perform in the long run, be economically cheaper, have high accuracy, better sensitivity, increased limits of detection and ensure rapid detection of food contaminants in real time. In addition, an ideal nano-biosensing detection system must be accessible to common people.

References

1. Zhao X., Lin, C.W., Wang, J., Oh, D.H., Advances in Rapid Detection Methods for Foodborne Pathogens. *J. Microbiol. and Biotechnol.* 24, 3, 297-312, 2014
2. Nyachuba, D.G., Foodborne illness: is it on the rise? *Nutr. Reviews. 68*, 5, 257-269, 2010.
3. Eleftheriadou, M., Pyrgiotakis, G., Demokritou, P., Nanotechnology to the rescue: using nano-enabled approaches in microbiological food safety and quality. *Current Opinion Biotechnol.* 44, 87–9, 2017.
4 Cushen, M., Kerry, J., Morris, M., Cruz-Romero, M., & Cummins, E., Nanotechnologies in the food industry – Recent developments, risks and regulation. *Trends in Food Science & Technology*, 24, 1, 30–46, 2012.
5. Sharma, R., Ragavan, K.V., Thakur, M.S., Raghavarao, K., Recent advances in nanoparticle based aptasensors for food contaminants. *Biosens. Bioelectron.* 74, 612–627, 2015.
6. Yasmin, J., Ahmed, M.R., Cho, B.K., Biosensors and their Applications in Food Safety: A Review. *J. of Biosystems Eng.* 41, 3, 240-254, 2016.
7. Uyttendaele, M., Franz, E., & Schlüter, O., Food Safety, a Global Challenge. *Int. J Env. Res. Pub. Health.* 13(1), 67, 2015.
8. World Health Organization (WHO). *WHO estimates of the global burden of foodborne diseases.* 2015 (https://www.who.int/news-room/fact-sheets/detail/food-safety).
9. Dhiman, S., Mukherjee, G. (2021) Recent Trends and Advancements in Biosensor Research for Food Safety. In: Prasad R., Kumar V., Singh J., Upadhyaya C.P. (eds.) *Recent Developments in Microbial Technologies. Environmental and Microbial Biotechnology*. Springer, Singapore. https://doi.org/10.1007/978-981-15-4439-2_5
10. Parish, M.E, Beuchat, L.R., Suslow, T.V., Harris, L.J., Garrett, E.H., Farber, J.N., Busta, F.F., Methods to Reduce/Eliminate Pathogens from Fresh and Fresh-Cut Produce. *Compre. Rev. Food Sci. Food Safety*, 2, 161-173, 2003.

11. Ruder, A.M., Potential Health Effects of Occupational Chlorinated Solvent Exposure. *Annals of the New York Academy of Sciences.* 1076, 1, 207–227, 2006.
12. Kou, L., Luo, Y., Wu, D., Liu, X., Effects of mild heat treatment on microbial growth and product quality of packaged fresh-cuttable grapes. *J Food Sci.*, 72, 8, S567-S573, 2007.
13. Karaca, H., Velioglu, Y.S., Ozone Applications in Fruit and Vegetable Processing. *Food Rev. Int.* 23, 1, 91–106, 2007.
14. Olaimat, A.N., Holley, R.A., Factors influencing the microbial safety of fresh produce: A review. *Food Microbiology*, 32, 1, 1–19, 2012.
15. Newell, D.G., Koopmans, M., Verhoef, L., Duizer, E., Aidara-Kane, A., Sprong, H., Opsteegh, M., Langelaar, M., Threfall, J., Scheutz, F *et al.*, Food-borne diseases—The challenges of 20 years ago still persist while new ones continue to emerge. *Int. J Food Microbio*l. 139, S3-S15, 2010.
16. Singh, T., Shukla, S., Kumar, P., Wahla, V., Bajpai, V.K. Rather, I.A., Application of Nanotechnology in Food Science: Perception and Overview. *Front. Microbiol.* 8,1501, 2017.
17. Bajpai, V. K., Kamle, M., Shukla, S., Mahato, D. K., Chandra, P., Hwang, S. K., … Han, Y.-K., Prospects of using nanotechnology for food preservation, safety, and security. *J Food and Drug Anal.* 1-14, 2018.
18. Nile, S.H., Baskar, V., Selvaraj, D. *et al.*, Nanotechnologies in Food Science: Applications, Recent Trends, and Future Perspectives. *Nano-Micro Lett.* 12, 45, 2020.
19. Mathivanan, S., Perspectives of Nano-Materials and Nano Biosensors in Food Safety and Agriculture, in *Novel Nanomaterials*, ed. Karthikeyan Krishnamoorthy, IntechOpen, 2021.
20. Elghanian, R., Storhoff, J.J., Mucic, R.C., Letsinger, R.L., Mirkin, C.A., Selective colorimetric detection of polynucleotides based on the distance-dependent optical properties of gold nanoparticles. *Science.* 277, 1078-1081, 1997.
21. Kumar, N., Seth, R., Kumar, H., Colorimetric detection of melamine in milk by citrate-stabilized gold nanoparticles. *Anal. Biochem.* 456, 43-49, 2013.
22. Li, L., Li, B., Cheng, D., Mao, L., Visual detection of melamine in raw milk using gold nanoparticles as colorimetric probe. *Food Chem.* 122, 895-900, 2010.
23. Qiu, S., Lin, Z., Zhou, Y., Wang, D., Yuan, L., Wei, Y., *et al.*, Highly selective colorimetric bacteria sensing based on protein-capped nanoparticles. *Analyst.* 140, 1149-1154, 2015.
24. Miranda, O.R., Li, X., Garcia-Gonzalez, L., Zhu, Z-J., Yan, B., Bunz, U.H.F., Rotello V.M., Colorimetric bacteria sensing using a supramolecular enzyme-nanoparticle biosensor. *J. Am. Chem. Soc.* 133, 9650-9653, 2011.
25. Fu, J., Lai, H., Zhang, Z., Li, G., UiO-66 metal-organic frameworks/gold nanoparticles based substrates for SERS analysis of food samples. *Analytica Chimica Acta*, 1161, 338464, 2021.

26. Leverette, C.L., Villa-Aleman, E., Jokela, S., Zhang, Z., Liu, Y., Zhao, Y., Smith, S.A., Trace detection and differentiation of uranyl (VI) ion cast films utilizing aligned Ag nanorod SERS substrates. *Vibr. Spectrosc.* 50, 143-151, 2009.
27. Shanmukh, S., Jones, L., Driskell, J., Zhao, Y., Dluhy, R., Tripp, R.A., Rapid and sensitive detection of respiratory virus molecular signatures using a silver nanorod array SERS substrate. *Nano Lett.* 6, 2630-2636, 2006.
28. Fu, J., Park B., Siragusa, G., Jones, L., Tripp, R., Zhao, Y., Cho, Y.-J., An Au/Si hetero-nanorod-based biosensor for Salmonella detection. *Nanotechnol.* 19, 155502, 2008.
29. Lin, Y., Huaqiao, J., Zhonghong, F., A Rare Case Report: Triethylamine Poisoning and The Pharmaceutical Care. 1-6, 2020.
30. Hu, Q., He, J., Chang, J., Gao, J., Huang, J., & Feng, L., Needle-Shaped WO3 Nanorods for Triethylamine Gas Sensing. *ACS Applied Nano Materials.* 3, 9, 9046–9054, 2020.
31. He, Z., Zhang, L.,Peng, G., Wang, G., Liang, X., Chemical Redox-Modulated Etching of Plasmonic Nanoparticles for Nitrite Detection: Comparison Among Gold Nanosphere, Nanorod, and Nanotriangle. *J Anal and Testing.* 1-10, 2021.
32. Kim, G.-Y., Son, A., Development and characterization of a magnetic bead-quantum dot nanoparticles-based assay capable of *Escherichia coli* O157: H7 quantification. *Anal. Chim. Acta.* 677, 90-96, 2010.
33. Bruno, J.G., Phillips, T., Carrillo, M.P., Crowell, R., Plastic adherent DNA aptamer-magnetic bead and quantum dot sandwich assay for Campylobacter detection. *J. Fluoresc.* 19: 427-435, 2009.
34. Sajwan, R.K., Lakshmi, G.B.V.S., Solanki P.R., Fluorescence tuning behaviour of carbon quantum dots with gold nanoparticles via novel intercalation effect of aldicarb. *Food Chem.* 340, 127835, 2021.
35. Frey, N.A., Peng, S., Cheng, K., Sun, S., Magnetic nanoparticles: synthesis, functionalization, and applications in bioimaging and magnetic energy storage. *Chem. Soc. Rev.* 38, 2532-2542, 2009.
36. Mornet, S., Vasseur, S., Grasset, F., Duguet, E., Magnetic nanoparticle design for medical diagnosis and therapy. *J. Mater. Chem.* 14, 2161-2175, 2004.
37. Varshney, M., Li, Y., Interdigitated array microelectrode-based impedance biosensor coupled with magnetic nanoparticle antibody conjugates for detection of *Escherichia coli* O157:H7 in food samples. *Biosens. Bioelectron.* 22, 2408-2414, 2007.
38. Lee, I.S., Lee, N., Park, J., Kim, B.H., Yi, Y.-W., Kim, T., *et al.* Ni/NiO core/shell nanoparticles for selective binding and magnetic separation of histidine-tagged proteins. *J. Am. Chem. Soc.* 128, 10658-10659, 2006.
39. Luo, Y., Nartker S., Wiederoder, M., Miller, H., Hochhalter D., Drzal, L.T., Alocilja, E.C., Novel biosensor based on electrospun nanofiber and magnetic nanoparticles for the detection of *E. coli* O157:H7. *IEEE Trans. Nanotechnol.* 11, 676-681, 2012.

40. Ravindranath, S.P., Mauer, L.J., Deb-Roy, C., Irudayaraj, J., Biofunctionalized magnetic nanoparticle integrated mid-infrared pathogen sensor for food matrixes. *Anal. Chem.* 81, 2840-2846, 2009.
41. Ma, Y., Liu, H., Wu, J., Yuan, L., Wang, Y., Du, X., Wang, R., Marwa, P.W., Petlulu, P., Chen, X., Zhang, H., The adverse health effects of bisphenol A and related toxicity mechanisms. *Environ Res.* 176, 108575, 2019.
42. Huang, L., Liu, S., Ren, S., Zhang, M., Wang, T., Wang, X., Gao, Z., Magnetic Relaxation Switch Biosensors Based on Self-Assembly of Polystyrene Microspheres and Magnetic Nanoparticles for Detection of Bisphenol A. *ACS Applied Nano Materials.* 4, 6, 5963-5971, 2021.
43. Wu, K., Liu, J., Saha, R., Su, D., Krishna, V. D., Cheeran, M. C.-J., & Wang, J.-P. (2020). Magnetic Particle Spectroscopy (MPS) for Detection of Influenza A Virus Subtype H1N1. *ACS Appl. Mater. Interfaces.* 12, 13686-13697, 2020.
44. Lee, S.-Y., Chen, F., & Lee, T. Y., Tryptamine-functionalized magnetic nanoparticles for highly sensitive detection of Salmonella typhimurium. *The Analyst.* 146, 8, 2559-2566, 2021.
45. Spitalsky, Z., Tasis, D., Papagelis, K., Galiotis, C., Carbon nanotube-polymer composites: chemistry, processing, mechanical and electrical properties. *Progress Polym. Sci.* 35, 357-401, 2010.
46. Zhou, L., Li, D-J., Gai, L., Wang, J.-P., Li, Y.-B., Electrochemical aptasensor for the detection of tetracycline with multi-walled carbon nanotubes amplification. *Sens. Actuators B Chem.* 162, 201-208, 2012.
47. Zhao, G., Zhan, X., Dou, W., A disposable immunosensor for Shigella flexneri based on multiwalled carbon nanotube/sodium alginate composite electrode. *Anal. Biochem.* 408, 53-58, 2011.
48. Tang, D., Tang, J., Su B., Chen, G., Ultrasensitive electrochemical immunoassay of staphylococcal enterotoxin B in food using enzyme-nanosilica-doped carbon nanotubes for signal amplification. *J. Agric. Food Chem.* 58, 10824-10830, 2010.
49. Lim, M.C., Kim, Y.R., Analytical Applications of Nanomaterials in Monitoring Biological and Chemical Contaminants in Food. *J. Microbiol. Biotechnol.* 26, 9, 1505–151, 2016.
50. Novoselov, K. S., Geim, A. K., Morozov, S. V., Jiang, D., Katsnelson, M. I., Grigorieva, I. V., … Firsov, A. A., Two-dimensional gas of massless Dirac fermions in graphene. *Nature*, 438, 7065, 197-200, 2005.
51. Schedin, F., Geim, A.K., Morozov, S.V., Hill, E.W., Blake, P., Katsnelson, M.I., Novoselov, K.S., Detection of individual gas molecules adsorbed on graphene. *Nat. Mater.* 6: 652-655, 2007.
52. Ghosh, S., Calizo, I., Teweldebrhan, D., Pokatilov, E. P., Nika, D. L., Balandin, A. A., … Lau, C. N., Extremely high thermal conductivity of graphene: Prospects for thermal management applications in nanoelectronic circuits. *Applied Physics Letters.* 92, 15, 151911, 2008.
53. Wan, Y., Lin, Z., Zhang, D., Wang, Y., Hou, B., Impedimetric immunosensor doped with reduced graphene sheets fabricated by controllable

electrodeposition for the non-labelled detection of bacteria. Biosens. *Bioelectron.* 26, 1959-1964, 2011.

54. Tanwar, S., Mathur, D., Graphene-based nanocomposites as sensing elements for the electrochemical detection of pesticides: a review. *J Solid State Electrochem.* 25, 2145-2159, 2021.
55. Kaushal, S., Pinnaka, A. K., Soni, S., & Singhal, N. K., (2021). Antibody assisted graphene oxide coated gold nanoparticles for rapid bacterial detection and near infrared light enhanced antibacterial activity. *Sensors and Actuators B: Chemical.* 329, 129141, 2021.
56. Zhao, X., Li, Z., Chen, C., Xie, L., Zhu, Z., Zhao, H., Lan, M., $MnFe_2O_4$ nanoparticles-decorated graphene nanosheets used as an efficient peroxidase minic enable the electrochemical detection of hydrogen peroxide with a low detection limit. *Microchem. Journal.* 166, 106240, 2021.
57. Yoo, K.H., Lee, S.H., Kim, H.J., Sung, K.W., Jung, H.L., Cho, E.J., Park, H.K., Kim, H.A., Koo, H.H., The impact of post-thaw colony-forming units-granulocyte/macrophage on engraftment following unrelated cord blood transplantation in pediatric recipients. *Bone Marrow Transplant.* 39, 515-521, 2007.
58. Verma M.S., Rogowski J.L., Jones L., Gu F.X., Colorimetric biosensing of pathogens using gold nanoparticles. *Biotechnol. Adv.* 33, 666-680, 2015.
59. Lee, S.Y., Lee, J., Lee, H.S., Chang, J.H., Rapid pathogen detection with bacterial-assembled magnetic mesoporous silica. *Biosens. Bioelectron.* 53, 123-128, 2014.
60. Russell, C., Welch, K., Jarvius, J., Cai, Y.X., Brucas, R., Nikolajeff, F., Svedlindh, P., Nilsson, M., Gold nanowire based electrical DNA detection using rolling circle amplification. *ACS Nano* 8, 1147-1153, 2014.
61. Khunrattanaporn, N., Rijiravanich, P., Somasundrum, M., Surareungchai, W., Highlysensitiveelectrochemical detection of genomic DNA based on stem loop probes structured for magnetic collection and measurement via metalised hollow polyelectrolyte shells. *Biosens. Bioelectron.* 73, 181-187, 2015.
62. Quintela, I.A., de los Reyes, B.G., Lin, C.S., Wu, V.C.H., Simultaneous direct detection of Shiga-toxin producing Escherichia coli (STEC) strains by optical biosensing with oligonucleotide-functionalized gold nanoparticles. *Nanoscale.* 7, 2417-2426, 2015.
63. Li, Y., Deng, J., Fang, L., Yu, K., Huang, H., Jiang, L., Liang, W., Zheng, J., A novel electrochemical DNA biosensor based on HRP-mimicking hemin/G-quadruplex wrapped Gox nano composites as tag for detection of *Escherichia coli* O157:H7. *Biosens. Bioelectron.* 63, 1-6, 2015.
64. Meng, K., Sun, W., Zhao, P., Zhang, L., Cai, D., Cheng, Z., Guo, H., Liu, J., Yang, D., Wang, S., Chai, T., Development of colloidal gold-based immunochromatographic assay for rapid detection of Mycoplasma suis in porcine plasma. *Biosens. Bioelectron.* 55, 396-399, 2014.

65. Ellchi, S., Leon, F., Vossier, L., Marchandin, H., Errachid, A., Coste, J., Jaffrezic Renault, N., Fournier-Wirth, C., (2014) Microconductometric immunosensor for label-free and sensitive detection of gram-negative bacteria. *Biosens. Bioelectron.* 54, 378-384, 2014.
66. Zhang, L., Huang, Y.J., Wang, J.Y., Rong, Y., Lai, W.H., Zhang, J.W., Chen, T., Hierarchical flowerlike gold nanoparticles labeled immunochromatography test strip for highly sensitive detection of *Escherichia coli* O157:H7. *Langmuir.* 31, 5537-5544, 2015.
67. Shukla, S., Lee, G., Song, X., Park, S., Kim, M., Immunoliposome-based immune magnetic concentration and separation assay for rapid detection of *Cronobacter sakazakii. Biosens. Bioelectron.* 77, 986-994, 2016.
68. Jiang, T., Song, Y., Wei, T.X., Li, H., Du, D., Zhu, M.J., Lin, Y.H., Sensitive detection of *Escherichia coli* O157:H7 using Pt-Au bimetal nanoparticles with peroxidase-like amplification. *Biosens. Bioelectron.* 77, 687-694, 2016.
69. Chen, Z.Z., Cai, L., Chen, M.Y., Lin, Y., Pang, D.W. Tang, H.W., Indirect immunofluorescence detection of *E. coli* O157:H7 with fluorescent silica nanoparticles. *Biosens. Bioelectron.* 66, 95-102, 2015.
70. De Oliveira, T.V., Soares, N.D.F., Coimbra, J.S.D., de Andrade, N.J., Moura, L.G., Medeiros, E.A.A., de Medeiros, H.S., Stability and sensitivity of polydiacetylene vesicles to detect *Salmonella. Sens. Actuators B Chem.* 221, 653-658, 2015.
71. Park, J., Shin, J.H., Park, J.-K., Pressed paper-based bipstick for detection of foodborne pathogens with multistep reactions. *Anal. Chem.* 88, 3781-3788, 2016.
72. Wang, L., Lyu, S., Gu, G., & Bolten, S., Selection of aptamers targeted to foodborne pathogenic bacteria *Vibrio parahaemolyticus. Food Sci. Nutrition.*1-8, 2020.
73. Rhouati, A., Marty, J.-L., Vasilescu, A., Metal Nanomaterial-Assisted Aptasensors for Emerging Pollutants Detection. *Nanotechnology and Biosensors.* 193-231, 2018.
74. Zhang, T., Tao, Q., Bian, X.-J., Chen, Q., Yan, J., Rapid Visualized Detection of *Escherichia Coli* O157:H7 by DNA Hydrogel Based on Rolling Circle Amplification. *Chinese J Anal. Chem.* 49, 3, 377-386, 2021.
75. Dehghani, Z., Nguyen, T., Golabi, M., Hosseini, M., Rezayan, A. H., Mohammadnejad, J., ... Vinayaka, A. C., Magnetic beads modified with Pt/Pd nanoparticle and aptamer as a catalytic nano-bioprobe in combination with loop mediated isothermal amplification for the on-site detection of *Salmonella Typhimurium* in food and fecal samples. *Food Control.* 107664, 2020.
76. Swierczewska, M., Liu, G., Lee, S., & Chen, X., High-sensitivity nanosensors for biomarker detection. *Chem. Soc. Rev.*, 41, 7, 2641-2655, 2012.
77. Sun, F., Zhang, J., Yang, Q., Wu, W., Quantum dot biosensor combined with antibody and aptamer for tracing foodborne pathogens, *Food Quality and Safety*, 5, 2021. https://doi.org/10.1093/fqsafe/fyab019

78. Conrad, R. C., Giver, L., Tian, Y., & Ellington, A. D., In vitro selection of nucleic acid aptamers that bind proteins. *Methods in Enzymology*. 336-367,1996.
79. Jayasena, S.D., Aptamers: An emerging class of molecules that rival antibodies in diagnostics. *Clin. Chem.* 45, 1628-1650, 1999.
80. Chen, J., Wu, X.M., Huang, Y.W., Zhao, Y.P., Detection of *E. coli* using SERS active filters with silver nanorod array. *Sens. Actuators B Chem.* 191, 485-490, 2014a.
81. Chen, J.H., Zhang, X., Cai, S.X., Wu, D.Z., Chen, M., Wang, S.H., Zhang J., A fluorescent aptasensor based on DNA-scaffolded silver-nanocluster for ochratoxin A detection. *Biosens. Bioelectron.* 57, 226-231, 2014b
82. Wang, Y., Sun, Q.Q., Zhu, L.L., Zhang, J.Y., Wang, F.Y., Lu, L.L., Yu, H.J., Xu, Z.A., Zhang, W., Triplex molecular beacons for sensitive recognition of melamine based on abasic-site-containing DNA and fluorescent silver nanoclusters. *Chem. Commun.* 51, 7958-7961, 2015.
83. Weng, C.-I., Chang, H.-T., Lin, C.-H., Shen, Y.-W., Unnikrishnan, B., Li, Y.-J., & Huang, C.-C. One-step synthesis of biofunctional carbon quantum dots for bacterial labeling. *Biosensors Bioelectronics.* 68, 1-6, 2015.
84. Tian, B., Bejhed, R.S., Svedlindh, P., Stromberg, M., Blu-ray optomagnetic measurement based competitive immunoassay for *Salmonella* detection. *Biosens. Bioelectron.* 77, 32-39, 2016.
85. Girigoswami, A., Ghosh, M.M., Pallavi, P., Ramesh, S., Girigoswami, K., Nanotechnology in *Detection of Food Toxins–Focus on the Dairy Products. Proceedings*, Abstract. 3, 1, 2021.
86. Hussein, H.S., Brasel, J.M., Toxicity, metabolism, and impact of mycotoxins on human and animals. *Toxicology*. 167, 101-134, 2001.
87. Turner, N.W., Subrahmanyam, S., Piletsky, S.A., Analytical methods for determination of mycotoxins: A review. *Anal. Chim. Acta.* 632, 168-180, 2009.
88. Das A.P., Kumar P.S., Swain S. (2014) Recent advances in biosensor-based endotoxin detection. Biosens. Bioelectron. 51, 62-75, 2014.
89. Lin, Y., Zhou, Q., Lin, Y., Tang, D., Chen, G., Tang, D., Simple and sensitive detection of aflatoxin B1 within five minute using a non-conventional competitive immunosensing mode. *Biosens. Bioelectron.* 74, 680-686, 2015.
90. Wang, B., Chen, Y., Wu, Y., Weng, B., Liu, Y., Lu, Z., Li, C.M., Yu, C. Aptamer induced assembly of fluorescent nitrogen-doped carbon dots on gold nanoparticles for sensitive detection of AFB 1. *Biosens. Bioelectron.* 78, 23-30, 2016a.
91. Wang, C.Q., Qian, J., Wang, K., Yang, X.W., Liu, Q., Hao, N., Wang, C.K., Dong, X.Y., Huang, X.Y., Colorimetric aptasensing of ochratoxin A using Au@Fe3O4 nanoparticles as signal indicator and magnetic separator. *Biosens. Bioelectron.* 77, 1183-1191, 2016b.
92. Bulbul, G., Hayat, A., Andreescu, S., A generic amplification strategy for electrochemical aptasensors using a non-enzymatic nanoceria tag. *Nanoscale.* 7, 13230-13238, 2015.

93. Rivas, L., Mayorga-Martinez, C.C., Quesada-Gonzalez, D., Zamora-Galvez, A., de la Escosura-Muniz, A., Merkoci, A., Label-free impedimetric aptasensor for ochratoxin-A detection using iridium oxide nanoparticles. *Anal. Chem.* 87, 5167-5172, 2015.
94. Lu, Z.S., Chen, X.J., Wang, Y., Zheng, X.T., Li, C.M., Aptamer based fluorescence recovery assay for aflatoxin B1 using a quencher system composed of quantum dots and graphene oxide. *Microchim. Acta.* 182, 571-578, 2015.
95. Qiang, L., Zhang, Y., Guo, X., Gao, Y., Han, Y., Sun, J., Han, L., A rapid and ultrasensitive colorimetric biosensor based on aptamer functionalized Au nanoparticles for detection of saxitoxin. *RSC Advances.* 10, 26, 15293-15298, 2020.
96. Verma, M., Chaudhary, M., Singh, A., Kaur, N., Singh, N., Naphthalimide-gold-based nanocomposite for the ratiometric detection of okadaic acid in shellfish. *J Materials Chem. B*, 8, 36, 8405-8413, 2020.
97. Wu, S.J., Duan, N., Shi, Z., Fang, C.C., Wang, Z.P., Simultaneous aptasensor for multiplex pathogenic bacteria detection based on multicolour upconversion nanoparticles labels. *Anal. Chem.* 86, 3100-3107, 2014.
98. Chen, J., Jackson, A.A., Rotello, V.M., Nugen, S.R., Colorimetric detection of *Escherichia coli* based on the enzyme-induced metallization of gold nanorods. *Small.* 12, 2469-2475, 2016.
99. Kuo, F.Y., Chang, B.Y., Wu, C.Y., Mong, K.K.T., Chen, Y.C., Magnetic nanoparticle-based platform for characterization of shiga-like Toxin 1 from complex samples. *Anal. Chem.* 87, 10513-10520, 2015.
100. Tang, D.P., Lin, Y.X., Zhou, Q., Lin, Y.P., Li, P.W., Niessner, R., Knopp, D., Low-cost and highly sensitive lmmuno sensing platform for aflatoxins using one-step competitive displacement reaction mode and portable glucometer-based detection. *Anal. Chem.* 86, 11451-11458, 2014.

14

Food Allergy and Food Intolerance

Sunita Aggarwal

Department of Microbiology, Institute of Home Economics, University of Delhi, New Delhi, India

Abstract

Food allergy is an inappropriate immune response that occurs immediately in a few individuals on the consumption of certain food. About 3-8% of people globally suffer from food allergy. The prevalence is more common among children as compared to adults and appears to be on the rise. Though allergic reactions have been reported to more than 170 foods, more than 90% of the allergic reactions are towards eight food types. Food allergic reactions occur on release of pharmacologically active mediators like histamine. IgE antibodies, non-IgE immune response, or a mix of both mediate the underlying mechanism. All the adverse reactions to food, however, are not due to allergy. The reactions similar to food allergy are also seen in cases of food intolerance, but there seems to be no involvement of immune response in latter. These may be reactions against food components and toxins or because of metabolic disorders.

The food allergy can be diagnosed by examining the medical history, diet elimination, skin and oral food challenge tests, and determination of allergen specific IgE. The food allergy management protocol includes avoidance of food allergens, education of the patient and caregivers, and medication. In some cases, oral allergen immune therapy and anti-IgE monoclonal antibody therapy are used to resolve the allergic symptoms. Proper labeling of foods can minimize accidental exposure to food allergens.

Keywords: Allergens, hypersensitivity, food allergy, food intolerance, anaphylaxis, diagnosis, therapy

Email: sunita.aggarwal@ihe.du.ac.in

Tejpal Dhewa, Anil Kumar Puniya and Anil Panghal (eds.) Nutritional Science and Technology: Concept to Application, (317–334) © 2023 Scrivener Publishing LLC

14.1 Introduction

Any abnormal or exaggerated immune reaction to substances prevalent in the environment is known as allergy or hypersensitivity. It is estimated that roughly 5-10% of the general population is hypersensitive to different environmental substances, namely pollen, fungi, insects, their parts, dust, and food, etc. These can cause a number of clinical manifestations like allergic rhinitis, asthma, angioedema, hives, and gastrointestinal problems.

Food allergy, also called food hypersensitivity, is quite common and has been reported to affect approximately 6-8% of children and around 3% of adults globally [1]. Generally, in children, the allergy resolves with age except in a few where high IgE level is associated with more severe symptoms and co-sensitization to a number of foods. Some allergies are more common in early childhood, while others develop late in life. The incidences of food allergies are increasing with time. It is reported that the prevalence rate of allergy to peanuts among children has almost doubled in 10 years [2]. The individuals may be hypersensitive to one or more foods.

The type and the time of appearance of symptoms in sufferers can vary. A severe allergic reaction called anaphylactic shock may develop in a few persons and may be fatal and need emergency medical attention.

In the upcoming sections, we will discuss food allergy, its definition and difference from food intolerance, causes and symptoms, the mechanism leading to allergic reactions, methods of diagnosis, and treatment and prevention of food allergy.

14.2 Food Allergy and Food Intolerance

A food allergy is an unusual reaction of the body to one or some particular foods mediated by the body's immune (defense) system. The response varies from slight abdominal discomfort to severe life-threatening anaphylactic reactions. Different body parts may be affected simultaneously, viz., mouth, eyes, skin, face, gastric tract by the allergic reactions. However, all the unpleasant responses to foods are not symptomatically allergic. These reactions may occur because of food intolerance.

Though the term "food allergy" and "food intolerance" (food sensitivities) are often used interchangeably and may produce many very similar symptoms, there is a marked difference between the two (Table 14.1). While food allergy is elicited by immunological reactions, food intolerance is basically caused by gastrointestinal reactions to food [3].

Table 14.1 Difference between food allergy and food intolerance.

Characteristics	Food allergy	Food intolerance
System involved	Immune system	Gastrointestinal system
Immune response	IgE antibodies/T cells/both	No antibodies/T cells involvement
Response time	Immediate	Hours to days
Dose of allergen	Small	Dose-dependent
Symptoms	Involves different parts of the body and include • Itchy skin and rashes • Trouble in breathing • Hives • Chest pain, etc.	Mainly gastrointestinal • Dry skin and acne • Bloating, gas, cramps • Constipation, acidity, • Heartburn, headache, irritation • Mood swings
Life-threatening reactions	Anaphylaxis	Does not occur
Adverse reactions	Even very less quantity can cause adverse reactions	Small quantity can be consumed without adverse reactions
Common cause	Eggs, nuts, milk, soya, peanut, shellfish, fish	Beans, gluten, lactose, fructose, processed food

There is no involvement of the immune system in food intolerance. The absence of sufficient digestive enzymes or reactions to food additives and added preservatives in food are the leading causes of food intolerance. The prevalence of food intolerance is much higher in comparison to food allergy. Around 15-20% of the population experiences food intolerance to certain foods, e.g., lactose and gluten intolerance [4]. No life-threatening symptoms appear during food intolerance and often the reaction takes time – hours to days – to develop.

Moreover, the severity of the reactions in food intolerance depends on the frequency and amount of the offending food consumed. Generally, a large quantity of food with frequent consumption is required to produce the symptoms. The food can be tolerated if taken in a small amount and not too often. Food intolerance, in some cases, may damage the intestinal lining and thereby interfere with nutrient absorption. In contrast, in case of food allergy, the reaction occurs immediately, i.e., within minutes and often can result in a life-threatening response that may be fatal if no immediate action is taken. A very small amount of food can result in an allergic response, and it recurs every time that food is consumed.

14.3 Food Allergens

Any food item can be a potent source of food allergens. However, as reported by FALCPA (Food Allergen Labeling and Consumer Protection Act, 2004), there are eight foods that causes over 90% of the reported food allergies [5]. These food allergens, also called as "Big 8" include (Figure 14.1):

i. Eggs
ii. Milk
iii. Tree nuts like walnut, almond, pistachios, cashews, pine nuts, etc.
iv. Shellfish, including crustaceans and mollusks like crayfish, crab, lobster, prawns, shrimp, squid, scallops
v. Fish
vi. Peanuts
vii. Wheat
viii. Soya

Figure 14.1 Major food allergens.

Generally, babies and young children (approximately 75%) are more susceptible to foods like milk, eggs, peanuts, tree nuts, wheat, and soya. Still, they lose these allergies with age except to tree nuts and peanuts, which persist lifelong. An inverse relationship has been noted between the severity of food allergies and the possibility of being resolved. In adults, the cause of most allergic reactions are peanuts, shellfish, and fish. Other food items that may cause allergic reactions less commonly in some people include celery, fruits and vegetables (like peach, banana, avocado, kiwi, passion fruit, garlic, celery), gluten, mustard, sesame and lin-seeds, meat and pine nuts. Over 170 foods can trigger allergic reactions in sensitive individuals [5]. Many, like sesame, are emerging food allergens. Geographical differences exist in the prevalence of food allergens, e.g., rice allergy is common in Japan and East Asia. Also, the methods of food processing and preparation alter its allergenicity.

As per the guidelines of the Codex Alimentarius Commission (CAC) on food labeling, the food and ingredients that are responsible for most of the allergic reactions should always be declared on food labels [6]. These include the following foods and their products causing food hypersensitivity globally:

i. Milk, egg, fish, and their products
ii. Tree nuts, peanuts, soya, and their products
iii. Shellfish, fish, and their products
iv. Gluten-containing grains or cereals likes wheat, rye, barley, oat, etc., and hybridized strains and their products.
v. Sulphites with concentration ≥10 mg/kg

Nations have the freedom to make alterations (add or delete) in the above list while declaring the national list of food allergens to be mentioned on labels based on the type of food allergens prevalent in their area. The foods that commonly produce allergic manifestaions are:

1. Milk and eggs – Milk, eggs and their products cause the most common childhood allergies among toddlers and babies. The breastfeeding mothers of such babies should avoid consuming food containing milk and eggs. Egg allergy to egg white is generally more common than to egg yolk. These allergies are rarely seen in adults as 70-90% of the children affected usually grow out of them. Cooked/baked egg products, e.g., biscuits, cakes, etc., usually are less or not allergic,

probably because of a change in egg allergens. The best way to curtail these allergies is to have a milk and egg-free diet.
2. Tree nuts and peanuts – Allergy to nuts like walnut, peanuts, cashew, pistachios, pine nuts, almonds, etc. Their food products (nut butter, oils) are quite common and frequently result in severe fatal allergic reactions. It is necessary for these patients to always keep adrenaline shots ready with them for emergency conditions. Usually, patients allergic to one or two nut types may be at risk of becoming allergic to other nuts. Though in some affected children, the condition may resolve, nuts and peanuts allergy is otherwise a lifelong allergic condition.
3. Shellfish – Shellfish allergy is commonly associated with protein tropomyosin and usually persists lifelong. In some individuals, shellfish allergy may develop even on inhaling vapors on cooking shellfish. Shellfish allergy many times may be confused with seafood-borne poisoning by microbial contaminants, but they are distinct.
4. Fish – Unlike other food allergies, fish allergy may have a later onset, i.e., it may appear in about 40% of people as an adult. Just like shellfish allergy, fish allergy may also be confused with foodborne infections by contaminating microbes.
5. Wheat – Allergy to wheat proteins is quite common among children and may get resolved by 10-11 years of age. Occasionally, it may produce severe fatal anaphylactic reactions. Because of similar digestive symptoms, it may be confused with celiac and non-celiac gluten sensitivities. Unlike a wheat allergy, these reactions are not life-threatening. Moreover, their treatment involves excluding all the grains containing gluten protein, whereas, in wheat allergy, only wheat consumption has to be avoided.
6. Soy – Soya beans and soy products like soymilk and soy sauce commonly trigger an allergic reaction in children under age of three. Sometimes soy allergy may be present along with cow's milk allergy. Allergic reactions may be associated with anaphylactic response.

Besides the consumption of food, inhalation of airborne food particles and food vapors can also result in food allergies. Commonly reported cases include reactions to peanut, legume, fish, milk, and seafood. These usually cause respiratory symptoms followed by anaphylaxis in a few cases. Due to cross-reactivity among food allergens, oral allergy syndrome may happen in people sensitive to pollens, e.g., ragweed and birch pollen sensitivity result in the appearance of itching in the mouth on the consumption of melon and apple peel.

14.4 Types of Food Allergy

Allergies caused by foods are grouped into three types depending on the underlying mechanism involved and the time and kind of symptoms that appeared [7, 8]:

i. IgE mediated allergy (Immediate and immediate delayed onset)
ii. Non-IgE mediated (delayed onset)
iii. Mixed IgE and non-IgE mediated (delayed onset)

14.4.1 IgE Mediated Food Allergies

IgE mediated allergic reactions are caused by the production of IgE antibodies by the immune system. These are the most common type of allergic reactions that have an immediate onset, i.e., those which develop within one to two hours of food ingestion. In some individuals, this immediate reaction may be followed by the late-phase reactions wherein the symptoms are ongoing and prolonged for a day, e.g., allergic dermatitis and respiratory symptoms involve early as well as late responses. IgE-mediated allergic reactions' clinical manifestations include allergic urticaria, angioedema, gastrointestinal and respiratory symptoms. The release of a large amount of mediators from sensitized mast cells and basophils in IgE mediated reactions are responsible for life-threatening anaphylaxis responses in a few individuals.

14.4.2 Non-IgE Mediated Food Allergies

These responses are unclear and probably triggered by different immune cells, particularly T cells. IgE antibodies are not involved in eliciting these allergic responses. These have delayed onset, i.e., their development occur in several hours and usually involve the skin and gastrointestinal system. These are associated with allergic contact dermatitis, enteropathy, allergic proctocolitis, and enterocolitis induced by food allergens [9]. The symptoms are mainly limited to gut and skin, such as gut wall inflammation, diarrhea, constipation, and emesis. It primarily affects infants who show lethargy, vomiting, starvation, and poor growth. Exposure to food proteins also results in mucosal edema within the intestine, blood-ridden diarrhea, and dehydration. The most common foods involved are cow's milk, rice, wheat, soya, oat, fruits, and vegetables.

14.4.3 Mixed IgE and Non-IgE Mediated Food Allergies

These arise due to combined IgE and cell-mediated immune responses. The onset is delayed and is characterized by abdominal pain, diarrhea, nausea, vomiting, dysphasia, edema, obstruction, eosiniophilic inflammation, and weight loss. Children may suffer with vomiting, difficulty in feeding, and have improper growth in early childhood. Allergy to eggs and milk in infants and children may be IgE or non-IgE mediated.

14.5 Mechanism of Food Allergy

The body's immune system is involved in eliciting allergic reactions to food. Food components, particularly proteins, are recognized by the immune cells, and antibodies belonging to class IgE (Immunoglobulin E) are produced. The generation of immune response involves two phases [10]:

i. Sensitization Phase – In this phase, the immune cells (antigen-presenting cells) recognize the food allergens for the first time. Antigen-presenting cells (APC) sensitize the allergen-specific Type-2 helper T cells by presenting antigen, which in turn sensitizes allergen-specific B cells. Activated B-lymphocytes then produce IgE antibodies against the allergen. These allergen specific IgE antibodies bind to FcεRI (high-affinity receptors for IgE antibodies) present on the surface of other immune cells, i.e., mast cells and basophils where they can reside for months. Both the mast cells (present in tissues) and basophils (present in circulation) elicit the inflammatory response.
ii. Effector phase – On later exposure to the same food allergens, the IgE antibodies on mast cells and basophils bind with these specific allergens and cross-link IgE receptors. The cross-linked IgE receptors then trigger an allergic response by activating mast cells and basophils. Signals are transmitted from the cell surface to the interior of the cell and result in the release of different chemical mediators-both preformed and newly synthesised- into

the surrounding tissues. Preformed chemicals are present within the granules of mast cells and basophils, which undergo degranulation on activation. The main chemicals released include histamine (preformed), leukotrienes, and prostaglandins (newly synthesized). These chemicals are responsible for causing allergic reactions. The allergic reactions occur immediately (immediate response) and subside but may be followed by late-phase reactions in many. Histamine is the most important preformed chemical (an inflammatory mediator), which binds with the histamine receptors present on other body cells (Figure 14.2). Histamine binding to cell surface results in:

- Increase vascular permeability
- Vasodilation
- Smooth muscle contraction and
- Mucus secretion and nerve stimulation

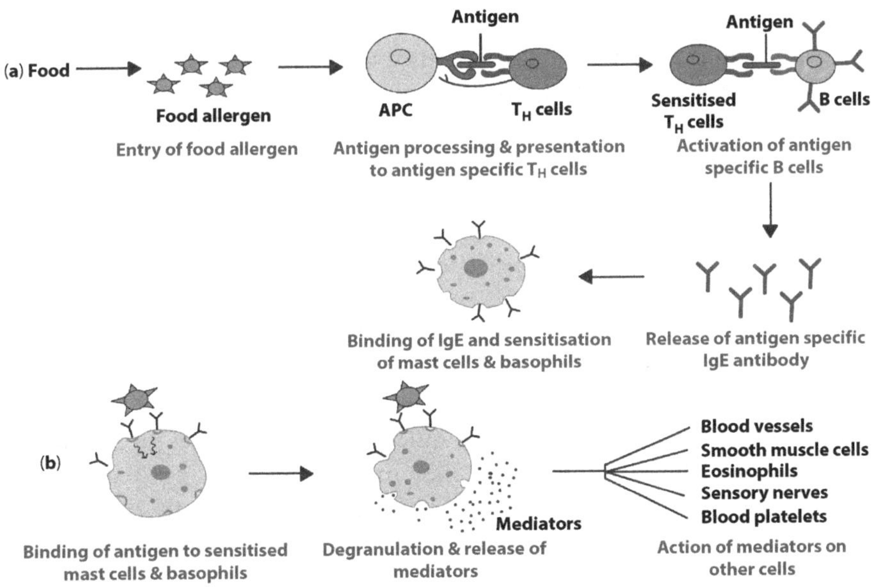

Figure 14.2 Mechanism of allergic reaction – (a) Sensitization phase (b) Effector phase.

The release of histamine in most food allergies is mainly restricted to skin, mouth and throat and results in quick appearance of skin (swelling and itching), and gastrointestinal (diarrhea, vomiting) symptoms. These symptoms may be accompanied by severe life-threatening allergic reactions often called anaphylaxis (dizziness, loss of consciousness, respiratory problems, shock, etc.) within a few minutes of consuming troublesome food. The onset of anaphylaxis requires immediate medical attention.

14.6 Risk Factors Involved in Food Allergy

The exact reason why some individuals become allergic to otherwise harmless food is unknown. Immunopathology and clinical manifestations of food allergy seem to depend on the interaction of genetic and environmental factors. The chances of having a food allergy are increased due to the following factors [11]:

i. Family history – Persons with a family history of allergies like asthma, eczema, or other allergic reactions are more prone to develop food allergies.
ii. Age – children with eczema at an early age have higher chances of food allergy.
iii. Food additives – The presence of additives and preservatives like sulphites, benzoates, and benzoic acids in food may enhance or worsen the symptoms in people with allergies.
iv. Diet, vitamin D deficiency, diagnostic methods and criteria, geographic variation, etc. also influences the prevalence range for food allergy, reported varying from 1-10% [2].

14.7 Symptoms of Food Allergy

The symptoms of food allergy arise quickly, within minutes to 3-4 hours, on consuming the causative food. They may involve different parts of the body, including the gastrointestinal tract, skin, respiratory tract, eyes, and cardiovascular system (Table 14.2). The symptoms may persist for days to weeks. The severity of the symptoms varies depending on the individual's sensitivity and the amount of food consumed.

The allergic person may experience mild symptoms like any other gastroenteritis problem, such as abdominal pain, nausea, or vomiting, to more severe symptoms like erythema (redness), angio-edema (swelling), urticaria (itching) and allergic dermatitis within minutes of food consumption [12]. Respiratory manifestations include allergic rhinitis, dyspnea and asthma and usually occur along with skin and gastrointestinal responses, than occurring on a stand-alone basis. A fatal, life-threatening response, i.e., anaphylaxis, may develop in a few patients, resulting in chest tightening, swollen tongue, shortness of breath, dizziness, difficulty in speaking and swallowing, and ultimately collapse if immediate medical attention is not provided. This occurs due to a release of very large amount of mediators from mast cells and basophils.

Table 14.2 Symptoms of food allergy.

Location	Symptoms
Skin and Face	• Itchy, red skin with or without rashes (hives) • Itchiness and swelling of whole face, eyelids, lips, and nose
Gastrointestinal tract	• Abdominal pain, stomach cramps • Nausea, vomiting • Swollen throat, vocal cord, and tongue • Itchiness and tingling in the mouth
Respiratory tract	• Allergic rhinoconjunctivitis (running nose, sneezing, red itchy watering eyes) • Asthma (difficulty in breathing, swallowing and speaking, coughing, wheezing) • Shortness of breath (dyspnea)
Cardiovascular system	• Low blood pressure, abnormal heartbeat, chest pain • Dizziness, faintness, and unconsciousness

In contrast to IgE-mediated food allergic reactions, non-IgE-mediated allergic reactions take several days to develop and are usually restricted to the skin and digestive tract. The symptoms generally include dry, cracked, itchy and red skin (eczema), along with vomiting, abdominal pain, and heartburn. Many patients with a food allergy may suffer from diarrhea or constipation also. These non-IgE mediated allergic reactions may result in diarrhea and reflex in babies. They cry inconsolably despite not being hungry or sick (colic in babies). Some babies may experience both swelling and constipation (mixed IgE and non-IgE symptoms). In rare cases, pneumonia may occur in affected children; pulmonary infiltrate and gastrointestinal blood loss, blood in stool, and vomiting may occur. Accurate diagnosis of non-IgE mediated allergy to food is difficult as IgE class of antibody does not cause it; rather, other immune cells are involved in raising it.

14.8 Diagnosis of Food Allergy

Different diagnostic tests can be used in combination to detect food allergy [13, 14]. These include family history, dietary review, skin test, oral allergen challenge and blood tests. Along with the clinical history of having allergic symptoms to food, a positive *in-vivo* and/or *in-vitro* tests are required for correct diagnosis of the food allergy [1, 15]. Non-IgE mediated allergic reactions, however, have a complicated diagnosis.

Diagnostic tests used to detect food allergy are:

i. Family and medical history – It is shown that people with siblings or parents with allergies are more sensitive to food allergies. If the medical history and physical examination do reveal the presence of food allergy, that needs to be substantiated further by performing *in-vivo* and *in-vitro* tests.
ii. Dietary Review – Diet information, food eaten, its timing, and symptoms should be reviewed in detail to find out the suspected culprit of food allergy.
iii. Skin Prick Test (SPT) – Skin test is done on the forearm or back of the allergic patients. Small amount of different food allergens are pricked into the skin using a fine needle

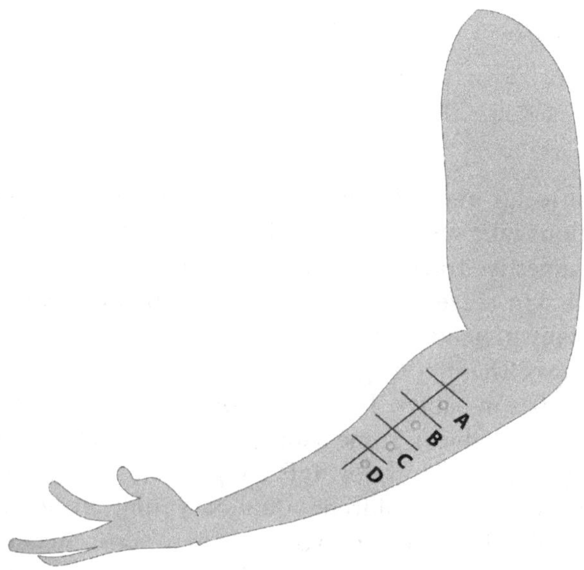

Figure 14.3 Skin prick test to diagnose food allergy.

(Figure 14.3). The reactions (weal and flare) that appeared on the pricked site after 15-20 minutes of introducing the allergen are monitored and compared with buffer (negative) and histamine (positive) controls for interpretation. The appearance of edema and erythema more than the positive control on the test site within a few minutes of testing is considered positive. The test is, however, applicable for IgE-mediated allergic reactions. Atopy Patch Test (APT) can be performed to diagnose the food allergy more specifically, the late reaction.

iv. Oral food challenges - It can also be used for allergies other than IgE-mediated ones. Suspected food is given to the patient to eat under medical supervision.

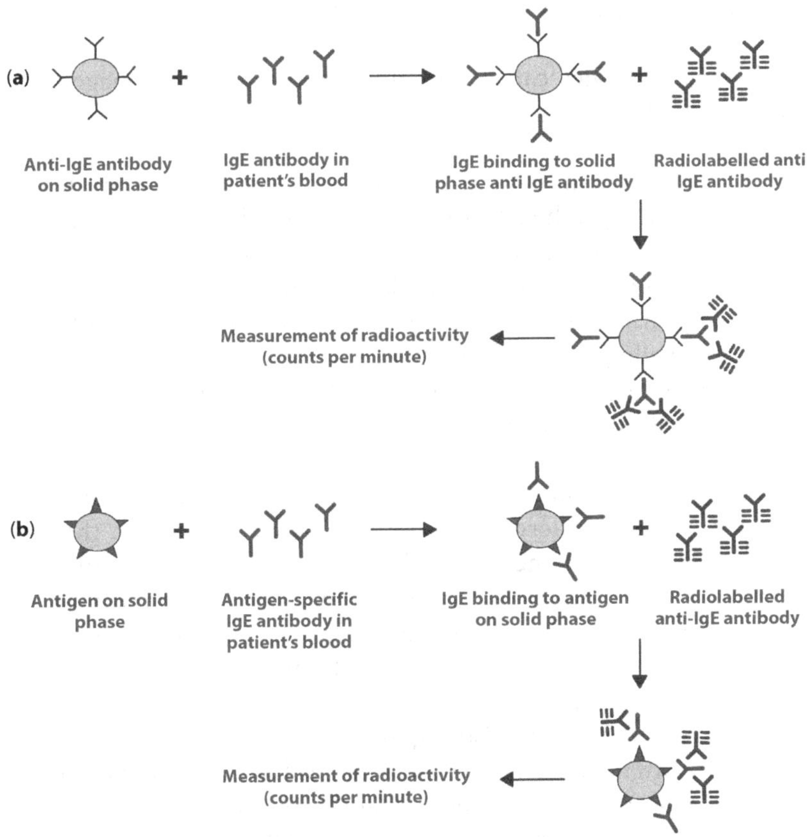

Figure 14.4 Blood tests (a) RIST (b) RAST.

The amount of food is gradually increased, and adverse reaction so developed is recorded. DBPCFC (double-blind, placebo-controlled food challenge) which is the appropriate test to diagnose all types of food allergies is however, rarely performed.

v. Blood Test – *In-vitro* test can be performed in the laboratory to detect the level of total and allergen-specific IgE antibodies in patients' blood by performing Radio Immuno Sorbent Test (RIST) and Radio Allergo Sorbent Test (RAST). RAST scores more than the predictive value indicate a food allergy. The test, however, can detect only IgE mediated allergy and not the non-IgE mediated one (Figure 14.4).

14.9 Treatment of Food Allergy

The following strategies can be adopted for the proper management of food allergies [15, 16].

i. Food avoidance – In the absence of permanent therapy for food allergy, the effective practice is total avoidance of the offending food. Sensitive people must also avoid inhalation of or touching the foods containing the allergen. The responsible food has to be diagnosed and avoided till the food allergy is resolved. Avoidance of food is a bit difficult in allergic children. To protect allergic patients from accidental exposure to food allergens-

  - Food companies are required to specifically mention the allergy-causing ingredients along with their source on the food label. As per FALCPA requirements, these should be written either in parenthesis following ingredient name, e.g., whey (milk) or can be given in a "contain" statement immediately next to the ingredient list.
  - Companies are also required to prevent "cross contact" with food allergens. If the multiple foods are produced in the processing unit using shared production line or equipment, the advisory statement be put up on the label as "may contain" or "produced in a unit also using allergen".

- FDA takes up many measures from providing guidance to establishing regulations to conduction of inspections to ensure consumers' protection from accidental entry of allergens in food.
- The food label should be read carefully by the allergic person or caregivers for the presence of an explicitly known labeled allergen or any hidden allergens in the form of "natural flavor", "spices", "may contain", or "cross contact". Allergic reactions can be avoided by avoiding such foods containing causative food allergens.

ii. Drugs – Drugs are used for treating food allergy and anaphylactic manifestations. Epinephrine or adrenaline is the drug of choice for alleviating anaphylactic reactions, improving blood circulation, and relaxing airways. It is available in two doses as an auto injector. A dose of 0.15 mg is recommended for individuals weighing less than 15 kg and 0.30 mg for those having weight > 30 kg. The drug might need to be repeated. Other drugs used for symptomatic relief include corticosteroids, anti-histamines, and prostaglandin synthesis inhibitors. Antihistamines, e.g., diphenhydramine, commonly used for food allergy, work by blocking histamine action on blood vessels and sensory nerve endings. Steroids generally take time to relieve symptoms and can be taken orally or through injection. The role of anti-IgE monoclonal antibody therapy is under investigation, though its use has been licensed in certain countries.

iii. Education – Patients, their parents, teachers, colleagues, employers, and caregivers need to be educated regarding the food to be avoided, early symptoms, treatment procedures to be followed and immediate medical measures to be taken in case of emergency. Relevant information, written material, and an action plan must be provided to them for better and long-term allergy management.

iv. Desensitization – In some patients, the allergy can be resolved permanently by giving increasing doses of the offending allergens. Many countries have now accepted this treatment (oral immune therapy) [17].

14.10 Food Intolerance

Food producers add thousands of food additives like artificial colorants, flavoring agents, antioxidants, sweetners, preservatives, etc., to enhance the flavor, appeal, and shelf life of food. Many of these food additives may produce an adverse reaction (food intolerance) in a few people. Examples include nitrates present in processed meat, sulphites in wine, jam, jellies, and monosodium glutamate. These agents are responsible for chest tightness, headache, hives, diarrhea, and nausea. The absence of metabolic enzymes in sensitive individuals to digest certain foods properly is another reason for food intolerance. The most common food intolerances are to lactose, gluten, and fructose.

Lactose intolerance occurs because of the poor release of enzyme lactase required to break milk sugar lactose to absorb monosaccharides easily. The persistence of lactose in GI tests leads to diarrhea, bloating, and stomachache. Gluten intolerance is also caused by non-digestion of gluten, a protein present in cereals like wheat, barley. Patients with gluten intolerance experience symptoms like abdomen pain, gas bloating, brain fog, headache, etc., on consuming gluten-laden food.

Fructose intolerance results from the malabsorption of fructose – a sugar present in fruits and vegetables. The reason is the lack of a protein involved in fructose absorption. Patients with this intolerance suffer fullness, diarrhea, bloating, and abdominal pain due to fructose fermentation in the gut.

The diagnosis of food intolerance is a bit complicated because of the similarity of symptoms with food allergy and chronic digestive problems, e.g., irritable bowel syndrome (IBS). The diagnosis may be based on the patient's recall of food taken and the timing and type of symptoms, along with ruling out the possibility of food allergy by SPT and RAST. The best way to diagnose food intolerance is to sequentially eliminate food components in a diet. The treatment includes the exclusion of or reduction in the amount of causative food components [18].

14.11 Conclusion

An unpleasant response can occur in some individuals on the consumption of a few foods. These abnormal responses are either food allergies or food intolerances. While food allergy is elicited by the immune system, food intolerances involve the gastric system. Though more than 170 food products are known that can cause hypersensitive reactions, more than 90% of food allergies are caused by eight foods, commonly called the "Big 8". Incidences of food allergies are higher in infants and children as compared to adults, but most children outgrow their allergies. The incidences of food allergies are on the rise. The occurrence of food intolerance is much higher (15-20%) as compared to food allergy (3-8%). Food intolerances occur because of the absence of metabolic enzymes or reactions to food additives. On the basis of the immune response against food allergens, food allergies are categorized as IgE mediated, non-IgE mediated, or a mix of both, the IgE being the commonest one. The allergic symptoms involve the skin and gastrointestinal system in association with or without the respiratory system. In a few cases, a life-threatening anaphylactic response, requiring immediate medical attention may develop. Proper diagnosis of food allergy includes medical history along with various *in-vivo* and *in-vitro* tests. As permanent treatment is not available for food allergy, the elimination of that food type from the diet is the best approach. Use of medication, education to patients and caregivers, and adopting desensitization procedures are also used for food allergy management. Food companies are also required to mention the allergy-causing ingredients on food labels to avoid any accidental exposure of sensitized individuals as per the guidelines and regulations.

References

1. Walsh, J. and O'Flynn, N. (2011). Diagnosis and assessment of food allergy in children and young people in primary care and community settings: NICE clinical guideline. *British Journal of General Practice* 2011; 61 (588): 473-475. DOI: https://doi.org/10.3399/bjgp11X583498
2. Alessandro, F. and Fierro, V. 2017. Food allergy. https:// www2017.worldallergy.org/education-and-programs/education/allergic-disease-resource-center/professionals/food-allergy
3. American Academy of Allergy, Asthma and immunology. https://www.aaaai.org/conditions-and-treatments/library/allergy-library/food-intolerance
4. Medical News Today. https://www.medicalnewstoday.com/articles/2639652.
5. Food Allergies. US Food and Drug Administration. https://www.fda.gov/food/food-labeling-nutrition/food-allergies
6. WHO. Food allergies (2006). https://www.who.int/foodsafety/fs_management/No_03_allergy_June06_en.pdf
7. NHS. https://www.nhs.uk/conditions/food-intolerance/
8. Waserman, S., Bégin, P. and Watson W. IgE-mediated food allergy. *Allergy Asthma Clin Immunol.* 2018 Sep 12;14(Suppl 2):55. doi: 10.1186/s13223-018-0284-3. PMID: 30263035; PMCID: PMC6156835.

9. Labrosse R, Graham F, Caubet JC. Non-IgE-Mediated Gastrointestinal Food Allergies in Children: An Update. *Nutrients.* 2020 Jul 14;12(7):2086. doi: 10.3390/nu12072086. PMID: 32674427; PMCID: PMC7400851.
10. Kuby Immunology. Sixth Edition. Thomas J. Kindt, Barbara A. Osborne, Richard A. Goldsby. 2007, W. H. Freeman and Company.
11. Mayo Clinic. https://www.mayoclinic.org/diseases-conditions/food-allergy/symptoms-causes/syc-20355095
12. Cleveland Clinic. https://my.clevelandclinic.org/health/diseases/10009-food-problems-is-it-an-allergy-or-intolerance
13. Sackeyfio, A., Senthinathan, A., Kandaswamy, P., Barry, P. W., Shaw, B. and Baker, M. (2011). Diagnosis and assessment of food allergy in children and young people: summary of NICE guidance. *BMJ, 342.*
14. Sicherer, S.C., and Sampson, H.A. 2006. Food allergy. *J Allergy Clin Immunol.* 117, issue 2 (2). S470-S475.
15. Waserman, S. and Watson, W. Food allergy. *Allergy Asthma Clin Immunol.* 2011 Nov 10;7 Suppl 1(Suppl 1):S7. doi: 10.1186/1710-1492-7-S1-S7. PMID: 22166142; PMCID: PMC3245440.
16. Onyimba, F., Crowe, S.E., Johnson, S. and Leung, J. 2021. Food allergies and intolerances: A clinical approach to the diagnosis and management of adverse reactions to food. *Clinical Gastroenterology and Hepatology.* 19, Issue 11, 2230-2240.e1, https://doi.org/10.1016/j/cgh.2021.01.025.
17. Nurmatov, U., Dhami, S., Arasi, S., Pajno, G.B., Fernandez-Rivas, M., Muraro, A., Roberts, G., Akdis, C., Alvaro-Lozano, M., Beyer, K., Bindslev-Jensen, C., Burks, W., du Toit, G., Ebisawa, M., Eigenmann, P., Knol, E., Makela, M., Nadeau, K.C., O'Mahony, L., Papadopoulos, N., Poulsen, L.K., Sackesen, C., Sampson, H., Santos, A.F., van Ree, R., Timmermans, F. and Sheikh, A. Allergen immunotherapy for IgE-mediated food allergy: a systematic review and meta-analysis. *Allergy.* 2017 Aug;72(8):1133-1147. doi: 10.1111/all.13124. Epub 2017 May 11. PMID: 28058751.
18. Food intolerance. NHS. https://www.nhs.uk/conditions/food-intolerance/

15

Molecular Nutrition and Nutrient–Gene Interactions

S. Hamsa[1], Aarti Yadav[2], Ruby Tiwari[1]* and Tejpal Dhewa[3]

[1]Department of Genetics, University of Delhi, South Campus, New Delhi, India
[2]Department of Microbiology, Shaheed Rajguru College of Applied Sciences for Women, New Delhi, India
[3]Department of Nutrition Biology, Central University of Haryana, Mahendergarh, Haryana, India

Abstract

A well-balanced and proper diet comprising essential nutrients is crucial for maintaining a healthy lifestyle. Over several decades, huge studies have been done on deficiencies and metabolic problems related to general or specific nutritional inadequacy. Molecular nutrition deals with the integration of nutritional science and molecular biology to explore how nutrition alters metabolism at cellular, molecular and tissue level. However, in past years extensive research has been carried out to perceive the significant role of various nutrients not only at the cellular level but also at the genome level for better human health. Recent advances in genomics and post-genomic technologies like proteomics, metabolomics, and transcriptomics have enriched our knowledge of nutrient–gene interactions. These interactions are broadly studied as Nutrigenomics; which aims to understand how nutritional factors influence gene expression and metabolism, and Nutrigenetics; which identifies the influence of genetic factors on nutritional requirements and their downstream responses. In the last decade, nutrigenomics has been utilized as the basis for treating several chronic disorders and diseases like obesity, cancer, diabetes, and cardiovascular diseases. Diet-related diseases have risen in India alarmingly in the past few years with the rise in urban population, changes in lifestyle and unhealthy food habits. This field has also paved the way for identifying biomarkers in the early stages, especially for diet-related metabolic disorders. Recent research also focuses on modifying common Single Nucleotide Polymorphisms (SNPs) known to alter nutritional requirements. This chapter provides a holistic view of

Corresponding author: ruby12tiwari@gmail.com

Tejpal Dhewa, Anil Kumar Puniya and Anil Panghal (eds.) Nutritional Science and Technology: Concept to Application, (335–354) © 2023 Scrivener Publishing LLC

the recent advances in nutritional genomics and its significance in providing new challenges and opportunities towards individualized health by following appropriate dietary guidelines.

Keywords: Nutrition, diet, lifestyle, nutrigenetics, nutrigenomics, personalized medicine

15.1 Introduction

A balanced diet comprises various foods crucial for fulfilling the basic nutrient requirements of the human body. According to World Health Organization (WHO) a healthy diet should include 400g of fruits and vegetables, a diet with more unsaturated fats, less sugar and salt intake, and nuts and whole grains should be considered an important part of the diet (WHO website). Over the years and with the onset of Industrial Revolution, lifestyle change has led to unhealthy food habits with improper nutrition. Functional food ingredients like n-3 fatty acids, polyphenols, sterols and fiber which were prominent in the Paleolithic and the Neolithic period, have been replaced by processed foods, vegetable oils, saturated fats and high cholesterol foods [1]. Several researchers have conducted studies to explore the association between a healthy diet and the prevention of diseases. A recent survey of dietary data from 195 countries showed a strong association between sub-optimal intake of proper nutrition and morbidities due to dietary factors [2]. With the nutrition transition, the Indian diet in recent years comprises more carbohydrates and less proteins, fruits and vegetables [3].

Nutrition research focuses on understanding basic nutrient requirements for reducing or preventing the severity of diseases or disorders through dietary means. In the last decade, nutrition research has revolutionised due to various factors like the density of data, scale expansion, accessibility, meta-analysis and personalization for population-based analysis [4]. Taking advantage of the latest technologies, techniques and databases, researchers have focused on understanding molecular insights regarding detailed mechanisms of action between the interaction of nutrients and diet-related disorders described as molecular nutrition. Though molecular nutrition helps to understand the basic mechanisms, exploring the interaction of genes at the genome level was considered a prerequisite to preventing diseases at an individual level and paving the way towards personalized medicine [5]. Nutrient–gene interactions are studied as Nutrigenetics and nutrigenomics, which are mainly concerned with the effect of nutrients at a molecular and cellular level. Nutritional science in recent years has focused on not only gene interactions but also integrated the knowledge from transcripts, proteins and metabolites combined with bioinformatics, epidemiology and clinical interventions [6].

This chapter elaborates on the basic principles and advances made at the molecular and genome level of important diseases and disorders like cancer, diabetes, cardiovascular diseases, Vitamin D deficiency, obesity and metabolic disorders through the integration of pre-existing knowledge and application of omics technologies, both at an individual and population level. We would be discussing briefly evolving topics such as nutrigenomics and nutrimetabolomics. Finally, we conclude with how this information has been applied to clinical trials and has impacted public health globally.

15.2 Molecular Nutrition: Understanding Basic Mechanisms

The success of the Human Genome Project, unlimited access to various tools and databases, and empowered consumers have improved our understanding of molecular nutrition [7]. Nutrients are considered as signalling or bioactive compounds that interact with various intracellular and extracellular molecules at molecular, cellular and genomic levels [8]. Unravelling the fundamental mechanisms of how various macronutrients like fats and lipids, carbohydrates, proteins and micronutrients like Vitamin A, and Vitamin D interact with various metabolic pathways receptors has helped us gain more profound knowledge at the cellular and genomic level [9].

a. Fats

Cholesterol is a necessary cellular molecule modulating several signalling pathways and synthesising various hormones, but its excess leads to Type II diabetes and cardiovascular diseases [10]. A basic idea about cholesterol synthesis and absorption is crucial since balancing the two is necessary to maintain cholesterol levels. Two major receptors which maintain cholesterol homeostasis are Sterol regulated element binding protein (SREBP) and Liver X receptor (LXR). Several compensatory mechanisms are involved in shifting the pathways in either direction. Low-Density Lipoprotein (LDL): High-Density Lipoprotein (HDL) cholesterol ratio is a standard to measure the risk of cardiovascular diseases in a population. Food supplements such as soy-based products and plant sterols reduce the LDL: HDL ratio by reducing total cholesterol by 10% [11]. The mechanism of cholesterol synthesis and excretion varies among individuals due to variations in post-transcriptional changes. Generally, an increase in cholesterol in the diet leads to low synthesis. However, in some cases, individuals consuming a high-cholesterol intake diet did not lead to reduced cholesterol synthesis, causing a high increase in plasma cholesterol concentration [12, 13]. Probiotics consumed in the form of yoghurt, cheese and functional foods lower the cholesterol concentration by producing bile salt hydrolases (BSH) which release primary bile salts from the conjugated cholesterol by hydrolyzing amide and by binding and increasing the activity of LXR receptor, which further leads to downregulation of Niemann-Pick C1-Like 1 (NPC1L1) protein which is responsible for cholesterol absorption [14].

b. Carbohydrates

Carbohydrates comprise sugars, oligosaccharides, and polysaccharides and contribute to 80% of the calorie intake in humans. There are two types of carbohydrates-digestible absorbed as monosaccharide provides glucose to the body, and indigestible ones, which are fermented to short-chain fatty acids by gut microbes known as dietary fiber [15]. High intake of dietary sugars has been associated with overweight, obesity, diabetes, fatty liver disease and cardiovascular diseases. High sucrose content (glucose and fructose) leads to an increase in levels of glucose-dependent insulinotropic peptide (GIP) which leads to higher appetite, body weight and insulin resistance [16]. Metabolism of carbohydrates produces Acetyl CoA, which combines with oxygen in mitochondria to release ATP molecules. Excessive carbohydrate consumption leads to high Acetyl CoA which generates high amounts of free radicals, considered the most latent mechanism of insulin resistance where insulin can no longer control blood glucose levels [17]. High fiber diet (dietary fiber) impact glucose and lipid metabolism by producing Short Chain Fatty Acids like acetate propionate and butyrate by fermentation of gut microbiota [18]. SCFAs are inhibitors of histone deacetylases like butyrate prevents type I diabetes by affecting the acetylation of histones H3 and H4 and inhibiting p38/ERK pathway by inducing beta cell proliferation [19].

c. Vitamins and minerals

Vitamins and minerals play an important role in various fundamental cellular processes crucial for brain and muscle function [20]. The recommended levels of vitamins are <1g/day for normal functioning of the body. Vitamins are to be taken as supplements exogenously as they cannot be synthesized in the body. Most of the vitamins are important to boost the immune system, both innate and adaptive, by the production of antibodies against the pathogen [21]. A list of vitamins and their roles are discussed in the table (Table 15.1). Recently it was reported that water-soluble vitamins Vitamin B and C help slow the digestion of starch by inhibiting the human pancreatic alpha–amylase (HPA) activity through conformational scrambling at the binding site of HPA. Thus, starch with vitamins can help maintain blood glucose levels [22]. Vitamin C and D have potential anti–tuberculosis properties [23]. Vitamins C, D and E have been known to reduce the side effects of chemotherapy and radiation used for treating various cancer [24]. In high amounts, vitamin C has protective effects on ovarian, pancreatic, colorectal, brain and blood cancer [25]. Apart from vitamins, several minerals comprising macro and micro minerals are required in trace amounts by the human body, and understanding the

Table 15.1 Role and diseases caused due to deficiency of vitamins.

Name of vitamin	Role	Diseases associated with deficiency
Water soluble		
Vitamin B	For a generation of energy acts as cofactors and is essential for the functioning of neurons, acts as cofactors	Loss of coordination in the body, loss of appetite, anemia, lethargy
Vitamin C	Aids in healing wounds, is antioxidant, and helps in protein metabolism	Scurvy, Gum and dental problems, Anemia
Fat soluble		
Vitamin A	Formation and functioning of membranes	Increases the susceptibility to infections, and causes diseases in the eyes, respiratory as well reproductive system
Vitamin D	Aids in absorption of calcium and maintains bones	Leads to soft bones
Vitamin E	Acts as antioxidant, important for reproductive functions and reduces toxicity from blood	Leads to problems in the kidney, decrease fertility and bone health
Vitamin K	Essential for clotting blood	Haemorrhage

mechanisms of absorption of minerals is the key step in maintaining their homeostasis [26].

Translating basic molecular nutrition science research into practical applications at a population and the individual level was the major challenge faced in this field [27]. Approaches to further study molecular nutrition started with genotype–phenotype correlation, quantification of the disease phenotype through biomarkers and various functional assays [28], then application systems-level biology to correlate gene-diet interactions at a population level using bioinformatics and knowledge management tools [29] and finally development of databases to store data integrated from multiple omics and association studies from different countries [30].

15.3 Nutrigenetics

Nutrigenetics refers to the interaction between the genotypic composition and the metabolic state of the individual. It majorly deals with genetic variation, which includes Single Nucleotide Polymorphism (SNP), Insertions and Deletions (INDELs), Copy Number Variations (CNVs) and its effect on the phenotype caused by the diet [31]. Several population studies, from candidate gene studies through high throughput sequencing to Genome Wide Association studies (GWAS) in various diseases, have explored the relationship between gene-nutrient-disease associations. An example to illustrate nutrigenetics: A subset of the population in Northeast India infected with liver disease was found to have a SNP in the CYP2E1 gene (C->T), which is a Cytochrome p50 gene involved in generating oxidative stress and hepatotoxicity. The indigenous foods of Northeast India contain a high number of nitrites which further trigger the risk of hepatitis and liver disease in individuals having mutant allele [32]. Recently GWAS studies have identified a candidate gene GRIN2A, an important gene of the Central Nervous System (CNS) encoding subunit of N-methyl-D-aspartate glutamate receptor, and the association of this gene to coffee drinkers, and Parkinson's disease was analyzed. Heavy coffee drinkers with genotype rs4998386_CC were at 18% risk of Parkinson's disease, while those with genotypers4998386_TC were at 59% risk compared to the CC genotype [33]. Nutrigenetics has been widely studied for cancer prevention and control. Various gene polymorphisms and the influence of food and bioactive compounds on the individual with the genotype have been characterized for prostate, colorectal, breast and lung cancer (Table 15.2). One of the common biomarkers for colorectal cancer is the MTHFR gene in which an individual having a 677TT genotype and high folate, Vitamin B6, and B12 diet intake has less risk of developing

Table 15.2 Reports of some nutrigenetics studies conducted for understanding gene polymorphism – diet and disease interaction.

Gene polymorphisms	Diet/nutrient/ metabolite	Disease/disorder	References
UGTA1, UGTA7	Crucifers/ sulforaphane	Colorectal cancer	[37]
Folate pathway genes MHFTR, TYMS, MTRR, DFR, BHMT	Folic acid	Colon cancer and coronary artery disease	[38, 39]
SEPP1, SEP15, GPX1	Selenium	Prostate cancer	[40, 41]
ABCG1, ABCG5, ABCG8	Fatty acids	High cholesterol, obesity, CVD	[42, 43]
APOB, APOA1, LIPC, PPARG2, APOE	Fatty acids	CVD, plasma lipids	[44, 43]
LPL, LXRA, TNF, CYP71A1, FABP2	Fatty acids	CVD, plasma lipid	[43]
Fok-1	Vitamin-D	Vitamin-D deficiency, diabetes	[45]
CBS, CTH	Vitamin- B_6, Vitamin B_{12}	Hypertension, diabetes	[46]
COMT, TXN	Vitamin-E	Obesity	[47]
RBP1, SLC23A1, SLC23A2	Vitamin-A, Vitamin-C	Glaucoma	[48]
SLC23A2/SLC23A1	Vitamin-C	Cancer, coronary syndrome, Inflammatory Bowel Disease	[49]
ALOX5	Omega 3 fatty acids	Asthma	[50]
Paraoxonase-1 PON1	Dietary zinc, copper, fructose corn syrup	Autism	[51]

(Continued)

Table 15.2 Reports of some nutrigenetics studies conducted for understanding gene polymorphism – diet and disease interaction. (*Continued*)

Gene polymorphisms	Diet/nutrient/ metabolite	Disease/disorder	References
CETP	Lipids	Type -2 diabetes, CVD, pancreatic cancer	[52]
ACE, AGT, BDKRB2	Micronutrients (Sodium, potassium, calcium, magnesium)	Blood pressure and other related diseases	[53]
GSTM1, GSTT1, TNF, IL6, PNPLA3	Fatty acids	Non-alcoholic fatty liver disease (NAFLD)	[54, 55]
CARTPT	Cholesterol, fats, sodium	Cardiac disease	[56]

cancer than the ones having less intake of these foods [34]. WHO projected that cardiovascular diseases (CVDs) are the leading cause of death globally, and by 2030 CVDs will be the single leading cause of death (WHO website). Thus, understanding the nutrigenetics of CVDs, including hypertension, heart disease, heart failure, stroke, and heart disease, is of utmost importance in nutritional science. Several gene polymorphisms like apolipoprotein E (APOE), fatty acid desaturases (FADS), peroxisome proliferator-activated receptors (*PPARs*), 5-lipoxygenase (*5-LO*) and the interaction with a specific diets mainly containing unsaturated fatty acids and high Omega3/Omega6 ratio have been shown to lower the risk of CVDs. Oxidative stress occurs due to an imbalance between the production of reactive oxygen species (ROS) and the degradation of ROS due to antioxidants. Variations in genes coding for antioxidant enzymes, their interactions with the diet and their role in disease development due to oxidative stress have been explored through the field of nutrigenetics [35]. Mediterranean diet has been extensively studied due to the presence of huge antioxidant enzymes and unsaturated fatty acids, which have high protective effects on individuals with the susceptible genotype [36]. This field has evolved over the years. Recent reports of other gene polymorphism-diet and the disease interaction have been provided in Table 15.2.

15.4 Nutrigenomics

The goal of nutrigenomics is to understand the impact of diet on the expression of genes of various metabolic pathways through omics approaches. Nutrigenetics and nutrigenomics are complementary to each other, and together, they provide a holistic picture of the metabolism of an individual [57]. A web-based application NutriGenomeDB is a manually curated database developed based on various nutrigenomic experiments carried out in the past [58]. An example to illustrate nutrigenomics: [59] studied the expression of genes in subjects having a High carbohydrate diet (HC) and a High Protein diet (HP). Differential expression of genes was observed in both cases, of which HP breakfast resulted in downregulation of PPRC1 gene, which is involved in causing obesity; thus high protein diet prevents the risk of obesity.

Most of the studies in nutrigenomics pertain to non-communicable chronic diseases like cancer, CVDs, diabetes, obesity and vitamin deficiency. Nearly 30-40% of cancers are influenced by diet [60]. Bioactive compounds having anti-cancerous properties like warfarin, revesterol, sulforaphane, carnesic acid cause downregulation in genes coding for DNA replication, microtubule assembles, signal transduction, necleoporin and kinesin [61]. Flavonoid-containing diets are known to reduce the risk of breast cancer. Recently, it was discovered that flavonoids inhibit aromatase (CYP91A1-CYP450 isoenzyme, which to steroid precursors to estrogen, promotes breast cancer [62]. Other phytochemicals like EECG target apoptotic genes, caspase 8,9 and Bcl2 genes; quercetin targets cyclin, cadherin and c-myc genes; revesterol targets MDR genes, Bax2, Bcl2 and p53 [63]. Pomegranates, *Gingko biloba*, milk thistle and soy have antioxidant properties which ameliorate oxidative stress. They target AMPK/PGCIa/SIRT1 genes involved in mitochondrial biogenesis leading to thermogenesis. Induction of thermogenesis increases energy expenditure and further reduces the risk of obesity and related diseases like hypertension, IR, hyperglycemia and CVDs [64]. Curcumin, caffeic acid, quercetin, genistein, anthocyanins, and gingerols diminish oxidative and inflammatory stress. Curcumin and flavanols have lipid regulatory properties and maintain cholesterol homeostasis. Phenols exhibit antioxidant properties by activating Nrf2 and AICT, which stimulate antioxidant enzymes and inhibit NADPH oxidase [65]. Vitamin D deficiency can lead to impaired immune system, asthma and other respiratory illness, cancer and CVDs. Since individuals show a very personalized response to Vitamin D, they are categorized as low, medium and high based on Vitamin D index.

Earlier Vitamin D index was based on 24 Vitamin D target genes, but nowadays, transcription expression of 702 genes is considered [66]. Thus, exploring the response of genes to vitamin D is crucial for precision medicine.

Antioxidants also protect from neuroinflammation, which causes neurodegenerative diseases like Alzheimer's, Parkinson's and multiple sclerosis. They repress the release of tumour necrosis factor, inhibit the production of iNOS, activate NADPH oxidase and manage ROS detoxification through MKKK and modulation of foxO gene family genes [67]. Association of diabetes has been correlated with gene expression changes in protein kinase B, duodenal homeobox, insulin receptor and glucokinase [68]. Nutrigenomics has expanded its wings in periodontal research [69], autism [70], understanding relations between gut microbe and human health [71] and retinal disease [72]. Nutrigenomics, Overweight/Obesity and Weight Management (NOW) have observed that weight management associated with nutrigenomics can help in the improvement of dietary fat intake in a population [73]. Gut microbiota consists of beneficial bacteria which mediate immune responses by Toll-like receptor signalling (TRL). Human breast milk, a gold standard for infant health, has favoured the growth of these bacteria [74] and has also been implicated in Inflammatory Bowel Syndrome [75]. These findings may assist in devising food supplements and nutraceuticals based on how the bioactive compounds regulate the expression of various genes in an individual.

15.5 Nutriepigenomics

Epigenetic mechanisms refer to gene expression and regulation changes without altering the DNA sequence. This includes histone modifications like acetylation or phosphorylation, DNA methylation and microRNA mediated modifications [76]. Various natural bioactive compounds have been known to impact gene expression in a metabolic pathway through epigenetic modifications. Several epigenetic modifications like hypomethylation, hypermethylation, histone modifications and non-coding RNA dysregulation are a major causal phenomena for cancer [77]. Bioactive compounds like green tea polyphenol epigallocatechin-3-gallate, oleuropin from olive oil and sulforaphane from crucifers have been shown to prevent cancer through epigenetic modifications [78–80]. Improper diet has adverse effects on the crucial genes involved in important biological processes in our body, like high methionine, and low folic acid diets have been shown to increase global methylation rates, thus causing epigenetic silencing of netrin-1 gene. Netrin-1 gene is involved in the maintenance of neuron cells in the brain, and its silencing causes learning and memory defects [81]. Candidate gene approaches and Epigenome Wide Association Studies (EWAS) have led to the identification of 18 differential hypomethylated genes associated with inflammatory responses and could be studied further as biomarkers [82]. Several dietary compounds like caffeic acid, quercitol, luteolin, resveratrol, curcumin, etc, regulate the epigenetic modulation of histone deacetylases (HDACs) in the heart leading to a reduction in heart failure and CVDs [83].

Similarly, phenyl isothiocyanate from crucifers has antitumor effects. They inhibit the activity of DNMTs and HDACs, which repress the activity of CDH1, a tumour suppressor gene involved in preventing breast cancer [84]. Folic acid supplementation reduces the risk of autism, Alzheimer's and schizophrenia through epigenetic modifications [85, 86]. These modifications also play an inevitable role in mediating the offspring's response to various diseases. An inflammatory diet in the prenatal stage, like fruits, vegetables, plant-based proteins, and healthy fats, decides the epigenetic signature of offspring, which may have long-term effects on neurodegenerative diseases [87]. Several genes involved in metabolic pathways and drug actions like CYP genes and UGTs, drug resistance, RNA binding proteins and microRNAs, and drug transporters like Multidrug resistance 1 (MDR) have been reported to undergo epigenetic modifications which can be used as drug targets in pharmacogenomics [88]. Therapeutic approaches regulating microRNA through antisense, microRNA sponges and CRISPR Cas9 are promising tools for clinical applications [89]. Since epigenetic variations are also dependent on sex, age and lifestyle both nutrigenetics and nutrigenomics need to be integrated along with epigenomics for precision medicine.

15.6 Nutrimetabolomics

Nutrimetabolomics explores the impact of diet on the metabolic status of the individual, revealing differences in the metabolic signature of healthy and infected groups [90]. Unravelling the metabolic flux and the metabolome is done through targeted or non-targeted metabolic profiling. Various tools and techniques like GC-MS, LC-MS, NMR, and FT-IR have been utilized separately or in combination to reveal the complexity of metabolomics [91]. Recently a new tool has been developed, Quantitative Dietary Fingerprinting (QDF), which combines Solid Phase Extraction (SPE) with UHPLC MS/MS to quantify 350 dietary markers of phase1 and phase2 metabolites simultaneously [92]. Untargeted metabolomics revealed the presence of adrenic acid (omega-6 PUFA) in subjects with high insulin resistance (IR) and obesity as compared to other control subjects. Thus, adrenic acid can be used as a potential biomarker for both IR and obese [93]. Metabolic syndrome (MetS) is associated with diabetes, IR, and obesity which is generally due to lifestyle and diet changes. Potential biomarkers for MetS like glucose, lactate, uric acid, glutamine, citrate associated with carbohydrate dysfunctioning, N-glycoproteins and Lys phospholipids for inflammation, uric acid and GSG/GSSG ratios for oxidative stress, TMAO for gut biota dysbiosis can be developed as a part of nutrimetabolomics [94]. Association of diet – dietary patterns and plasma metabolite concentration is a requisite for personalized medicine approach in subgroups. Patients with colorectal cancer were observed to have high phosphotidylcholine concentration in their plasma as compared to the control. Carnivore dietary pattern was positively correlated with an increase in phosphatidylcholine concentrations as compared to the western or prudent patterns. Thus, patients with high metabolite phosphatidylcholine should avoid carnivore diet since they risk developing colorectal cancer [95]. Similar associations have been for prostate cancer – homocysteine levels - whole grain rye through NMR-based Metabolomics [96]. A database called Food Ontology Biomarker Ontology (FOBI) has been developed, which explores or integrates information about various biomarkers (metabolites) associated with a specific diet. This understanding may help in strategically selecting a specific diet based on individual's disease state and metabolome [97].

15.7 Conclusion and Future Perspectives

The outcome of all the research on nutritional health is targeted advice for personalized nutrition based on diet, phenotype and genotype to cure and prevent chronic diseases. Nutritional science emphasizes integrating an individual's genetic makeup and then device strategies for improving an individual's health. Several dietary guidelines have been created to improve the overall health of an individual for a person having high cholesterol, dietary guidelines recommend more of unsaturated fatty acids along with lifestyle changes. The amalgamation of all these studies is finally important to establish personalized nutrition on an individual's health. Though enormous research and databases are available in nutrition health studies, still benefits to public health is limiting. Various challenges faced in precision medicine are mechanisms of drug activity, optimization of diet for a specific individuals, and implicating adverse effects of dietary components in an individual [98]. Rather than exploring at the individual level, the dietary requirements need to be analyzed at the subgroup level based on genetic variants, differential gene expression, epigenome studies and metabolic profiling. With an enormous knowledge base, novel scientific and clinical research, applying research to improve public health, personalized nutrition will play a major role in leading a good, happy and healthy life.

References

1. Jew, S., AbuMweis, S.S., Jones, P.J., Evolution of the human diet: linking our ancestral diet to modern functional foods as a means of chronic disease prevention. *J. Med. Food.*, 12, 5, 925-934, 2009.
2. Afshin, A., Sur, P.J., Fay, K.A., Cornaby, L., Ferrara, G., Salama, J.S., Mullany, E.C., Abate, K.H., Abbafati, C., Abebe, Z., Afarideh, M., Health effects of dietary risks in 195 countries, 1990–2017: a systematic analysis for the Global Burden of Disease Study 2017. *Lancet*, 393, 10184, 1958-1972, 2019.
3. Sharma, M., Kishore, A., Roy, D., Joshi, K., A comparison of the Indian diet with the EAT-Lancet reference diet. *BMC Public Health.*, 20, 1-13, 2020.
4. Van Ommen, B., Bouwman, J., Dragsted, L.O., Drevon, C.A., Elliott, R., De Groot, P., Kaput, J., Mathers, J.C., Müller, M., Pepping, F. and Saito, J., Challenges of molecular nutrition research 6: the nutritional phenotype database to store, share and evaluate nutritional systems biology studies. *Genes. Nutr.*, 5, 189–203, 2010.
5. Bashiardes, S., Godneva, A., Elinav, E., Segal, E., Towards utilization of the human genome and microbiome for personalized nutrition. *Curr. Opin. Biotechnol.*, 51, 57-63, 2018.
6. Norheim, F., Gjelstad, I.M., Hjorth, M., Vinknes, K.J., Langleite, T.M., Holen, T., Jensen, J., Dalen, K.T., Karlsen, A.S., Kielland, A., Rustan, A.C., Molecular nutrition research—the modern way of performing nutritional science. *Nutrients*, 412,1898-1944, 2012.
7. Gillies, P.J., Nutrigenomics: the Rubicon of molecular nutrition. *J. Am. Diet. Assoc.*, 103, 12, 50-55, 2003.
8. Gebauer, S.K., VandenHeuvel, J.P., Kris-Etherton, P.M., Gillies, P.J., Integration Of Molecular Biology And Nutrition: The Role Of Nutritional Genomics In Optimizing Dietary Guidance In Lipids. *Future. Lipidol.*, 2, 165-171, 2007.
9. Jump, D.B., and Clarke, S.D., Regulation of gene expression by dietary fat. *Annu. Rev. Nutr.*, 19, 63-90, 1999.
10. Maxfield, F.R., Van Meer G. Cholesterol, the central lipid of mammalian cells. *Curr. Opin. Cell. Biol.*, 22, 422–429, 2010.

11. Hermansen, K., Dinesen, B., Hoie, L.H., Morgenstern, E. and Gruenwald, J., Effects of soy and other natural products on LDL: HDL ratio and other lipid parameters: A literature review. *Adv. Therapy.*, 20, 50, 2003.
12. McNamara, J.R., and Schaefer, E.J., Automated enzymatic standardized lipid analyses for plasma and lipoprotein fractions. *Clinica Chimica Acta.*, 166, 1, 1-8, 1987.
13. Afonso M.S., Machado R.M., Lavrador M.S., Quintao E.C.R., Moore K.J., Lottenberg, A.M., Molecular pathways underlying cholesterol homeostasis. *Nutrients*, 10, 760, 2018.
14. Ishimwe, N., Daliri, E.B., Lee, B.H., Fang, F., Du, G., The perspective on cholesterol-lowering mechanisms of probiotics. *Mol. Nutr. Food. Res.*, 59, 1, 94–105, 2015.
15. Asp N.G., Nutritional classification and analysis of food carbohydrates. *Am. J. Clin. Nutr.*, 59, 3, 679S–681S, 1994.
16. Prinz, P., The role of dietary sugars in health: molecular composition or just calories?. *Eur. J. Clin. Nutr.*, 73, 1216–1223, 2019.
17. Skuratovskaia, D., Komar, A., Vulf, M., Litvinova, L., Mitochondrial destiny in type 2 diabetes: the effects of oxidative stress on the dynamics and biogenesis of mitochondria. *Peer J.*, 8, e9741, 2020.
18. Tremaroli V. and Backhed F., Functional interactions between the gut microbiota and host metabolism. *Nature.*, 489, 242–249, 2012.
19. Portune, K.J., Benítez-Páez, A., Del Pulgar, E.M.G., Cerrudo, V., Sanz, Y., Gut microbiota, diet, and obesity-related disorders—The good, the bad, and the future challenges. *Mol. Nutr. Food. Res.*, 61, 1, 1600252, 2017.
20. Tardy, A.L., Pouteau, E., Marquez, D., Yilmaz, C., Scholey, A., Vitamins and minerals for energy, fatigue and cognition: a narrative review of the biochemical and clinical evidence. *Nutrients.*, 12, 1, 228, 2020.
21. Aslam, M.F., Majeed, S., Aslam, S., Irfan, J.A., Vitamins: Key role players in boosting up immune response—A mini review. *Vitam. Miner.*, 6, 1, 2376-1318, 2017.
22. Borah, P.K., Sarkar, A., Duary, R.K., Water-soluble vitamins for controlling starch digestion: Conformational scrambling and inhibition mechanism of human pancreatic α-amylase by ascorbic acid and folic acid. *Food. Chem.*, 288, 395-404, 2019.
23. Tyagi G., Singh P., Varma-Basil M., Bose M., Role of Vitamins B, C, and D in the fight against tuberculosis. *Int. J. Mycobacteriol.*, 6, 4, 328, 2017.
24. Jain, A., Tiwari, A., Verma, A., Jain, S.K., Vitamins for cancer prevention and treatment: an insight. *Curr. Mol. Med.*, 17, 5, 321-340, 2017.
25. Zhao, C.N., Li, Y., Meng, X., Li, S., Liu, Q., Tang, G.Y., Gan, R.Y., Li, H.B., Insight into the roles of vitamins C and D against cancer: Myth or truth?. *Cancer Lett.*, 431, 161-170, 2018.
26. Gharibzahedi, S.M.T. and Jafari, S.M., The importance of minerals in human nutrition: Bioavailability, food fortification, processing effects and nanoencapsulation. *Food. Sci. Technol. Res.*, 62, 119-132, 2017.

27. Williams, C.M., Ordovas, J.M., Lairon, D., Hesketh, J., Lietz, G., Gibney, M., van Ommen B., The challenges for molecular nutrition research 1: linking genotype to healthy nutrition. *Genes. Nutr.*, 3, 41–49, 2008.
28. Van Ommen, B., Keijer, J., Kleemann, R., Elliott, R., Drevon, CA., McArdle, H., Gibney, M., Müller, M., The challenges for molecular nutrition research 2: quantification of the nutritional phenotype. *Genes. Nutr.*, 3, 51–59, 2008.
29. Daniel, H., Drevon, C.A., Klein, U.I., Kleemann, R., van Ommen, B., The challenges for molecular nutrition research 3: comparative nutrigenomics research as a basis for entering the systems level. *Genes. Nutr.*, 3, 3-4, 101-106, 2008.
30. Van Ommen, B., Keijer, J., Heil, S.G., Kaput J., Challenging homeostasis to define biomarkers for nutrition related health. *Mol Nutr Food Res.*, 53, 7, 795-804, 2009.
31. Fenech, M., El-Sohemy, A., Cahill, L., Ferguson, L.R., French, T.A.C., Tai, E.S., Milner, J., Koh, W.P., Xie, L., Zucker, M., Buckley, M., Nutrigenetics and nutrigenomics: viewpoints on the current status and applications in nutrition research and practice. *Lifestyle. Genom.*, 4, 2, 69-89, 2011.
32. Deka, M., Bose, M., Baruah, B., Bose, P.D., Medhi, S., Bose, S., Saikia, A., Kar, P., Role of CYP2E1 gene polymorphisms association with hepatitis risk in Northeast India. *World J Gastroenterol.*, 16, 38, 4800, 2010.
33. Hamza, T.H., Chen, H., Hill-Burns, E.M., Rhodes, S.L., Montimurro, J., Kay, D.M., Tenesa, A., Kusel, V.I., Sheehan, P., Eaaswarkhanth, M., Yearout, D., Genome-wide gene-environment study identifies glutamate receptor gene GRIN2A as a Parkinson's disease modifier gene via interaction with coffee. *PLoS. Genet.*, 7(8):e1002237, 2011.
34. Ulrich, C.M., Nutrigenetics in Cancer Research—Folate Metabolism and Colorectal Cancer. *J. Nutr.*, 135, 11, 2698–2702, 2005.
35. Da Costa, L.A., Badawi, A., El-Sohemy, A., Nutrigenetics and modulation of oxidative stress. *Ann. Nutr. Metab.*, 60, Suppl. 3, 27-36, 2012.
36. Caradonna, F., Consiglio, O., Luparello, C., Gentile, C., Science and healthy meals in the world: nutritional epigenomics and nutrigenetics of the Mediterranean diet. *Nutrients*. 12(6), 1748, 2020
37. Tang, K.S., Chiu, H.F., Chen, H.H., Eng, H.L., Tsai, C.J., Teng, H.C., Huang, C.S., Link between colorectal cancer and polymorphisms in the uridine-diphosphoglucuronosyltransferase 1A7 and 1A1 genes. *World. J. Gastroenterology.*, 11, 21, 3250-4, 2005.
38. Jokić, M. and Kapitanović, S., Nutrigenetics and colon cancer. *Period. Biol.*, 111, 1, 51-54, 2009.
39. Singh, P.R. and Lele, S.S., Folate gene polymorphisms MTR A2756G, MTRR A66G, and BHMT G742A and risk for coronary artery disease: a meta-analysis. *Genet. Test. Mol. Biomarkers*, 16, 6, 471-5, 2012.
40. Kadkol, S. and Diamond, A.M., The Interaction between Dietary Selenium Intake and Genetics in Determining Cancer Risk and Outcome. *Nutrients.*, 12, 8, 2424, 2020.

41. Steinbrecher, A, Méplan, C, Hesketh, J, Schomburg, L, Endermann, T, Jansen, E, Akesson, B, Rohrmann, S, Linseisen, J., Effects of selenium status and polymorphisms in selenoprotein genes on prostate cancer risk in a prospective study of European men. *Cancer Epidemiol. Biomarkers Prev.*, 19, 11, 2958-68, 2010.
42. Viturro, E., de Oya, M., Lasunción, M.A., Gorgojo, L., Benavente, M., Cano, B., Garces, C., Cholesterol and saturated fat intake determine the effect of polymorphisms at *ABCG5/ABCG8* genes on lipid levels in children. *Genet. Med.*, 8, 594–599, 2006.
43. Abdullah, M.M., Jones, P.J., Eck, P.K., Nutrigenetics of cholesterol metabolism: observational and dietary intervention studies in the post-genomic era. *Nutr. Rev.*, 73, 8, 523-543, 2015.
44. Rudkowska, I., Dewailly, E., Hegele, R.A., Boiteau, V., Dubé-Linteau, A., Abdous, B., Giguere, Y., Chateau-Degat, M.L., Vohl, M.C., Gene–diet interactions on plasma lipid levels in the Inuit population. *Br. J. Nutr.*,109, 5, 953-961, 2013.
45. Neyestani, T.R., Djazayery, A., Shab-Bidar, S., Eshraghian, M.R., Kalayi, A., Shariátzadeh, N., Khalaji, N., Zahedirad, M., Houshiarrad, A., Chamari, M., Asadzadeh, S., Vitamin D Receptor Fok-I polymorphism modulates diabetic host response to vitamin D intake: need for a nutrigenetic approach. *Diabetes care.*, 36, 3, 550-556, 2013.
46. Lucock, M., Yates, Z., Martin, C., Choi, J.H., Boyd, L., Tang, S., Naumovski, N., Roach P. and Veysey M., Hydrogen sulphide-related thiol metabolism and nutrigenetics in relation to hypertension in an elderly population. *Genes. Nutr.*, 8, 2, 221-229, 2013.
47. Mansego, M.L., De Marco, G., Ivorra, C., Lopez-Izquierdo, R., Morcillo, S., Rojo-Martínez, G., González-Albert, V., Martinez, F., Soriguer, F., Martín-Escudero, J.C., Redon, J., The nutrigenetic influence of the interaction between dietary vitamin E and TXN and COMT gene polymorphisms on waist circumference: a case control study. *J. Transl. Med.* 13, 286, 2015.
48. Zanon-Moreno, V., Ciancotti-Olivares, L., Asencio, J., Sanz, P., Ortega-Azorin, C., Pinazo-Duran, M.D., Corella, D., Association between a SLC23A2 gene variation, plasma vitamin C levels, and risk of glaucoma in a Mediterranean population. *Mol. Vis.*, 17, 2997, 2011.
49. Wasilewska, J., Wasilewski, T., Zwierz, K., Epigenetic and nutrigenetic aspects of vitamin C in clinical practice including pregnancy. *Advances in Hygiene & Experimental Medicine*, 73, 2019.
50. Lang, J.E., Mougey, E.B., Allayee, H., Blake, K.V., Lockey, R., Gong, Y., Hossain, J., Killen, K., Lima, J.J., Nutrigenetic response to omega-3 fatty acids in obese asthmatics (NOOA): rationale and methods. *Contemp. Clin. Trials.*, 34, 2, 326-335, 2013.
51. Dufault, R., Lukiw, W.J., Crider, R., Schnoll, R., Wallinga, D., Deth, R., A macroepigenetic approach to identify factors responsible for the autism epidemic in the United States. *Clin. Epigenetics.*, 4, 1, 1-12, 2012.

52. Campos-Perez, W., Perez-Robles, M., Torres-Castillo, N., Rodríguez-Reyes, S.C., De la Cerda Trujillo, L.F., Navarro-Muñiz, E., Lopez-Lizárraga, C.R., Llamas-Covarrubias, I.M. and Martinez-Lopez, E., Physical inactivity and excessive sucrose consumption are associated with higher serum lipids in subjects with Taq1B CETP polymorphism. *J. Hum. Nutr. Diet.*, 33, 3, 299-307, 2020.
53. Giovanella, J., Wollinger, L.M., Capra, L., Dresch, F., Genro, J.P., Contini, V., Diet-gene interaction: effects of polymorphisms in the ACE, AGT and BDKRB2 genes and the consumption of sodium, potassium, calcium, and magnesium on blood pressure of normotensive adult individuals. *Mol. Cell. Biochem.*, 476, 1211–1219, 2021.
54. Damavandi, N. and Zeinali, S., Association of xenobiotic-metabolizing enzymes (GSTM1 and GSTT 1), and pro-inflammatory cytokines (TNF-α and IL-6) genetic polymorphisms with non-alcoholic fatty liver disease. *Mol. Biol. Rep.*, 48, 1225–1231, 2021.
55. Haigh, L., McPherson, S., Mathers, J.C., Anstee, Q.M., Nutrigenetics-based intervention approach for adults with non-alcoholic fatty liver disease (NAFLD): study protocol for a randomised controlled feasibility trial. *BMJ Open.*,11, 4, e045922, 2021.
56. Mahmoudi-Nezhad, M., Farhangi, M.A., Kahroba H., Cocaine and amphetamine-regulated transcript prepropeptide gene (CARTPT) polymorphism interacts with Diet Quality Index-International (DQI-I) and Healthy Eating Index (HEI) to affect hypothalamic hormones and cardio-metabolic risk factors among obese individuals. *J. Transl. Med.*, 18, 1, 1-10, 2020.
57. Marcum, J.A., Nutrigenetics/nutrigenomics, personalized nutrition, and precision healthcare. *Curr. Nutr. Rep.*, 1-8, 2020.
58. Martín-Hernández, R., Reglero, G., Ordovás, J.M., Dávalos, A., Nutrigenomedb: a nutrigenomics exploratory and analytical platform. *Database.* 2019.
59. Van Erk, M.J., Blom, W.A., van Ommen, B., Hendriks, H.F., High-protein and high-carbohydrate breakfasts differentially change the transcriptome of human blood cells. *Am. J. Clin. Nutr.*, 84, 5, 1233-1241, 2006.
60. Ardekani, A.M. and Jabbari, S., Nutrigenomics and cancer. *Avicenna J. Med. Biotechnol.* 1, 1, 9, 2009.
61. Martín-Hernández, R., Reglero, G., Dávalos, A., Data mining of nutrigenomics experiments: identification of a cancer protective gene signature. *J. Funct. Foods.*, 42, 380-386, 2018.
62. Jenzer, H. and Sadeghi-Reeves, L., Nutrigenomics-associated impacts of nutrients on genes and enzymes with special consideration of aromatase. *Front. Nutr.*, 7, 2020
63. Braicu, C., Zanoaga, O., Zimta, A.A., Tigu, A.B., Kilpatrick, K.L., Bishayee, A., Nabavi, S.M. and Berindan-Neagoe, I., Natural compounds modulate the crosstalk between apoptosis-and autophagy-regulated signaling pathways:

Controlling the uncontrolled expansion of tumor cells. *Semin. Oncol.*, 80, 218-236, 2022.

64. Reguero, M., Gómez de Cedrón, M., Wagner, S., Reglero, G., Quintela, J.C., Ramírez de Molina, A., Precision Nutrition to Activate Thermogenesis as a Complementary Approach to Target Obesity and Associated-Metabolic-Disorders. *Cancers.*, 13, 4, 866, 2021.
65. Toma, L., Sanda, G.M., Niculescu, L.S., Deleanu, M., Sima, A.V., Stancu, C.S., Phenolic compounds exerting lipid-regulatory, anti-inflammatory and epigenetic effects as complementary treatments in cardiovascular diseases. *Biomolecules.*, 10, 4, 2020.
66. Carlberg, C., Nutrigenomics of vitamin D. *Nutrients.* 11, 3, 676, 2009.
67. Virmani, A., Pinto, L., Binienda, Z., Ali, S., Food, nutrigenomics, and neurodegeneration—neuroprotection by what you eat. *Mol. Neurobiol.*, 48, 2, 353-362, 2013.
68. Rana, J.S., Liu, J.Y., Moffet, H.H., Jaffe, M., Karter, A.J., Diabetes and prior coronary heart disease are not necessarily risk equivalent for future coronary heart disease events. *J. Gen. Intern. Med.*, 31, 387-393, 2016.
69. Behura, S.S., Rajesh, E., Patra, A., Babu N.A., Nutrigenomics and its Role in Periodontal Health–A Review. *Indian J. Forensic Med. Toxicol.*, 14, 4, 1409, 2020.
70. Jaureguiberry, M.S. and Venturino, A., Nutritional and environmental contributions to autism spectrum disorders: Focus on nutrigenomics as complementary therapy. *Int. J. Vitam. Nutr. Res.*, 17, 1-19 2020.
71. Yin, R., Kuo H.C., Hudlikar, R., Sargsyan, D., Li, S., Wang, L., Wu R., Kong, A.N., Gut microbiota, dietary phytochemicals, and benefits to human health., *Curr. Pharmacol. Rep.*, 5, 5, 332-344, 2019.
72. Neelam, K., Dey, S., Sim, R., Lee, J., Au Eong, K.G., Fructuslycii: A Natural Dietary Supplement for Amelioration of Retinal Diseases. *Nutrients.*, 13, 1, 246, 2021.
73. Horne, J.R., Gilliland, J.A., O'Connor, C.P., Seabrook, J.A., Madill, J., Change in Weight, BMI, and Body Composition in a Population-Based Intervention Versus Genetic-Based Intervention: The NOW Trial. *Obesity.*, 28, 8, 1419-1427, 2020.
74. Toscano, M., De Grandi, R., Grossi, E., Drago, L., Role of the human breast milk-associated microbiota on the newborns' immune system: A mini review. *Front. Microbiol.*, 8, 2100, 2017.
75. Ferguson, L.R., Inflammatory bowel disease: why this provides a useful example of the evolving science of nutrigenomics. *J. R. Soc. N. Z.*, 50, 2, 299-315, 2020.
76. Anderson, O.S., Sant, K.E., Dolinoy D.C., Nutrition and epigenetics: an interplay of dietary methyl donors, one-carbon metabolism and DNA methylation. *J. Nutr. Biochem.*, 23, 8, 853-859, 2012.

77. Joseph, P.V., Abey, S.K., Henderson, W.A., Emerging role of nutri-epigenetics in inflammation and cancer. *Oncol. Nurs. Forum.*, NIH Public Access, 43, 6, 784, 2016.
78. Li, F., Qasim, S., Li, D., Dou, Q.P., Updated review on green tea polyphenol epigallocatechin-3-gallate as a cancer epigenetic regulator. *Semin. Cancer Biol.*, 2021.
79. Su, X., Jiang, X., Meng, L., Dong, X., Shen, Y., Xin, Y., Anticancer activity of sulforaphane: the epigenetic mechanisms and the Nrf2 signaling pathway. *Oxid. Med. Cell. Longev.*, 2018.
80. Hamed, M.M., Handoussa, H., Hussein, N.H., Eissa, R.A., Abdel-Aal, L.K., El Tayebi, H.M., Oleuropin controls miR-194/XIST/PD-L1 loop in triple negative breast cancer: New role of nutri-epigenetics in immune-oncology. *Life Sciences.*,11935, 2021.
81. Kalani, A., Chaturvedi, P., Kalani, K., Kamat, P.K., Chaturvedi, P., Tyagi, N., A high methionine, low folate and vitamin B6/B12 containing diet can be associated with memory loss by epigenetic silencing of netrin-1. *Neural. Regen. Res.*, 14, 7,1247, 2019.
82. Gonzalez-Jaramillo, V., Portilla-Fernandez, E., Glisic, M., Voortman, T., Ghanbari, M., Bramer, W., Chowdhury, R., Nijsten, T., Dehghan, A., Franco, O.H., Nano, J., Epigenetics and inflammatory markers: a systematic review of the current evidence. *Int. J. Inflamm.*, 6273680, 2019.
83. Evans, L.W. and Ferguson, B.S., Food bioactive HDAC inhibitors in the epigenetic regulation of heart failure. *Nutrients*, 10, 8, 1120, 2018.
84. Zhang, T., Zhang, W., Hao, M., Phenethylisothiocyanate reduces breast cancer stem cell-like properties by epigenetic reactivation of CDH1. *Oncol. Rep.*, 45, 1, 337-348, 2021.
85. Nardone, S. and Elliott, E., The interaction between the immune system and epigenetics in the etiology of autism spectrum disorders. *Front. Neurosci.*, 10, 329, 2016.
86. Li, W., Liu, H., Yu, M., Zhang, X., Zhang, Y., Liu, H., Wilson, J.X., Huang, G., Folic acid alters methylation profile of JAK-STAT and long-term depression signaling pathways in Alzheimer's disease models. *Mol. Neurobiol.*, 53, 9, 6548-6556, 2016.
87. Gabbianelli, R. and Damiani, E., Epigenetics and neurodegeneration: role of early-life nutrition. *J. Nutr. Biochem.*, 57, 1-13, 2018.
88. Cacabelos, R., Epigenomic networking in drug development: from pathogenic mechanisms to pharmacogenomics. *Drug Dev. Res..*, 75, 6, 348-365, 2014.
89. Cacabelos, R., Metabolomics of drug resistance in cancer: the epigenetic component. *Metabolomics*: Open Access, 4, 5, 2015.
90. Lorach, R., Garcia-Aloy, M., Tulipani, S., Vazquez-Fresno, R., Andres-Lacueva, C., Nutrimetabolomic strategies to develop new biomarkers of intake and health effects. *J. Agric. Food. Chem.*, 60, 36, 8797-8808, 2012.

91. Rangel-Huerta, O.D. and Gil, A., Nutrimetabolomics: An update on analytical approaches to investigate the role of plant-based foods and their bioactive compounds in non-communicable chronic diseases. *Int. J. Mol. Sci.*,17, 12, 2072, 2016.
92. González-Domínguez, R., Urpi-Sarda, M., Jáuregui, O., Needs, P.W., Kroon P.A. and Andrés-Lacueva C., Quantitative dietary fingerprinting (QDF)—a novel tool for comprehensive dietary assessment based on urinary nutrimetabolomics. *J. Agric. Food. Chem.*, 68, 7, 1851-1861, 2019.
93. Marco-Ramell, A., Tulipani, S., Palau-Rodriguez, M., Gonzalez-Dominguez, R., Miñarro, A., Jauregui, O., Sanchez-Pla, A., Macias-Gonzalez, M., Cardona, F., Tinahones, F.J., Andres-Lacueva, C., Untargeted profiling of concordant/discordant phenotypes of high insulin resistance and obesity to predict the risk of developing diabetes. *J. Proteome. Res.*, 17, 7, 2307-2317, 2018.
94. Hernandez-Baixauli, J., Quesada-Vázquez, S., Mariné-Casadó, R., Gil Cardoso, K., Caimari, A., Del Bas, J.M., Escoté, X. and Baselga-Escudero, L., Detection of early disease risk factors associated with metabolic syndrome: a new era with the NMR metabolomics assessment. *Nutrients*, 12, 3, 806, 2020.
95. Geijsen, A.J., Kok, D.E., van Zutphen, M., Keski-Rahkonen, P., Achaintre, D., Gicquiau, A., Gsur, A., Kruyt, F.M., Ulrich, C.M., Weijenberg, M.P., de Wilt, J.H., Diet quality indices and dietary patterns are associated with plasma metabolites in colorectal cancer patients. *Eur. J. Nutr.*, 1-14, 2021.
96. Moazzami, A.A., Zhang, J.X., Kamal-Eldin, A., Åman, P., Hallmans, G., Johansson, J.E., Andersson, S.O., Nuclear magnetic resonance–based metabolomics enable detection of the effects of a whole grain rye and rye bran diet on the metabolic profile of plasma in prostate cancer patients. *J. Nutr.*, 141, 12, 2126-2132, 2011.
97. Castellano-Escuder, P., González-Domínguez, R., Wishart, D.S., Andrés-Lacueva, C., Sánchez-Pla, A., FOBI: An ontology to represent food intake data and associate it with metabolomic data. *Database*. 2020.
98. Bashiardes, S., Abdeen, S.K., Elinav, E., Personalized nutrition: Are we there yet?. *J. Pediatr. Gastroenterol. Nutr.*, 69, 6, 633-638, 2019.

16

Food Metabolism and Chronic Diseases

Ruby Tiwari[1]*, Aarti Yadav[2], S. Hamsa[1] and Tejpal Dhewa[3]

[1]Department of Genetics, University of Delhi, South Campus, New Delhi, India
[2]Department of Microbiology, Shaheed Rajguru College of Applied Sciences for Women, New Delhi, India
[3]Department of Nutrition Biology, Central University of Haryana, Mahendergarh, Haryana, India

Abstract

Our body performs a biochemical process and transforms energy stored in the food to utilize it as fuel. This process is broadly referred to as food metabolism. The human diet consists of macronutrients, i.e., proteins, carbohydrates, and fats, that provide energy and various micronutrients (vitamins and minerals) required in small amounts by the body. Literature studies have shown that protein, lipid, and carbohydrate metabolism abnormalities are associated with a chronic condition. Failure of the metabolic process and too much or little essential metabolites leads to a metabolic disorder that causes chronic diseases. The majority of known metabolic disorders such as obesity, hypertension, metabolic syndrome, neurodegeneration, cancer, and type-2 diabetes have reached epidemic. The changes in nutritional level, nutrients processing affect chronic conditions beyond those considered to be specific metabolic disorders. Hence, the management of these chronic conditions is multifaceted, and alteration in food metabolism becomes one of the modifiable risks for most chronic diseases. The knowledge of detailed mechanisms and biochemistry behind metabolic processes will help evaluate the condition with a better disease prognosis. Besides, modern biotechnological approaches have uncovered and increased our understanding of chronic diseases linked to metabolic disorders. This chapter attempts to summarize and better understand the metabolic processes involved and the defects in these processes that lead to chronic conditions which will help in prevention and cure of disease.

***Keywords*:** Food metabolism, chronic disease, metabolic derangements, metabolites, diet, food nutrition

**Corresponding author*: ruby12tiwari@gmail.com

Tejpal Dhewa, Anil Kumar Puniya and Anil Panghal (eds.) Nutritional Science and Technology: Concept to Application, (355–382) © 2023 Scrivener Publishing LLC

16.1 Introduction

The last century has shown improvement in life expectancy, which has increased from 45 years to 77 years presently [1]. This significant increase is mainly due to improved healthcare, sanitation, better hygiene, development of vaccines and antibiotics [2]. However, evidence has determined that the average lifespan is much greater than the achieved improvement. This directly correlates with the rise in the number of people affected with chronic diseases- respiratory diseases, cardiovascular diseases, malignancy, type 2 diabetes, and more [3]. Data shows 71% of global deaths are due to chronic diseases occurring due to alterations in metabolism. The metabolic disorder includes impaired glucose tolerance, insulin resistance, and obesity, increasing the likelihood of cardiovascular disease, type 2 diabetes mellitus, and cancer. The complex interaction in food metabolism within the human contributes to chronic organ damage [4].

Chronic diseases that result from genetic, physiological, environmental, and behavioural factors include cardiovascular diseases, cancers, chronic respiratory disorders, osteoporosis, diabetes, Alzheimer's disease, and chronic kidney disease. The increasing prevalence of these diseases demands proper management strategies that usually focus on nutritional, physical activity, dietary, and lifestyle modifications. The food metabolism role is crucial and is reported to be associated with most chronic diseases as a single condition or in comorbid states [5, 6]. National Center for Chronic Disease Prevention and Health Promotion (NCCDPHP) defines chronic disease as a condition that is long-lasting, usually for a year or more. Such conditions require continuous medical attention or limited day-to-day activity, or both. Chronic diseases such as cardiovascular, diabetes, and cancer are the leading cause of death in the United States, costing the nation $3.8 trillion in the annual health sector. According to NCCDPHP, major risk behaviours behind such conditions are the use of tobacco or secondhand smoke exposure, lack of physical exercise, poor nutrition, and excessive drinking of alcohol.

According to the WHO technical report series, a modifiable determinant of chronic disease is food nutrition and is largely preventable. The National Library of Medicine defines metabolism as the sequence of events undertaken to break down complex molecules for sustained energy production. A defect in the metabolic process causes the body to synthesize either too much or little essential metabolites [7]. Our bodies need carbohydrates, fats, protein, minerals, and vitamins to perform numerous functions like amino acids. The brain needs sodium, potassium, and calcium to generate electrical impulses and lipids for the maintenance of the nervous system [8]. A metabolic disorder can range from a missing vitamin or enzyme important for a chemical reaction to a disease in the pancrease, liver, or other organs involved in metabolism [9]. All food must pass through the gastrointestinal tract for absorption of food and intake of energy that can be modified by hormone action [10]. Studies show that children born with low birth weight are at greater risk for developing metabolic diseases such as obesity, excess weight, hypertension, etc. [11]. Studies have shown that bioactive compounds in food possess several benefits like anti-cancerous, anti-allergens, anti-fungal, and anti-inflammatory properties. These compounds in food, once consumed, are broken down into various metabolites via numerous biological processes in the human body [12]. This signifies that any food that a human body consumes does not directly reach the bloodstream in its complex form. Rather, it is broken into several compounds and absorbed after digestion [13]. Hence, food metabolism, including dosage and availability, plays an important role while proposing respective health effects [14]. This chapter largely talks about various food metabolic pathways and associated chronic diseases in humans.

16.2 Food Metabolism and Associated Diseases

Carbohydrate metabolism

Carbohydrates are organic molecules that include both simple sugars glucose and fructose, and complex sugars comprising starch, cellulose and glycogen. Carbohydrate is broken down to generate ATP via glycolysis, the kerb cycle, and oxidative phosphorylation [15, 16]. In normal conditions, monosaccharides are stored in the skeletal muscle and liver cells, but in obesity, excess glucose gets converted to triglycerides which are then stored in lipid droplets in mature adipocytes [17, 18]. In the lipotoxicity state, other liver tissues, pancreatic beta cells and skeletal muscles accumulate triacylglycerol. Hence, this leads to metabolic deregulation and an enhanced rate of insulin resistance and type II diabetes [19, 20]. Increasing rates of diabetes type II and insulin resistance and several other related metabolic abnormalities lead to cardiovascular disease [20, 21]. This has focussed interest on identifying major nutritional determinants of metabolic derangements. Since, sucrose is considered to be the prime determinant of Type II diabetes, foods with low glycaemic index (GI) are favoured and encouraged [22]. Glycemic response (GR) is a tool created in 1981 to select food for diabetic persons, and carbohydrate-rich food greatly impacts GR. GR is defined as the appearance of glucose in the blood after eating and depends on a number of factors like the entry rate of glucose into the circulation, absorption of glucose amount, the disappearance rate from the circulation due to uptake by cells, and hepatic glucose release regulation [23, 24].

The most important carbohydrate is glucose, and 70% of glucose uptake is by muscle tissue. Glucose is directly converted to glycogen through glycogenesis (70%) or enters the glycolytic pathway to produce energy (90%), and only 10% go through anaerobic glycolysis [18]. Insulin and glucagon are central regulators of glucose metabolism [25]. Insulin resistance is defined as 200 units or more of insulin requirement per day to produce an expected biological effect and attain glycemic control to prevent ketosis, i.e., IR condition needs higher than normal insulin concentrations to achieve normal metabolic response [26]. Insulin resistance (IR) not only causes impairment of glucose tolerance but is also an underlying factor of series of diseases in a combination known as the "metabolic syndrome" or "Syndrome X" [27]. The two hormones, Gastric inhibitory polypeptide (GIP) and glucagon-like peptide-1 (GLP-1) stimulate the secretion of insulin from pancreatic cells [25, 28]. Glucose and fatty acid metabolism have been observed to play a crucial role in IR. Ingestion of glucose in the evening leads to higher plasma glucose concentration than in the morning [25]. This diurnal variation has been attributed to the following:

1. Gene expression changes are involved in fatty acid oxidation occurring in muscle and adipose tissue during lipogenesis. The former has a higher expression in the morning while the latter in the evening. The free fatty acids in the evening derived from previous meals impair insulin function [29].
2. Glucose metabolism in the liver is through gluconeogenesis which uses glutamine as the predominant precursor to produce glucose in the final step. Glucose is then stored as glycogen in the liver. The deficiency of several enzymes involved in this process like aldolase B, pyruvate carboxylase, fructose-1,6-bisphosphatase, Phosphoenolpyruvate carboxykinase (PEPCK) causes hypoglycaemia and lactic acidosis [30].

A moderate amount of sugar is required for energy balance, and one should restrict sugar-containing energy-dense foods as these promote overweight and obesity [31, 32]. It has been reported that diets enriched with whole grain and dietary fibre are protective against diabetes and pre-diabetic states as they promote satiety and weight loss [21, 33]. Some studies also report that diets with a low GI protect against carbohydrate

Table 16.1 Few metabolic disorders associated with carbohydrate metabolism.

Deficient enzyme	Disorder	Clinical manifestation	References
Fructose-1-phosphate-aldolase	Fructosemia	Aversion to sweets and fruits. Abdominal pain, vomiting, and hypoglycaemia. If left untreated, it can harm the liver and kidneys	[37]
Galactose-1-phosphate uridyl transferase	Galactosemia	Jaundice, inflamed and swollen liver, loss of weight; hypoglycaemia, renal collapse, ovarian failure in women, retarded mental and physical development, cataract in the eyes	[37]
L-xylulase	Pentosuria	Xylulose levels in the urine significantly increased and completely harmless	[37]
Hepatic fructokinase	Fructosuria	Significant increase in fructose levels in the urine and blood. It's extremely rare and completely harmless	[37]
Muscle glycogen phosphorylase	McArdle disease (Glycogen storage disorder (GSD) type V)	Glycogen levels increase moderately, muscle cramps	[37]
Amylo 1,6-glucosidase (debranching enzyme; muscle and liver)	Cori's disease (GSD type III)	Increase in branched glycogen, liver enlargement, hyperlipemia, hypoglycemia, hyperuricemia, ketosis	[37]
Glucose-6-phosphatase (liver and kidney)	Von Gierke's disease (GSD type I)	Hyperlipemia, inability to thrive, elevated glycogen, enormous liver enlargement, severe hypoglycemia, ketosis, hyperuricemia	[37]

metabolism-related disorders [21]. A study analyzed that person consuming a diet composed of high protein and unsaturated fatty acids; a moderate amount of fibre-rich carbohydrates; low in total fat, but relatively more fibre-rich carbohydrates sustained weight reduction and improvement in metabolic players [34]. Weight loss or obesity is needed to be managed to improve insulin sensitivity which can be achieved by the intake of a carbohydrate-rich diet provided by whole grain and dietary fibres [35]. Studies have also suggested that high sucrose and high fructose (corn syrup) have unpleasant effects on humans and contribute to overweight and obesity, leading to metabolic disorder [36].

Finally, it is necessary to emphasize the consumption of conventional and functional foods having low GI and high dietary fibre [21]. However, evidence for the benefit of these foods is still limited. Some reports were undertaken to study the extent of magnitude and nature of dietary carbohydrates that influence IR, GI and other related cardiovascular diseases in individuals with type II diabetes [21]. Some metabolic disorders associated with carbohydrates are discussed in Table 16.1.

Lipids and fats metabolism

Metabolism of lipids involves the synthesis of functional and structural lipids (glycolipids, phospholipids, sphingolipids, cholesterol, prostaglandins) and their degradation for the production of energy. This suggests that lipids are in a dynamic equilibrium state where some lipids are continuously being oxidized to meet the immediate energy requirement of the body, while others are being synthesized and stored. This dynamic equilibrium determines the body's fat mass [38]. Leptin is one of the important hormones that regulate lipid metabolism, and the majority of the total energy reserves (90%) stored as triglycerides (TG) in adipocytes can be hydrolyzed following hormonal stimulation and thus releases fatty acids (FA) [39]. Two possible fates of FA are B-oxidation that produces energy (ATP), or re-esterification again into neutral lipids TGs for storage [39]. Another essential plasma lipid, along with TG is cholesterol. The human body tightly controls the homeostasis of cholesterol, and its deregulation is linked to atherosclerosis and cardiovascular diseases [40]. Excessive intake of food or genetic factors results in the accumulation of cholesterol in the heart's arterial wall leading to atherosclerosis, vessel dysfunction, and blocked blood flow. This further leads to heart attack and stroke (CVDs), which are responsible for the majority of deaths in the western world [40, 41].

Cholesterol is hydrophobic and forms a complex with protein called the lipoprotein complex, which is then transported to various body parts.

The Lipoprotein complex composes of a hydrophilic phospholipid layer, apolipoproteins, and a hydrophobic cholesterol ester core. Lipoproteins have been classified into seven categories:

1. Chylomicrons
2. chylomicron remnants
3. very-low-density lipoproteins (VLDL),
4. intermediate-density lipoproteins (IDL),
5. low-density lipoproteins (LDL),
6. high-density lipoproteins (HDL), and
7. lipoprotein

Chylomicrons function to shuttle cholesterol and triglycerides via the chylous fluid to the liver. Its size varies with the amount of consumption of dietary fat. Another critical transport protein called microsomal transport protein (MTP) is essential for the movement of lipids to apoproteins from the endoplasmic reticulum in intestinal epithelial cells. The defect and lack of MTP lead to abetalipoproteinemia, an inability to produce chylomicrons. A chylomicron remnant, cholesterol-enriched-VLDL particles are synthesized in the liver, and this also utilizes MTP for the uptake and transfer of cholesterol and TG. IDL forms from VLDL particles once triglycerides are removed from peripheral tissue. IDL transforms into LDL particles and carries the majority of circulating cholesterol. The LDL molecule easily binds and penetrates the arterial wall and forms a tendency to atherogenesis [42].

Lipid metabolism involves exogenous lipoprotein pathways, which start with ingested dietary lipids as triglycerides-rich chylomicrons. This is metabolized by lipoprotein lipase in adipose and muscle tissue. Finally, the liver rapidly takes a byproduct- Chylomicron remnant through various receptors, thus completing the pathway. A pathway is initiated endogenously with VLDL formulation in the liver. VLDL is metabolized by lipoprotein lipase into ester-enriched LDL molecules and absorbed through LDL receptors which further gets resorbed in the liver. Mature HDL is formed when lipids and cholesterol are deposited into Apo AI via ATP-binding cassette 1 (ABCA1). Mature HDL molecules help in depositing cholesterol to the liver [42]. Patients with Insulin-resistant and chronic kidney disease (CDK) are associated with higher catabolic rate of Apo A-1 [43].

Additionally, progressive kidney disease is linked to low levels of lipoprotein lipase, dysfunction of endothelial cells, structural changes to IDL and LDL, and decreased receptor mediating lipoprotein re-uptake [44, 45].

It has been shown that CKD leads to impaired lipid metabolism and the formation of excess small LDL molecules, increased circulating atherogenic lipoproteins leading to atherosclerosis, decreased lipoprotein lipase activity, and defect in reverse cholesterol transport. Statins are a class of drugs that reduces cholesterol synthesis and increases LDL receptor expression in hepatocytes, optimizing the removal of circulating LDL. Thus, lipids, essential membrane constituents, need to have a tight balance between synthesis, uptake, and export. It is linked with several chronic diseases and cannot be degraded in higher organisms [46].

Similarly, an alteration in fat metabolism results in increased free fatty acid flux from fatty acid tissues (adipose tissue) to the non-adipose organ, leading to the accumulation of hepatic triglyceride, contributing to abnormal glucose metabolism and insulin sensitivity in muscle and the hepatic tissues [47]. Insulin resistance (IR) is common in metabolic syndrome and is a systemic disease affecting the nervous system, muscles, pancreas, kidney, heart, immune system, and liver [48]. The metabolic syndrome causes advanced fibrotic liver disease, and such abnormalities are common in liver disease. *Hepatogenous diabetes*, describes hyperglycemia that develops in advanced cirrhosis. The liver is a major player in nutrient and hormone metabolism. Lately, insulin metabolism has been shown to directly impact the pathogenesis of liver disease ranging from fatty liver to non-alcoholic steatohepatitis. Insulin binding to its receptor initiates a cascade of downstream signalling events, and the mechanism involved in IR either involves insulin binding IRS protein or translocation of GLUT4 (Glucose Transporter 4) [48].

Protein metabolism

Proteins are composed of amino acids (AAs) and are well-balanced in a healthy individual. AAs serve various functions in our body, are pivotal for life of the cell, and are the only nitrogen source in mammals [49]. This nitrogen is utilized in synthesizing precursors such as purines and pyrimidines and regulates numerous biochemical processes. Thus, AAs are biochemical totipotent molecules that convert into energy, carbohydrates, and lipids as the body demands. This corresponds to a higher demand for AAs in hypercatabolic conditions wherein there is increased energy expenditure in the resting state [49]. This increased catabolism and impaired anabolic/catabolic stimulation of protein refers as "hypercatabolic syndrome" (HS), which causes an imbalance in body's metabolism [50]. Hyper-catabolism results in biochemical consequences and signs such as anaemia, sarcopenia, infection hypalbuminemia, etc. [49]. HS induces IR and decreases cytoplasmic and mitochondrial cell synthesis of protein, impairing cell metabolism. HS reduces appetite leading to reduced nutrient availability, including AAs [50, 51]. It has been shown that 50% of chronic condition patients show altered protein metabolism associated with muscle wasting in older patients, increasing life-threatening complications leading to respiratory failure and pulmonary edema [49, 52, 53]. These can also result in cardiac dysfunction, renal insufficiency, and ventricular arrhythmias [53]. Protein metabolism plays a notable role in albumin, prothrombin synthesis, and ammonia detoxification [54]. It has been studied that 30% of chronic heart failure patients exhibited reduced serum albumin, which increases morbidity and mortality [55].

Zinc, a trace metal, has an important role in the structural maintenance and activation of approximately 300 enzymes and proteins in the body that contributes to growth, immune response, aging, and carcinogenesis [56–61]. Zinc deficiency is most often observed in patients suffering from chronic liver conditions, which strikes protein metabolism [54]. In the urea cycle, a zinc enzyme-ornithine transcarbamylase is involved and its activity is simultaneously decreased during zinc deficiency. Thus, it reduces the hepatic capacity to metabolize ammonia [54, 62, 63]. As the disease progresses, skeletal muscle mass decreases due to decreased ability to metabolize ammonia in the liver, resulting in hyperammonemia [54]. Branched-chain amino acids (BCAA) increase protein synthesis by phosphorylation of mTOR and further activation of protein kinase S6 and eukaryotic transcription factor 4E binding protein 1 in human muscle [64]. The metabolism of BCAA plays an important role during physical exercise and fasting by decreasing the catabolism of these amino acids. Dysregulation of branched chain amino acids (BCAA) catabolism leads to heart failure [65], Insulin Resistance [66], non-alcoholic fatty liver disease [67], and others. Therefore, proper dietary intake providing sufficient AAs matching the organ demand preserves the organ integrity, ultimately, patient survival. The identification and treatment of protein metabolic impairment, muscle mass maintenance, and protein metabolism are relevant parameters to study for better prognosis with hypercatabolic inflammatory disease [49]. Some common food metabolic pathways and associated chronic diseases are discussed in Table 16.2.

Table 16.2 Common food metabolic pathways and associated chronic diseases.

Metabolic pathway	Key regulator/ biomarker	Chronic diseases	References
Phosphorus	Fibroblast Growth Factor 23 (FGF23)	Chronic Kidney disease, Cardiovascular disease	[68, 69]
Glucose	Suppression of tumorigenicity 2 (sST2)	Type 2 diabetes, Inflammation, obesity and impaired glucose tolerance, Insulin resistance, Cardiovascular disease	[70]
	Concentration of breath acetone, β-hydroxybutyrate	Lipolysis	[71]
	C/EBPα binding sites and FTO promoter	Risk for diabetes	[72]
	C-peptide, Pro-insulin, Fasting plasma glucose (FPG), oral glucose tolerance test (OGTT) glycated haemoglobin (HbA1c)	Obesity, type 2 diabetes mellitus (T2DM)	[73, 74]
Cholesterol	Campesterol, sitosterol and cholestanol, lathosterol and desmosterol	Cardiovascular disease, diabetes mellitus, hyperlipidemia, intestinal disease, kidney disease, liver disease, metabolic syndrome	[75]

(*Continued*)

Table 16.2 Common food metabolic pathways and associated chronic diseases. (*Continued*)

Metabolic pathway	Key regulator/ biomarker	Chronic diseases	References
	The oxidation products of cholesterol	Peroxisomal disorders, and lysosomal storage disorders	[76]
Phenolic compound	4-*O*-methylgallic acid (4OMGA) and isoferulicacid	Obesity, diabetes	[77]
Vitamin D	Vitamin D receptor (VDR), Retinoid X receptor (RXR)	Obesity, diabetes mellitus, insulin resistance, cardiovascular diseases including hypertension	[78]
Insulin	Class I glucose transporters, GLUT1–GLUT4	Impaired glucose homeostasis, impacts lipid metabolism	[79]
Triglycerides	Triglyceride/HDL-C, apoB-containing lipoproteins	Cardiovascular disease	[80]
	HDL-triglyceride HDL-cholesterol, circulating cholesterol ester transfer protein	Cardiovascular disease	[81]

(*Continued*)

Table 16.2 Common food metabolic pathways and associated chronic diseases. (*Continued*)

Metabolic pathway	Key regulator/ biomarker	Chronic diseases	References
Polyunsaturated fatty acids	Plasma/serum fatty acids, circulating *n*-3 PUFAs, especially DPA and DHA	Obesity, hypertension, dyslipidemia (elevated triglyceride (TG) and low high-density lipoprotein cholesterol concentrations), and impaired glucose tolerance	[82]
Thyroid hormone	5'-D2, 5′-deiodinase type 2; TRH, thyrotropin-releasing hormone; TSH, thyroid-stimulating hormone	End-stage renal disease	[83]

16.3 Nutrient Intake Goals for Preventing Diet-Related Chronic Disease

The alteration in genetic information that responds to food metabolism, nutrients and bioactive compounds refers to as nutrigenomics. It has been noted that genes and diet both regulate an individual's health and susceptibility to a particular disease. Thus, identifying candidate genes that regulate diet and cause chronic conditions is of utmost importance [84]. The information gained from such studies can help develop diagnostic tools and strategies to maintain health [85, 86]. Contemporary lifestyle has dramatically reduced physical activity, which remains the key element in weight control [87]. Regular physical exercise helps in maintaining a healthy weight and averts the likelihood of chronic diseases such as type 2 diabetes, cardiovascular disease, osteoarthritis, colon cancer, etc. [88–90]. Traditional plant-based diets, such as fruit, vegetables, nuts, and whole grains, have been associated with low rates of CVD and mortality [91, 92]. Numerous epidemiologic and clinical works have examined this, and most found protective effects of the plant-based diet against chronic diseases

[93–95]. Polyphenols have a crucial role in averting diseases related to lifestyle. Flavonoids are major polyphenols and are found in a wide variety of plant foods [96]. Green tea has been shown to reduce obesity due to caffeine and theanine. It has also been studied that when high-fat and high sucrose-fed mice were fed with green tea polyphenols, the weight of the liver and lipid content got reduced [97, 98]. Phytonutrients found in plants and vegetables have been shown to reduce the risk of cancer [99], and food such as legumes, and nuts minimizes the possibility of many chronic conditions [100]. Studies have pointed towards including plant-based food, including whole grain as a source of carbohydrate and unsaturated fats, fruits, and vegetables that plays a paramount role in preventing chronic disease and deserve more attention in dietary recommendations [101, 102]. It is highly recommended to consume fruits and vegetables as they protect against chronic diseases due to bioactive components (vitamins, minerals, fiber, and phytochemicals) and are also studied to play a role in the modulation of inflammation [103].

Eating a healthy diet has long been recognized to lower the chances of CVD, renal stones, diabetes, cancers, dental disease, and birth defects.

1. Replacing saturated and trans fats with unsaturated fats as trans-fats have adverse effects on blood lipids [104].
2. Ensure higher consumption of vegetables, fruits, and potassium and folic acid [105, 106].
3. Increase the intake of cereal products in high-fiber and whole-grain form [107].
4. Avoid and reduce the intake of sugar and beverages that are sugar-based [108].
5. Limit excessive calorie and sodium intake [109].

Following are the list of some of the dietary intake, their content, and benefits for maintaining a healthy lifestyle.

Fruits and vegetables:

- Cranberry contains antioxidants, vitamin C, citric, quinic and malic acids, and other phytochemicals) which benefit the urinary tract and cardiovascular system [110].
- Grape has resveratrol provides several health benefits [111].
- Pomegranate is composed of antioxidants, and phenolic compounds, which induce cyclooxygenase-2 (COX-2) inhibition and subsequently reduce Nitric Oxide (Ino),

Table 16.3 Dietary intake to maintain a healthy lifestyle.

Diet/functional food	**Metabolism affected**	**Disease/ disorder**	**References**
Green tea	Lipid transport and metabolism	Atherosclerosis and CVDs	[4, 123–125]
Flax seeds	Cholesterol homeostasis and lipid regulation	Cancer, Insulin Resistance, Dyslipidemia	[126–128]
Nuts (Almonds, Walnuts, Hazelnuts)	Glucose and fat metabolism	Insulin Resistance	[129–132]
Vegetable oils	Fat metabolism	Obesity, Coronary heart disease	[133]
Chia seeds	Fat metabolism	Obesity and related diseases	[134, 135]
Legumes	Glucose, fat, and protein	Inflammatory bowel disease, Indigestion, NAFLD, and hypertension	[136]
Camellia sinensis	Fat metabolism	Coronary heart disease, CVDs	[137, 138]
Barley	Glucose and lipid metabolism	Inflammatory bowel syndrome	[139]
Spices	Fat and lipid metabolism	Heart, liver, and skin diseases	[140, 141]
Vitamins	Fats, carbohydrate, protein	Cancer, CVDs, protein deficiency	[121, 142–144]
Whole grain cereals	Carbohydrate metabolism	Obesity and Type 2 diabetes	[145–149]

prostaglandin E-2 (PGE-2), inflammatory cytokines, reactive oxygen species (ROS) [112].

- Strawberry majorly contains flavonoids, flavonols, phenolic and ellagic acids, glutathione, and ascorbic acid, which help fight inflammation in chronic diseases [113].
- Whole grains comprise insoluble fiber, phenolic compounds, vitamin E, phytosterols [114–116] and provide better digestion and well-balanced microbiota [117].
- Tea and spices contain antioxidants having anti-inflammatory properties [118].
- Legumes (Soy, Nuts, and seeds) are highly rich in fiber, proteins, and minerals which play roles in cardiovascular protection [119].

Animal source food:

- Yoghurt is rich in probiotics (protein, calcium, nutrients) that improve glucose metabolism and cardiovascular protection [120, 121].
- Sardines contain proteins and omega-3 fatty acids that help in preventing heart diseases [122].

The detailed list is provided in Table 16.3.

16.4 Conclusion and Future Prospects

In today's scenario, the rise in consumption of unhealthy diet is leading to tremendrous increase in individual's body weight. The sedentary lifestyle of people results into metabolic alterations in an individual. Obesity is one of the significant risk factors for major diseases, including coronary heart disease, hypertension, cancers, and Type II diabetes [150]. Another area of concern is that plant-based food has been swiftly and steadily replaced by a high-fat, energy-dense diet that is one of the risk factors in the rise of lifestyle-associated chronic diseases. Intake of an empty calorie diet which is poor in nutrition, smoking and physical inactivity, are major causes for the onset of some of the most prevalent chronic conditions [151]. This burdens human and economic costs from CVD, diabetes, cancer etc. which are expected to rise in the near future without substantial reduction in the underlying risk factors and economical. It has been studied that genes and diet govern one's health and susceptibility to disease [152].

Therefore, research in the area to identify genes regulated by diet that cause or contribute towards chronic diseases would result in the development of strategies such as individualized intervention and diagnostic tools for maintaining health. Maintaining a proper diet, including whole grain food, legumes, fruits, vegetables and limiting full-fat dairy products, red meat, and sugar-based food and beverages helps reduce the risk of chronic diseases. Several dietary flavonoids have anti-osteoporotic anti-inflammatory, and anti-oxidative properties that are relevant and help in preventing lifestyle-related and chronic diseases [153–155]. Thus, strategies to cope with the ever-increasing rate of chronic diseases are multidimensional, and approaches to deal with such conditions demand nutritional and/or dietary interventions, modifications in lifestyle and regular physical activity. Identifying and treating metabolic impairment and appropriate therapies will be crucial to improving the prognosis of chronically ill patients. The role of food nutrition is primary and vital for many chronic diseases as diet is a modifiable risk factor and should be emphasized in research. Hence, multidimensional approaches are essential in managing chronic conditions, and dietary interventions are central to these strategies.

References

1. Christensen, K., Doblhammer, G., Rau, R., Vaupel, J.W., Ageing populations: the challenges ahead. *Lancet*, 374, 1196-1208, 2009.
2. US Government Printing Office, Centers for Disease Control and Prevention, Achievements in Public Health, 1900–1999: Control of InfectiousDiseases, *Morbidity and Mortality Weekly Report* 48(29), US GovernmentPrinting Office, Atlanta, GA, 621–629, 1999.
3. Eyre, H., Kahn, R., Robertson, R.M., ACS/ADA/AHA Collaborative Writing Committee, ACS/ADA/AHA Collaborative Writing Committee Members, Clark, N.G., Doyle, C., Hong, Y., Gansler, T., Glynn, T., Smith, R.A., Preventing cancer, cardiovascular disease, and diabetes: a common agenda for the American Cancer Society, the American Diabetes Association, and the American Heart Association. *Circulation*, 109, 25, 3244-3255, 2004.
4. Luvián-Morales, J., Varela-Castillo, F.O., Flores-Cisneros, L., Cetina-Pérez, L., Castro-Eguiluz, D., Functional foods modulating inflammation and metabolism in chronic diseases: a systematic review. *Crit. Rev. Food Sci. Nutr.*, 1-22, 2021.
5. World Health Organisation (WHO) Noncommunicable Diseases, [(accessed on 29 January 2019)]; 2018.
6. Ojo, O., Nutrition and chronic conditions. *Nutrients*, 459, 2019.

7. Rice, S.C., Nutrition and metabolism disorder. *Healtline*, (Reviewed by Olsen, 2018) https://www.healthline.com/health/nutrition-metabolism-disorders. 2018.
8. Morris, A.L., and Mohiuddin, S.S., Biochemistry, Nutrients, StatPearls Publishing, Treasure Island (FL), 2020.
9. Agana, M., Frueh, J., Kamboj, M., Patel, D.R., Kanungo, S., Common metabolic disorder (inborn errors of metabolism) concerns in primary care practice. *Ann. Transl. Med.*, 6, 24, 2018.
10. Begg, D.P. and Woods, S.C., The endocrinology of food intake. *Nat. Rev. Endocrinol.*, 9, 10, 584-597, 2013.
11. Jain, V. and Singhal, A., Catch up growth in low birth weight infants: striking a healthy balance. *Rev. Endocr. Metab. Disord.*, 13, 2, 141-147, 2012.
12. Samtiya, M., Aluko, R.E., Dhewa, T., Moreno-Rojas, J.M., Potential Health Benefits of Plant Food-Derived Bioactive Components: An Overview. *Foods*, 10, 4, 839, 2021.
13. Judge, A. and Dodd, M.S., Metabolism. *Essays Biochem.*, 64, 4, 607–647, 2020.
14. Hasler, C.M., Functional foods: benefits, concerns and challenges-a position paper from the American Council on Science and Health. *J. Nutr.*, 132, 12, 3772-3781, 2002.
15. Kowaltowski, A. and Abdulkader, F., Carbohydrate Metabolism. In *Where Does All That Food Go?*, 17-36, Copernicus, Cham, 2020.
16. Hemegowda, R., Sanjana, M.C., Sushma, R.E., Carbohydrate metabolism-A constant supply of energy, EPRA. *Int. J. Res. Dev.*, 4, 2, 151-156, 2019.
17. Sikaris, K.A., The clinical biochemistry of obesity. *Clin. Biochem. Rev.*, 25, 3, 165, 2004.
18. Jensen, J., Rustad, P.I., Kolnes, A.J., Lai, Y.C., The role of skeletal muscle glycogen breakdown for regulation of insulin sensitivity by exercise. *Front. Physiol.*, 2, 112, 2011.
19. Unger, R.H. and Zhou, Y.T., Lipotoxicity of beta-cells in obesity and in other causes of fatty acid spillover, *Diabetes*, 50, suppl 1, S118, 2001.
20. Ormazabal, V., Nair, S., Elfeky, O., Aguayo, C., Salomon. C, Zuñiga, F.A., Association between insulin resistance and the development of cardiovascular disease, *Cardiovasc. Diabetol.*, 17, 1, 1-14, 2018.
21. Mann, J., Dietary carbohydrate: relationship to cardiovascular disease and disorders of carbohydrate metabolism, *Eur. J. Clin. Nutr.*, *61*, 1, S100-S111, 2007.
22. Yudkin, J., Dietary fat and dietary sugar in relation to ischaemic heart disease and diabetes. *Lancet*, 4, 4–5, 1964.
23. Hermansen, M.L.F., Eriksen, N.M., Mortensen, L.S., Holm, L., Hermansen, K., Can the Glycemic Index (GI) be used as a tool in the prevention and management of type 2 diabetes?. *Rev. diabet. stud.*, 3, 2, 61, 2006.

24. Augustin, L.S., Kendall, C.W., Jenkins, D.J., Willett, W.C., Astrup, A., Barclay, A.W., Björck, I., Brand-Miller, J.C., Brighenti, F., Buyken, A.E., Ceriello, A., Glycemic index, glycemic load and glycemic response: an International Scientific Consensus Summit from the International Carbohydrate Quality Consortium (ICQC). *Nutr. Metab. Cardiovasc. Dis.*, 25, 9, 795-815, 2015.
25. Aronoff, S.L., Berkowitz, K., Shreiner, B., Want, L., Glucose metabolism and regulation: beyond insulin and glucagon. *Diabetes spectr.*, 17, 3, 183-190, 2004.
26. Freeman, A.M., Soman-Faulkner, K. and Pennings, N., Insulin resistance. *NCBI Bookshelf*, 2018.
27. Wilcox, G., Insulin and insulin resistance. *Clin. Biochem. Rev.*, 26, 2, 19, 2005.
28. Meier, J.J., Gallwitz, B., Nauck, M.A., Glucagon-Like Peptide 1 and Gastric Inhibitory Polypeptide. *BioDrugs*, 17, 2, 93-102, 2003.
29. Yoshino, J., Almeda-Valdes, P., Patterson, B.W., Okunade, A.L., Imai, S.I., Mittendorfer, B. and Klein, S., Diurnal variation in insulin sensitivity of glucose metabolism is associated with diurnal variations in whole-body and cellular fatty acid metabolism in metabolically normal women. *J. Clin. Endocrinol. Metab.*, 99, 9, E1666-E1670, 2014.
30. Adeva-Andany, M.M., González-Lucán, M., Donapetry-García, C., Fernández-Fernández, C., Ameneiros-Rodríguez, E., Glycogen metabolism in humans. *BBA clinical*, 5, 85-100, 2016.
31. Swinburn, B.A., Caterson, I., Seidell, J.C., James, W.P.T., Diet, nutrition and the prevention of excess weight gain and obesity. *Public Health Nutr.*, 7, 1a, 123-146. 2004.
32. TeMorenga, L., Mallard, S. and Mann, J., Dietary sugars and body weight: systematic review and meta-analyses of randomized controlled trials and cohort studies. *BMJ*, 346, 2013.
33. Buttriss, J.L. and Stokes, C.S., Dietary fibre and health: an overview. *Nutr. Bull.*, 33, 3, 186-200, 2008.
34. Salas-Salvadó, J., Bulló, M., Pérez-Heras, A., Ros, E., Dietary fibre, nuts and cardiovascular diseases. *Br. J. Nutr.*, 96, S2, S45-S51, 2006.
35. Brouns, F., Overweight and diabetes prevention: is a low-carbohydrate–high-fat diet recommendable. *Eur. J. Nutr.*, 57, 4, 1301-1312, 2018.
36. Rippe, J.M. and Angelopoulos, T.J., Sucrose, high-fructose corn syrup, and fructose, their metabolism and potential health effects: what do we really know?. *Adv. Nutr.*, 4, 2, 236-245, 2013.
37. Vangala, S.and Tonelli A., Biomarkers, metabonomics, and drug development: can inborn errors of metabolism help in understanding drug toxicity?. AAPS J., 9, 3, E284-E297, 2007.
38. Gyamfi, D., Awuah, E.O., Owusu, S., Lipid Metabolism: An Overview. *The Molecular Nutrition of Fats*, 17-32, 2019.
39. Vairappan, B., Cholesterol Regulation by Leptin in Alcoholic Liver Disease. In *Molecular Aspects of Alcohol and Nutrition*, 187-200, Academic Press, 2016.

40. Röhrl, C. and Stangl, H., Cholesterol metabolism—physiological regulation and pathophysiological deregulation by the endoplasmic reticulum. *Wiener Medizinische Wochenschrift*, 168, 11, 280-285, 2018.
41. Hegele, R.A., Plasma lipoproteins: genetic influences and clinical implications. *Nat. Rev. Genet.*,10, 2, 109–121, 2009.
42. Feingold, K.R. and Grunfeld, C., Introduction to lipids and lipoproteins. 2015.
43. Okubo, K., Ikewaki, K., Sakai, S., Tada, N., Kawaguchi, Y., Mochizuki, S., Abnormal HDL apolipoprotein AI and A-II kinetics in hemodialysis patients: a stable isotope study. *J. Am. Soc. Nephrol.* , 15, 4, 1008-1015, 2004.
44. Kaysen, G.A., Lipid and lipoprotein metabolism in chronic kidney disease. *J. Ren. Nutr.*, 19, 73-77, 2009.
45. Zhang, J., Bottiglieri, T., McCullough, P.A., The Central Role of Endothelial Dysfunction in Cardiorenal Syndrome. *Cardiorenal Med.*, 7, 104-117, 2017.
46. Sudhakaran, S., Bottiglieri, T., Tecson, K.M., Kluger, A.Y. and McCullough, P.A., 2018. Alteration of lipid metabolism in chronic kidney disease, the role of novel antihyperlipidemic agents, and future directions. Rev. *Cardiovasc. Med.*, 19, 3, 77-88.
47. Mittendorfer, B., Origins of metabolic complications in obesity: adipose tissue and free fatty acid trafficking. *Curr. Opin. Clin. Nutr. Metab. Care*, 14, 6, 535, 2011.
48. Bugianesi, E., McCullough, A.J., Marchesini, G., Insulin resistance: a metabolic pathway to chronic liver disease. *Hepatology*, 42, 5, 987-1000, 2005.
49. Pasini, E., Corsetti, G., Aquilani, R., Romano, C., Picca, A., Calvani, R., Dioguardi, F. S., Protein-amino acid metabolism disarrangements: The hidden enemy of chronic age-related conditions. *Nutrients*, 10, 4, 391, 2018.
50. Pasini, E., Aquilani, R., Dioguardi, F. S., D'Antona, G., Gheorghiade, M., Taegtmeyer, H., Hypercatabolic syndrome: molecular basis and effects of nutritional supplements with amino acids. *Am. J. Cardiol.*, 101, 11, S11-S15, 2008.
51. Tremblay, F., Lavigne, C., Jacques, H., Marette, A., Role of dietary proteins and amino acids in the pathogenesis of insulin resistance. *Annu. Rev. Nutr.*, *27*, 293-310, 2007.
52. Guigoz, Y., The Mini Nutritional Assessment (MNA) review of the literature--What does it tell us?. *J. Nutr. Health Aging*, 10, 6, 466-487, 2006.
53. Aquilani, R., Opasich, C., Viglio, S., Iadarola, P., Pasini, E., Nutrition in acute decompensation of patients with acute heart failure syndrome. In *Acute Heart Failure*, 876-882, Springer, London, 2008.
54. Katayama, K., Zinc and protein metabolism in chronic liver diseases. *Nutr. Res.*, 74, 1-9, 2020.
55. Liu, M., Chan, C.P., Yan, B.P., Zhang, Q., Lam, Y.Y., Li, R.J., Sanderson, J.E., Coats, A.J., Sun, J.P., Yip, G.W.K., Yu, C.M., Albumin levels predict survival in patients with heart failure and preserved ejection fraction. *Eur. J. Heart Fail.*, 14, 1, 39-44, 2012.

56. Prasad, A.S., Zinc: an overview. *Nutrition (Burbank, Los Angeles County, Calif.)*, 11, 1 Suppl, 93-99, 1995.
57. Prasad, A.S., Zinc and immunity. *Molecular and Cellular Effects of Nutrition on Disease Processes*, 63-69, 1998.
58. Vallee, B.L. and Falchuk, K.H., The biochemical basis of zinc physiology. *Physiol. Rev.* 73, 1, 79-118, 1993.
59. Leoni, G., Rosato, A., Perozzi, G., Murgia, C., Zinc proteome interaction network as a model to identify nutrient-affected pathways in human pathologies. *Genes Nutr.*, 9, 6, 436, 2014.
60. Powell, S.R., The antioxidant properties of zinc. *J. Nutr.*, 130, 5, 1447S-1454S, 2000.
61. Kitamura, H., Morikawa, H., Kamon, H., Iguchi, M., Hojyo, S., Fukada, T., Yamashita, S., Kaisho, T., Akira, S., Murakami, M., Hirano, T., Toll-like receptor–mediated regulation of zinc homeostasis influences dendritic cell function. *Nat. Immunol.*, 7, 9, 971-977, 2006.
62. Katayama, K., Ammonia metabolism and hepatic encephalopathy. *Hepatol. Res.*, 30, 73-80, 2004.
63. Riggio, O., Merli, M., Capocaccia, L., Caschera, M., Zullo, A., Pinto, G., Gaudio, E., Franchitto, A., Spagnoli, R., D'aquilino, E., Seri, S., Zinc supplementation reduces blood ammonia and increases liver ornithine transcarbamylase activity in experimental cirrhosis. *Hepatology*, 16, 3, 785-789, 1992.
64. Blomstrand, E., Eliasson, J., Karlsson, H.K. and Köhnke, R., Branched-chain amino acids activate key enzymes in protein synthesis after physical exercise. *J. Nutr.*, 136, 1, 269S-273S, 2006.
65. Sun, H. and Wang, Y., Branched chain amino acid metabolic reprogramming in heart failure, *Biochimica et Biophysica Acta (BBA)-Molecular Basis of Disease*, 1862, 12, 2270-2275, 2016.
66. Gannon, N.P., Schnuck, J.K., Vaughan, R.A., BCAA metabolism and insulin sensitivity–Dysregulated by metabolic status?. *Mol. Nutr. Food Res.*, 62, 6, 1700756, 2018.
67. Goffredo, M., Santoro, N., Tricò, D., Giannini, C., D'Adamo, E., Zhao, H., Peng, G., Yu, X., Lam, T.T., Pierpont, B., Caprio, S., A branched-chain amino acid-related metabolic signature characterizes obese adolescents with non-alcoholic fatty liver disease. *Nutrients*, 9, 7, 642, 2017.
68. Wolf, M., Update on fibroblast growth factor 23 in chronic kidney disease. *Kidney Int.*, 82, 7, 737-747, 2012.
69. Gutiérrez, O.M., Wolf, M., Taylor, E.N., Fibroblast growth factor 23, cardiovascular disease risk factors, and phosphorus intake in the health professionals follow-up study. *Clin. J. Am Soc. Nephrol.*, 6, 12, 2871-2878, 2011.
70. Cardellini, M., Rizza, S., Casagrande, V., Cardolini, I., Ballanti, M., Davato, F., Porzio, O., Canale, M.P., Legramante, J.M., Mavilio, M., Menghini, R., Soluble ST2 is a biomarker for cardiovascular mortality related to abnormal glucose metabolism in high-risk subjects. *Acta diabetologica*, 56, 3, 273-280, 2019.

71. Qiao, Y., Gao, Z., Liu, Y., Cheng, Y., Yu, M., Zhao, L., Duan, Y., Liu, Y., Breath ketone testing: a new biomarker for diagnosis and therapeutic monitoring of diabetic ketosis. *Biomed. Res. Int., 2014.*
72. Ren, W., Guo, J., Jiang, F., Lu, J., Ding, Y., Li, A., Liang, X., Jia, W., CCAAT/enhancer-binding protein α is a crucial regulator of human fat mass and obesity associated gene transcription and expression. *BioMed Res. Int., 2014.*
73. Jain, K.K., Biomarkers in Metabolic Disorders, In: The Handbook of Biomarkers. Humana Press, New York, NY, 2017.
74. Neutzsky-Wulff, A.V., Andreassen, K.V., Hjuler, S.T., Feigh, M., Bay-Jensen, A.C., Zheng, Q., Henriksen, K., Karsdal, M.A., Future detection and monitoring of diabetes may entail analysis of both β-cell function and volume: how markers of β-cell loss may assist. *J. Transl. Med.*, 10, 1, 1-16, 2012.
75. Mashnafi, S., Plat, J., Mensink, R.P., Baumgartner, S., Non-Cholesterol Sterol Concentrations as Biomarkers for Cholesterol Absorption and Synthesis in Different Metabolic Disorders: A Systematic Review. *Nutrients*, 11, 1, 124, 2019.
76. Mashima, R. and Maekawa, M., Lipid biomarkers for the peroxisomal and lysosomal disorders: their formation, metabolism and measurement. *Biomark Med.*, 12, 1, 83-95, 2018.
77. Hodgson, J.M., Chan, S.Y., Puddey, I.B., Devine, A., Wattanapenpaiboon, N., Wahlqvist, M.L., Lukito, W., Burke, V., Ward, N.C., Prince, R.L., Croft, K.D., Phenolic acid metabolites as biomarkers for tea-and coffee-derived polyphenol exposure in human subjects. *Br. J. Nutr.*, 91, 2, 301-305, 2004.
78. Park, J.E., Pichiah, P.B.T. and Cha, Y.S., Vitamin D and Metabolic Diseases: Growing Roles of Vitamin D. *J. Obes. Metab. Syndr.*, 27, 4, 223-232, 2018.
79. Roberts, C.K., Hevener, A.L., Barnard, R.J., Metabolic syndrome and insulin resistance: underlying causes and modification by exercise training. *Compr. Physiol.*, 3, 1, 1-58, 2013.
80. Krauss, R.M. and Siri, P.W., Metabolic abnormalities: triglyceride and low-density lipoprotein. *Endocrinol. Metab. Clin. North Am.*, 33, 2, 405-415, 2004.
81. Girona, J., Amigó, N., Ibarretxe, D., Plana, N., Rodríguez-Borjabad, C., Heras, M., Ferré, R., Gil, M., Correig, X., Masana, L., HDL triglycerides: a new marker of metabolic and cardiovascular risk. *Int. J. Mol. Sci.*, 20, 13, 3151, 2019.
82. Guo, X.F., Li, X., Shi, M., Li, D., n-3 Polyunsaturated Fatty Acids and Metabolic Syndrome Risk: A Meta-Analysis. *Nutrients*, 9, 7, 703, 2017.
83. Mullur, R., Liu, Y.Y., Brent, G.A., Thyroid hormone regulation of metabolism. *Physiol. Rev.*, 94, 2, 355-382, 2014.
84. Neeha, V.S. and Kinth, P., Nutrigenomics research: a review. *J. Food Sci. Technol.*, 50, 3, 415-28, 2013.
85. Aruoma, O.I., Hausman-Cohen, S., Pizano, J., Schmidt, M.A., Minich, D.M., Joffe, Y., Brandhorst, S., Evans, S.J., Brady, D.M., Personalized nutrition: Translating the science of nutrigenomics into practice: Proceedings from the

2018 American college of nutrition meeting. *J. Am. Coll. Nutr.*, 38, 4, 287-301, 2019.

86. Loos, R.J., From nutrigenomics to personalizing diets: are we ready for precision medicine?. 2019.
87. Butland, B., Jebb, S., Kopelman. P., McPherson, K., Thomas, S., Mardell, J., Parry, V., Tackling obesities: future choices-project report. London: Department of Innovation. Universities and Skills, 2007.
88. Willett, W.C., Koplan, J.P., Nugent, R., Dusenbury, C., Puska, P., Gaziano, T.A., Prevention of chronic disease by means of diet and lifestyle changes. *Disease Control Priorities in Developing Countries,* 2nd edition, 2006.
89. Passi, S.J., Prevention of non-communicable diseases by balanced nutrition: population-specific effective public health approaches in developing countries, *Curr. Diabetes Rev.*, 13, 5, 461-76, 2017.
90. Balakumar, P., Maung-U, K., Jagadeesh, G., Prevalence and prevention of cardiovascular disease and diabetes mellitus. *Pharmacol. Res.*, 113, 600-9, 2016.
91. Pallazola, V.A., Davis, D.M., Whelton, S.P., Cardoso, R., Latina, J.M., Michos, E.D., Sarkar, S., Blumenthal, R.S., Arnett, D.K., Stone, N.J., Welty, F.K., A Clinician's guide to healthy eating for cardiovascular disease prevention. Mayo Clinic Proceedings: Innovations, *Quality & Outcomes.*, 3, 3, 251-67, 2019.
92. Islam, S.U., Ahmed, M.B., Ahsan, H., Lee, Y.S., Recent Molecular Mechanisms and Beneficial Effects of Phytochemicals and Plant-Based Whole Foods in Reducing LDL-C and Preventing Cardiovascular Disease. *Antioxidants*, 10, 5, 784, 2021.
93. Miller, W.C., How effective are traditional dietary and exercise interventions for weight loss?. *Med. Sci. Sports Exerc.*, 31, 1129-34, 1999.
94. Caprara, G., Mediterranean-type dietary pattern and physical activity: The winning combination to counteract the rising burden of non-communicable diseases (NCDS). *Nutrients*, 13, 2, 429, 2021.
95. Cena, H., Calder, P.C., Defining a healthy diet: evidence for the role of contemporary dietary patterns in health and disease. *Nutrients.*, 12, 2, 334, 2020.
96. Rasouli, H., Farzaei, M.H., Khodarahmi, R., Polyphenols and their benefits: A review. *Int. J. Food Prop.*, 20, sup2, 1700-41, 2017.
97. Dinh, T.C., Phuong, T.N.T., Thuc, V.T.M., Bac, N.D., Van Tien, N., Show, P.L., Tao, Y., Ngoc, V.T.N., Ngoc, N.T.B., Jurgoński, A., Raj, D.B.T.G., The effects of green tea on lipid metabolism and its potential applications for obesity and related metabolic disorders-an existing update. *Diabetes Metab. Syndr.: Clin. Res. Rev.*, 13, 2, 1667-1673, 2019.
98. Bae, J., Kumazoe, M., Fujimura, Y., Tachibana, H., Diallyl disulfide potentiates anti-obesity effect of green tea in high-fat/high-sucrose diet-induced obesity. *J. Nutr. Biochem.*, 64, 152-61, 2019.

99. Van Breda, S.G. and de Kok, T.M., Smart combinations of bioactive compounds in fruits and vegetables may guide new strategies for personalized prevention of chronic diseases. *Mol. Nutr. Food Res.*, 62, 1, 1700597, 2018.
100. Albuquerque, T.G., Nunes, M.A., Bessada, S.M., Costa, H.S., Oliveira, M.B.P., Biologically active and health promoting food components of nuts, oilseeds, fruits, vegetables, cereals, and legumes. *Chemical Analysis of Food*, 609-656, Academic Press, 2020.
101. Hemler, E.C. and Hu, F.B., Plant-based diets for personal, population, and planetary health. *Adv Nutr*, 10, Supplement_4, S275-S283, 2019.
102. Clifton, P.M. and Keogh, J.B., A systematic review of the effect of dietary saturated and polyunsaturated fat on heart disease. *Nutr.Metab. Cardiovasc. Dis.*, 27, 12, 1060-80, 2017.
103. Prasad, S., B. Sung, B., B. B. Aggarwal, B. B., Age-associated chronic diseases require age-old medicine: Role of chronic inflammation. *Prev. Med.*, 54, S29–S37, 2012.
104. Islam, M.A., Amin, M.N., Siddiqui, S.A., Hossain, M.P., Sultana, F., Kabir, M.R., Trans fatty acids and lipid profile: A serious risk factor to cardiovascular disease, cancer and diabetes, *Diabetes Metab. Syndr.*, 13, 2, 1643-1647, 2019.
105. Czarnowska-Kujawska, M., Gujska, E., Michalak, J., Testing of different extraction procedures for folate HPLC determination in fresh fruits and vegetables. *J. Food Compost. Anal.*, 57, 64-72, 2017.
106. Stone MS, Martyn L, Weaver CM. Potassium intake, bioavailability, hypertension, and glucose control, *Nutrients*, 8, 7, 444, 2016.
107. Wolfram, T. and Ismail-Beigi, F., Efficacy of high-fiber diets in the management of type 2 diabetes mellitus. *Endocr. Pract.*,17, 1, 132-42, 2011.
108. Singh, J., Rasane, P., Kaur, S., Kumar, V., Dhawan, K., Mahato, D.K., Malhotra, S., Sarma, C., Kaur, D., Bhattacharya, J., Nutritional Interventions and Considerations for the development of low calorie or sugar free foods. *Curr. Diabetes Rev.*, 16, 4, 301-12, 2020.
109. Lopez-Jaramillo, P., Lopez-Lopez, J., Lopez-Lopez, C., Sodium intake recommendations: a subject that needs to be reconsidered. *Curr. Hypertens. Rev.*, 11, 1, 8-13, 2015.
110. Cunningham, D. G., Vannozzi, S. A., Turk, R., Roderick R., O'Shea, E., Brilliant., K., Cranberry phytochemicals and their health benefits. Washington, DC: ACS Publications, 2004.
111. Xia, E. Q., Deng, G. F., Guo, Y. J., H.-B. Li, H. B., Biological activities of polyphenols from grapes. *Int. J. Mol. Sci.*, 11, 2, 622–46, 2010.
112. Akhtar, S., Ismail, T., Layla, A., Pomegranate bioactive molecules and health benefits. In Bioactive molecules in food, ed. Merillon, J.M., Ramawat, K.G., 1253–79. Cham/Switzerland, Spring, 2019.
113. Hannum, S. M., Potential impact of strawberries on human health: A review of the science. *Crit. Rev Food Sci. Nutr.*, 44, 1, 1–17, 2004.

114. Neacsu, M., McMonagle, J., Fletcher, R., Scobbie, L., Duncan G., Cantlay, L., de Roos, B., Duthie, G., Russell, W., Bound phytophenols from ready-to-eat cereals: Comparison with other plantbased foods. *Food Chem.*, 141, 3, 2880–6, 2013.
115. Zhang, G. and Hamaker, B. R., Cereal carbohydrates and colon health. *Cereal Chem.*, 87, 4, 331–41, 2010.
116. Fardet, A. New hypotheses for the health-protective mechanisms of whole-grain cereals: What is beyond fibre?. *Nutr. Res. Rev.*, 23, 1, 65–134, 2010.
117. Zeng, H., Lazarova, D. L., M. Bordonaro, M., Mechanisms linking dietary fiber, gut microbiota and colon cancer prevention. *World J. Gastrointest. Oncol.*, 6, 2, 41–51, 2014.
118. Rubió, L., Motilva, M.J., Romero, M.P., Recent advances in biologically active compounds in herbs and spices: a review of the most effective antioxidant and anti-inflammatory active principles. *Crit. Rev. Food Sci. Nutr.*, 53, 9, 943-53, 2013.
119. Gebrelibanos, M., Tesfaye, D., Raghavendra, Y., Sintayeyu, B., Nutritional and health implications of legumes. *Int. J. Pharm. Sci. Res.*, 4, 4, 1269, 2013.
120. Watanabe, T., Watanabe-Kominato, K., Takahashi, Y., Kojima M., R. Watanabe, R., Adipose tissue-derived omentin-1 function and regulation. *Compr. Physiol.*, 7, 765–81, 2011.
121. Jafari, T., Faghihimani, E., Feizi, A., Iraj, B., Javanmard, S. H., Esmaillzadeh, A., Fallah, A. A., Askari, G., Effects of vitamin D-fortified low fat yogurt on glycemic status, anthropometric indexes, inflammation, and bone turnover in diabetic postmenopausal women: A randomized controlled clinical trial. *Clin. Nutr.*, 35, 1, 67–76, 2016.
122. Díaz-Rizzolo, D.A., Miro, A., Gomis, R., Prevention of type 2 diabetes through sardines consumption: an integrative review. *Food Rev. Int.*, 7, 1-9, 2021.
123. Bogdanski, P., Suliburska, J., Szulinska, M., Stepien, M., Pupek-Musialik, D., Jablecka, A., Green tea extract reduces blood pressure, inflammatory biomarkers, and oxidative stress and improves parameters associated with insulin resistance in obese, hypertensive patients. *Nutr. Res.*, 32, 6, 421-427, 2012.
124. Fukino. Y., Shimbo, M., Aoki, N., Okubo, T., Iso, H., Randomized controlled trial for an effect of green tea consumption on insulin resistance and inflammation markers. *J. Nutr. Sci. Vitaminol (Tokyo)*, 51, 5, 335-342, 2005.
125. Nogueira, L.P., Nogueira,Neto, J.F., Klein, M.R., Sanjuliani, A.F., Short-term Effects of Green Tea on Blood Pressure, Endothelial Function, and Metabolic Profile in Obese Prehypertensive Women: A Crossover Randomized Clinical Trial. *J. Am. Coll. Nutr.*, 36, 2, 108-115, 2017.
126. Hashemzadeh, A.A., Nasoohi, N., Raygan, F., Aghadavod, E., Akbari, E., Taghizadeh, M., Memarzadeh, M.R., Asemi, Z., Flaxseed oil supplementation improve gene expression levels of PPAR-γ, LP (a), IL-1 and TNF-α in type 2 diabetic patients with coronary heart disease. *Lipids*, 52, 11, 907-915, 2017.

127. Hutchins, A.M., Brown, B.D., Cunnane, S.C., Domitrovich, S.G., Adams, E.R., Bobowiec, C.E., Daily flaxseed consumption improves glycemic control in obese men and women with pre-diabetes: a randomized study. *Nutr. Res.*, 33, 5, 367-375, 2013.
128. Devarshi, P.P., Jangale, N.M., Ghule, A.E., Bodhankar, S.L., Harsulkar, A.M., Beneficial effects of flaxseed oil and fish oil diet are through modulation of different hepatic genes involved in lipid metabolism in streptozotocin-nicotinamide induced diabetic rats. *Genes Nutr.*, 8, 3, 329-342, 2013.
129. Lee, Y.J., Nam, G.E., Seo, J.A., Yoon, T., Seo, I., Lee, J.H., Im, D., Bahn, K.N., Jeong, S.A., Kang, T.S., Ahn, J.H., Nut consumption has favorable effects on lipid profiles of Korean women with metabolic syndrome. *Nutr. Res.*, 34, 9, 814-820, 2014.
130. Tey, S.L., Gray, A.R., Chisholm, A.W., Delahunty, C.M., Brown, R.C., The dose of hazelnuts influences acceptance and diet quality but not inflammatory markers and body composition in overweight and obese individuals. *J. Nutr.*, 143, 8, 1254-1262, 2013.
131. Acharjee, S., Zhou, J.R., Elajami, T.K., Welty, F.K., Effect of soy nuts and equol status on blood pressure, lipids and inflammation in postmenopausal women stratified by metabolic syndrome status. *Metabolism*, 64, 2, 236-243, 2015.
132. Casas-Agustench, P., López-Uriarte, P., Bulló, M., Ros, E., Cabré-Vila, J.J., Salas-Salvadó, J., Effects of one serving of mixed nuts on serum lipids, insulin resistance and inflammatory markers in patients with the metabolic syndrome. *Nutr. Metab. Cardiovasc. Dis.*, 21, 2, 126-135, 2011.
133. Orsavova, J., Misurcova, L., Ambrozova, J.V., Vicha, R., Mlcek, J., Fatty Acids Composition of Vegetable Oils and Its Contribution to Dietary Energy Intake and Dependence of Cardiovascular Mortality on Dietary Intake of Fatty Acids. *Int. J. Mol. Sci.*, 16, 6, 12871-12890, 2015.
134. Nieman, D.C., Cayea, E.J., Austin, M.D., Henson, D.A., McAnulty, S.R., Jin, F., Chia seed does not promote weight loss or alter disease risk factors in overweight adults. *Nutr Res.*, 29, 6, 414-418, 2009.
135. Vuksan, V., Jenkins, A.L., Brissette, C., Choleva, L., Jovanovski, E., Gibbs, A.L., Bazinet, R.P., Au-Yeung, F., Zurbau, A., Ho, H.V.T. and Duvnjak, L., Salba-chia (Salvia hispanica L.) in the treatment of overweight and obese patients with type 2 diabetes: A double-blind randomized controlled trial. *Nutr. Metab. Cardiovasc. Dis.*, 27, 2, 138-146, 2017.
136. Hermsdorff, H.H., Zulet, M.Á., Abete, I., Martínez, J.A., A legume-based hypocaloric diet reduces proinflammatory status and improves metabolic features in overweight/obese subjects. *Eur. J. Nutr.*, 50, 1, 61-69, 2011.
137. Singh, B.N., Rawat, A.K., Bhagat, R.M., Singh, B.R., Black tea: Phytochemicals, cancer chemoprevention, and clinical studies. *Crit. Rev. Food Sci. Nutr.*, 57, 7, 1394-1410, 2017.
138. Khan, N. and Mukhtar, H., Tea and health: studies in humans. *Curr. Pharm. Des.*, 19, 34, 6141-6147, 2013.

139. Aoe, S., Mio, K., Yamanaka, C., Kuge, T., Low molecular weight barley β-glucan affects glucose and lipid metabolism by prebiotic effects. *Nutrients*, 13, 1, 130, 2020.
140. Dog, T.L., A reason to season: the therapeutic benefits of spices and culinary herbs. *Explore: J. Sci. Heal*,2, 5, 446-449, 2006.
141. Nieman, D.C., Cialdella-Kam, L., Knab, A.M., Shanely, R.A., Influence of red pepper spice and turmeric on inflammation and oxidative stress biomarkers in overweight females: a metabolomics approach. *Plant Foods Hum. Nutr.*, 67, 4, 415-421, 2012.
142. Jung, H., Chen, C.O., Blumberg, J.B., Kwak, H.K., The effect of almonds on vitamin E status and cardiovascular risk factors in Korean adults: a randomized clinical trial. *Eur. J. Nutr.*, 57, 6, 2069-2079, 2018.
143. Krul-Poel, Y.H., Ter Wee, M.M., Lips, P., Simsek, S., Management of endocrine disease: the effect of vitamin D supplementation on glycaemic control in patients with type 2 diabetes mellitus: a systematic review and meta-analysis. *Eur. J. Endocrinol.*, 176, 1, R1-R14, 2017.
144. Mellenthin, L., Wallaschofski, H., Grotevendt, A., Völzke, H., Nauck, M., Hannemann, A., Association between serum vitamin D concentrations and inflammatory markers in the general adult population. *Metabolism*, 63, 8, 1056-1062, 2014.
145. Brownlee, I.A., Moore, C., Chatfield, M., Richardson, D.P., Ashby, P., Kuznesof, S.A., Jebb, S.A., Seal, C.J., Markers of cardiovascular risk are not changed by increased whole-grain intake: the WHOLEheart study, a randomized, controlled dietary intervention. *Br. J. Nutr.*, 104, 1, 125-134, 2010.
146. Kopf, J.C., Suhr, M.J., Clarke, J., Eyun, S.I., Riethoven, J.J.M., Ramer-Tait, A.E., Rose, D.J., Role of whole grains versus fruits and vegetables in reducing subclinical inflammation and promoting gastrointestinal health in individuals affected by overweight and obesity: A randomized controlled trial. *Nutr. J.*, 17, 1, 1-13, 2018.
147. Lambert-Porcheron, S., Normand, S., Blond, E., Sothier, M., Roth, H., Meynier, A., Vinoy, S., Laville, M., Nazare, J.A., Modulation of starch digestibility in breakfast cereals consumed by subjects with metabolic risk: impact on markers of oxidative stress and inflammation during fasting and the postprandial period. *Mol. Nutr. Food Res.*, 61, 12, 1700212, 2017.
148. Roager, H.M., Vogt, J.K., Kristensen, M., Hansen, L.B.S., Ibrügger, S., Mærkedahl, R.B., Bahl, M.I., Lind, M.V., Nielsen, R.L., Frøkiær, H., Gøbel, R.J., Whole grain-rich diet reduces body weight and systemic low-grade inflammation without inducing major changes of the gut microbiome: a randomized cross-over trial. *Gut*, 68, 1, 83-93, 2019.
149. Vitaglione, P., Mennella, I., Ferracane, R., Rivellese, A.A., Giacco, R., Ercolini, D., Gibbons, S.M., La Storia, A., Gilbert, J.A., Jonnalagadda, S., Thielecke, F., Whole-grain wheat consumption reduces inflammation in a randomized controlled trial on overweight and obese subjects with unhealthy

dietary and lifestyle behaviors: role of polyphenols bound to cereal dietary fiber. *Am. J. Clin. Nutr.,* 101, 2, 251-261, 2015.

150. Meldrum, D.R., Morris, M.A., Gambone, J.C., Obesity pandemic: causes, consequences, and solutions—but do we have the will?. *Fertil. Steril.,*107, 4, 833-9, 2017.
151. Cecchini, M., Sassi, F., Lauer, J.A., Lee, Y.Y., Guajardo-Barron, V., Chisholm, D., Tackling of unhealthy diets, physical inactivity, and obesity: health effects and cost-effectiveness. *Lancet*, 376, 9754, 1775-84, 2010.
152. Sales, N.M., Pelegrini, P.B., Goersch, M.C., Nutrigenomics: definitions and advances of this new science. *J. Nutr. Metab.*, 2014.
153. Costa, C., Tsatsakis, A., Mamoulakis, C., Teodoro, M., Briguglio, G., Caruso, E., Tsoukalas, D., Margina, D., Dardiotis, E., Kouretas, D., Fenga, C., Current evidence on the effect of dietary polyphenols intake on chronic diseases. *Food Chem. Toxicol.*, 110, 286-99, 2017.
154. Romagnolo, D.F., and Selmin, O.I., Mediterranean diet and prevention of chronic diseases. *Nutr. Today*, 52, 5, 208, 2017.
155. Alkhatib, A., Tsang, C., Tiss, A., Bahorun, T., Arefanian, H., Barake, R., Khadir, A., Tuomilehto, J., Functional foods and lifestyle approaches for diabetes prevention and management. *Nutrients,* 9, 12, 1310, 2017.

17

Nanomaterials in Food System

Anu Kumar[1]*, Bhanu Krishan[1,2], Sunny Dhiman[1] and Tejpal Dhewa[3]

[1]University Institute of Biotechnology, Chandigarh University, Gharuan, Mohali, Punjab, India
[2]Biotechnology Department, GGDSD College, Chandigarh, India
[3]Department of Nutrition Biology, Central University of Haryana, Mahendergarh, Haryana

Abstract
Incorporating nanomaterials into food systems is emerging as an innovative tool in the present scenario. It offers the advantage that the material exploited is eco-friendly and completely devoid of side effects. Moreover, shelf-life maintenance is of great concern in the case of food industries. So, this approach is significantly promising in this context and can ensure the longer shelf life of products. Additionally, such incorporation can focus more on improving food quality and traceability. The growing demand of consumers in terms of food quality is attracting researchers to identify new ways to improve food quality. Also, awareness of a healthy lifestyle encourages researchers to find eco-friendly approaches Therefore, the present chapter is focused on some innovative nanomaterials and their applications in food systems.

Keywords: Nanomaterials, nanotechnology, food packaging, eco-friendly

17.1 Introduction

The term Food is associated with complex systems involving biological, chemical and physical parameters such as microorganisms, reactions, enzymes, texture and colour. These parameters are foremost the crucial aspects to be maintained in order to preserve the food and to maintain its

**Corresponding author*: anu.uibt@cumail.in

Tejpal Dhewa, Anil Kumar Puniya and Anil Panghal (eds.) Nutritional Science and Technology: Concept to Application, (383–394) © 2023 Scrivener Publishing LLC

integrity by the food industry; it is essential to fulfill the demand of the customers by providing them with the desired food material with a longer shelf life, better nutrition and quality rich supply [1]. As food includes a wide range of sources such as vegetables, fruits, meat and fisheries, there are different strategies to maintain the probity by adequate preservation and packaging opted by the specific industry to keep the food secure from microbes, fungi and other deteriorating processes. Innovations, advancements and betterment widely pressure these practices to enhance the parameters involved in the food material's nutrient, storage, packaging, and health aspects [2]. Globalization has widely affected the input and output aspects of food integrity; with an increase in demand, these industries have to maintain an adequate supply, minimize spoilage, inhibition of foodborne pathogens, longer shelf life and better-quality product which is directly or indirectly affected by the processes taken up by the industries [3].

According to an estimation by WHO, around 600 million people get infected and 4 lakh 20 thousand deaths occur from foodborne diseases. Packaging serves an important purpose in protecting the food from any unfavourable conditions, namely spoilage, and maintaining the nutrition and overall quality of the food product [4, 5]. Packaging is of different types based upon the food material, such as tetra-packaging or aseptic processing, which includes sterile packaging made up of multiple tight layers of a mixture of paper, polyethene and aluminium, used for packaging milk products, fruit juices and other drinks. Paperboard trays have been widely used to pack eggs, fruit and drinks safely. Plastic bags are commonly used for the packaging of snacks and fruits which are generally aerated with preservatives such as nitrogen gas. Other packaging materials such as aluminium cans, plastic boxes and paper wrappers are widely used for packaging food materials [6, 7]. The materials used in these packaging materials are potential toxins for the environment as well as for the food material. These materials are generally coated with chemical additives to increase the shelf life and taste of the food material. As most of the material used in the packaging is built of plastic, metal, paper and glass, it takes about certain weeks to several years to degrade, generating a large amount of waste (Figure 17.1) [8].

As discussed earlier, these materials are incorporated with an additive; phthalates are used with plastics which gives plastic more flexibility, and perfluorinated chemicals are used in cardboard containers to contain liquid, bisphenol lining that coat aluminium cans. These chemicals not only serve as toxins for human health but also decrease food integrity as they exhibit very weak antimicrobial properties. Plastic has been known to be

Figure 17.1 Time taken by the food packaging material to get degraded.

the world's major problem being non-biodegradable, it breaks into micro-plastic, and in some cases, the plastic is burnt, and some of it gets recycled, but only a fraction is recovered from recycling [8, 9]. Most countries have taken the initiative to minimize the use of plastic, and similarly, the food industries have also developed modernization integrating recent technologies. The evolving industrialization involves the use of nanomaterials.

17.2 Nanomaterials as an Emerging Tool

Nanotechnology is defined as the manipulation and application of nanomaterials, which are small-scaled manufactured substances with at least one dimension less than 100nm with unique physical, optical, magnetic and chemical properties and a large surface area to volume ratio due to a relatively small size [10]. Nanomaterials can be natural or engineered and can be synthesized using green resources. Due to the small scale, these materials exhibit different quantum properties, which have the potential for greater applications in electronics, medicine, cosmetics, and other fields. These nanomaterials can be classified into different groups based on their shape, state and chemical structure; they can exist as a single substance, fused or aggregated [11]. Based on dimension and shape, nanomaterials are of 4 types, zero-dimensional with a size less than 100nm, such as nanorods, core shells and quantum dots etc. One-dimensional nanoparticles such as ceramics, nanowires, nanofibers have two dimensions in scale while the one is not in the scale. 2D has only two dimensions in scale, while the other two are not in scale, such as thin films, nanoplates and nano-coatings. 3D nanomaterials have dimensions over 100nm; these include fullerenes, nanoparticles, carbon nanobuds and honeycombs [10]. Based on the chemical constituents and properties, these nanoparticles exist as colloids, suspensions, composites, metallic nanomaterials and quantum dots [10, 11]. Nanoparticles of metal and metal oxides have been reported to exhibit antimicrobial and antioxidant properties. Ceramic nanoparticles have unique superconductive, ferromagnetic, and ferroelectric properties [10]. Quantum dots are good semiconductors with a high sensing ability; they have been used in various fields such as bio-labelling, sensors, lasers, light-emitting diodes and medicine [12]. Nanotechnology has emerged as a beneficial tool for therapeutics, cosmetic discipline, health care products, and food technology in the past four decades [13].

17.3 Nanomaterials in Food Industry

In the recent era, nanotechnology has been integrated with the food and beverage industry. Nanotechnology holds the foremost capability in food processing; the modern use of technology is a novel technique that supports various beneficial properties such as antimicrobial and antioxidant properties. Figures 17.2 and 17.3 shows the processes involved in the food industry and the integration of nanotechnology in these processes. The most favourable applications of nanomaterials in the food industry involve the following:

- Use nano-based sensors to detect toxins, pathogens or the food material's overall physiochemical or biological properties.
- Nanoparticles for their antimicrobial and antioxidant properties, suitable for controlling pathogens and their related diseases.
- Nanopolymers and nanocomposites in food packaging have been extensively used to prevent food materials from deterioration.
- Nanofilms are a replacement for synthetic chemicals because of nanofilms' antimicrobial and antioxidant properties.
- Edible coating with nanomaterials serves a beneficiary role in the storage processes of fruits and vegetables.

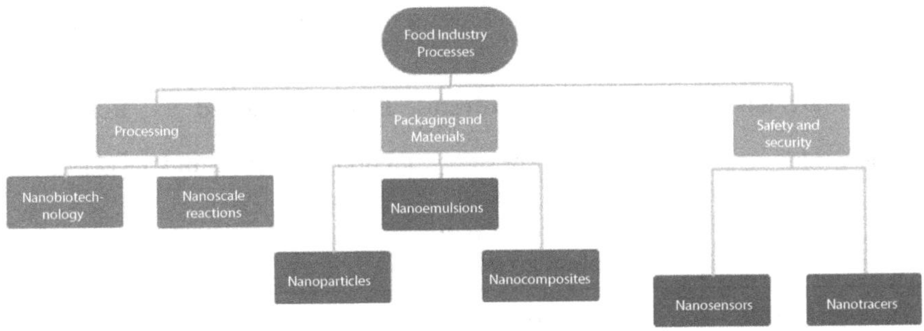

Figure 17.2 Processes involved in food industry and integration of nanotechnology into it.

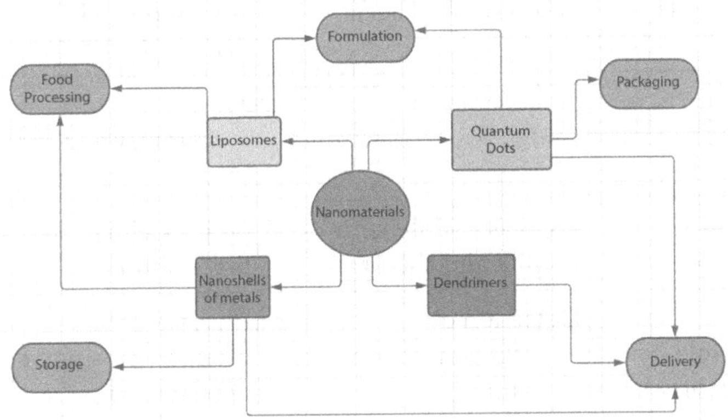

Figure 17.3 Applications of different nanomaterials in the fields of food industry.

17.3.1 Nanosensors

Nano-based sensors and probes are considered intelligent food packaging systems for detecting pathogens, additives, and adulterants. Nanosensors in agriculture have served a vital role in detecting analytes such as pesticides and fertilizers (namely DDT, carbofuran, residues of organophosphates and organochlorine) [14–16]. Due to high sensitivity and unique optical and electrochemical properties, these nanosensors are more convenient than traditional biosensors. Nanomaterials exploited in nanosensors have a high surface area to volume ratio, and due to this property, they can detect traces of specific pathogens in a food sample with a lower sample volume [14]. Further, the table below represents different nanomaterial-based sensors used in the food industry (Table 17.1).

Table 17.1 List of different nanomaterials used in nanosensors exploited in the food industry.

Nanomaterials	Application involved	Process involved	Reference
Zinc oxide	Reduce the flow of oxygen in packed containers, prevents microbial growth	Food packaging and preservation	[17]
Silicon Dioxide	Used as a food colourant, drying agent and anti-caking agent	Food packaging and preservation	[18]
Silver Zeolite	Antimicrobial agent	Preservation and disinfection	[19]
Graphene	Nanocomposites	Detection of contaminants in food	[20]
Titanium Dioxide	Used as a whitener in milk and cheese	Food packaging and preservation	[17]
Copper and Gold Nanoparticles	Detection of Pathogens in food samples	Surface water for production of fruits and vegetables	[21]

17.3.2 Nanoparticles

Nanoparticles have been widely integrated with food packaging to increase the shelf life of the food material due to their antimicrobial activity. These particles are used as reactive substances in packaging materials and are mostly incorporated in nanosensors to detect any change in the environment of the packed food material. Nanomaterials of metal and metal oxides, such as silver, Zinc oxides, and titanium oxide nanomaterial, have been reported to inhibit the growth of certain food-related pathogens [22]. Silver nanoparticles incorporated in polymeric films have been reported to increase the shelf life of food products. The silver nanoparticles react with dissolved oxygen and hydrogen ions and disrupt the cell wall and membrane of microorganisms. Also, silver nanoparticles can block ethylene which is released by fruit metabolism. With a good UV blocking ability, Titanium dioxide nanoparticles has been extensively used in food packaging as a UV blocker on food materials. Due to this property, these nanoparticles exhibit self-cleaning activity and antibacterial activity. Also, titanium dioxide nanoparticles have been used as pigment and coating additives in food packaging. Integration of titanium dioxide in pasteurized milk packaging was reported to reduce the effect of light on the quality of milk [23]. Zinc oxide being cheaper than silver nanoparticles, is used in packaging and provides blockade properties and durability to the packaging material [24].

17.3.3 Nanofilms and Nanocomposites

Nanofilms are nanoscaled thin layers of material with unique physiochemical properties. Polymeric nanofilms incorporated with nanoparticles of silver and titanium dioxide have been used as an efficient packaging material. Polymeric films supporting nanofilms can induce antimicrobial effects in packaging without migration [24, 25].

Nanocomposites are materials consisting of a mixture of polymers with inorganic clays or oxides at a nanoscale. In food packaging, nanocomposites of silver were reported to have efficient bioactivity and increased the shelf life of meat. Chitosan-based nanocomposites of silver (70% polyvinyl alcohol and 30% aqueous chitosan) were reported to have efficient antibacterial properties against most gram-positive and gram-negative bacteria [26]. Thus nanocomposites are mainly used for coating and packaging.

17.3.4 Liposomes

Engineered GT 1 Fluorescent labelled ganglioside liposomes with a size ranging from 120-130nm were extensively reported to detect the presence of exotoxins such as botulinum, cholera and tetanus [27]. Also, liposomes are used to increase the overall stability and availability of bioactive compounds present in food [28].

17.3.5 Nanotubes and Nanofibers

Nanotubes are generally hollow tubes of nanoscale; in the area of the food industry, these materials have been used to deliver protection to fresh food and in the safe transportation of food materials during storage. Carbon nanotubes with the strongest bactericidal properties have been used in safe food packaging [24]. Nanofibers on the other hand, exhibit similar antimicrobial properties as nanotubes, nanofibers have been used as a coating material in packaging and ensure better encapsulation of bioactive

Table 17.2 List of products manufactured by different companies integrating nanotechnology in the food system.

Product	**Manufacturer**	**Nanomaterial exploited**
Oat Chocolate Nutrition Drink mix	Toddler Health	Sun active iron (300nm)
Nano plastic wraps	Songsing Nanotechnology	Nano zinc crystals
AdNano	Evonik	Zinc oxide nanoparticles
Constantia multifilm N-coat	Constantia Multifilm	Nanocomposite polymer
Nano tea	Shenzen Become Industry & Trading Co	Nanoparticles of 160nm
Canola Active oil	Shemen	Nano self-assembled liquid micelles
Nanoceuticals Slim Shake Vanilla	RBC Lifescience	Nanodusters
Du Pont Light Stabilizer 210	Du Pont	Nanoparticles of titanium dioxide
Millar Beers	Nanocor	Nanocomposite Technology

substances. This material has been used in food preservation to increase nutritional quality and stability [29].

Further, the Table 17.2 enlists the food industries that have incorporated nanotechnology in food processing summarized by Blasco and Pico, 2011 [30].

17.4 Conclusion

Nanotechnology is an evolving revolution in the market; being more convenient than traditional technologies, nanotechnology serves more benefits in the food industry. It serves more efficient antioxidant and antimicrobial properties with the enhanced nutritional quality of the food product. Due to their small size and unique physiochemical properties, they are known to serve benefits to the food material and less toxins. Thus, nanomaterials can be used to detect the overall nutritional quality and incorporating nanosensors into the packaging material can be used to detect any amount of physical change in the environment of the food product. Nanomaterials can be used to enhance the nutritional potential of food products and can be used to enrich fruits and vegetables with the desired nutrient. Quantum dots, on the other hand, can be used to detect any additive present in the vegetable and to monitor the presence of any toxic pesticide or fertilizer in the food product. Therefore, these nanotechnologies incorporated in the food product can serve the consumers a better quality product with no or minimal chance of food getting infected by a pathogen or spoiled. This technology can change the food industries entirely if accepted widely and monitor the negative aspects of nanotechnology.

References

1. T. K. Singh, and K. R. Cadwallader, The shelf life of foods: An overview, p. 2-21 2003.
2. Q. Chaudhry, R. Watkins, and L. Castle, Nanotechnologies in food: What, why and how? p. 1-19, 2017.
3. E. K. Ling, and S. N. Wahab, Integrity of food supply chain: going beyond food safety and food quality. International Journal of Productivity and Quality Management, Vol. 29(2), p. 216-232, 2020.
4. A. Bahrami, Z.M. Baboli, K. Schimmel, S.M. Jafari, and L. William, Efficiency of novel processing technologies for the control of Listeria monocytogenes in food products, Trends Food Sci. Technol, Vol. 96, p. 61–78, 2020.
5. K. Hoelzer, A.I.M. Switt, M. Wiedmann, and K.J. Boor, Emerging needs and opportunities in foodborne disease detection and prevention: from tools to people, Food Microbiol, Vol. 75, p. 65–71, 2018.
6. R. Shaw. Food Packaging: 9 Types and Differences Explained, Assemblies Unlimited, https://www.assemblies.com/food-packaging-9-types-and-differences-explained/, 2013
7. Types of food packaging used by leading suppliers, York Saw and Knife, https://www.yorksaw.com/types-of-food-packaging/
8. Fadare, Oluniyi O., *et al.* Microplastics from consumer plastic food containers: Are we consuming it?. *Chemosphere* 253, 126787, 2020.
9. J. Hopewell, R. Dvorak, and E. Kosior, Plastics recycling: challenges and opportunities, Philos Trans R Soc, Vol. 364, p. 2115–2126, 2009.

10. T.A. Saleh, Nanomaterials: Classification, properties, and environmental toxicities, Environmental Technology and Innovation, p. 101067, 2020.
11. H. Gleiter, Nanostructured materials: basic concepts and microstructure, Acta Mater, Vol. 48, p. 1–29, 2000.
12. D. Vasudevan, R.R. Gaddam, A. Trinchi, and I. Cole, Core–shell quantum dots: Properties and applications, Journal of Alloys and Compounds, Vol. 636, p. 395-404, 2015.
13. S. Nafisi, and H. Maibach, Nanotechnology in cosmetics, Cosmetic science and technology: theoretical principles and applications, p. 337, 2017.
14. S.M. Rodrigues, P. Demokritou, N. Dokoozlian, C.O. Hendren, B. Karn, M.S. Mauter, and G.V. Lowry, Nanotechnology for sustainable food production: promising opportunities and scientific challenges, Environmental Science: Nano, Vol. *4*(4), p. 767-781, 2017.
15. B.S. Sekhon, Nanotechnology in agri-food production: an overview, Nanotechnology, science and applications, Vol. 7, p. 31, 2014.
16. V. Rai, S. Acharya, and N. Dey, Implications of nanobiosensors in agriculture, 2012.
17. M. Manikandan, P.N. Francis, S. Dhanuskodi, N. Maheswari, and G. Muralidharan, High performance supercapacitor behavior of hydrothermally synthesized CdTe nanorods, Journal of Materials Science: Materials in Electronics, Vol. 29(20), p. 17397-17404, 2018.
18. N. Jones, B. Ray, K.T. Ranjit, and A.C. Manna, Antibacterial activity of ZnO nanoparticle suspensions on a broad spectrum of microorganisms, *FEMS microbiology letters*, Vol. *279*(1), p. 71-76, 2008.
19. Y. Matsumura, K. Yoshikata, S.I. Kunisaki, and T. Tsuchido, Mode of bactericidal action of silver zeolite and its comparison with that of silver nitrate, Applied and environmental microbiology, Vol. *69*(7), p. 4278-4281, 2003.
20. A.K. Sundramoorthy, and S. Gunasekaran, Applications of graphene in quality assurance and safety of food, TrAC Trends in Analytical Chemistry, Vol. *60*, p. 36-53, 2014.
21. H. Zhang, J. Wang, and S. Ye, Predictions of acidity, soluble solids and firmness of pear using electronic nose technique, Journal of Food Engineering, Vol. *86*(3), p. 370-378, 2008
22. H. Bouwmeester, S. Dekkers, M.Y. Noordam, W.I. Hagens, A.S. Bulder, C. De Heer, and A.J. Sips, Review of health safety aspects of nanotechnologies in food production, Regulatory toxicology and pharmacology, Vol. 53(1), p. 52-62, 2009.
23. T. Moyssiadi, A. Badeka, E. Kondyli, T. Vakirtzi, I. Savvaidis, and M.G. Kontominas, Effect of light transmittance and oxygen permeability of various packaging materials on keeping quality of low fat pasteurized milk: chemical and sensorial aspects, International dairy journal, Vol. 14(5), p. 429-436, 2004.

24. M.A. Emamhadi, M. Sarafraz, M. Akbari,Y. Fakhri, N..T.T. Linh, and A.M. Khaneghah, Nanomaterials for food packaging applications: A systematic review, Food and Chemical Toxicology, p.111825, 2020.
25. E. Lloret, P. Picouet, and A. Fernández, Matrix effects on the antimicrobial capacity of silver based nanocomposite absorbing materials, *LWT-Food Science and Technology*, Vol. *49*(2), p. 333-338, 2012.
26. V.K. Pandey, S.N. Upadhyay, K. Niranjan, and P.K. Mishra, Antimicrobial biodegradable chitosan-based composite Nano-layers for food packaging, International journal of biological macromolecules, Vol. 157, p. 212-219, 2020.
27. A.K. Singh, S.H. Harrison, J.S. Schoeniger, Gangliosidesas receptors for biological toxins: development of sensitive fluoroimmunoassays using ganglioside-bearing liposomes, Analytical Chemistry, Vol. 72, p. 6019–6024, 2000, https://doi.org/10.1021/ac000 846l.
28. K.K Ajeeshkumar, P.A. Aneesh, N. Raju, M. Suseela, C.N. Ravishankar, and S. Benjakul, Advancements in liposome technology: Preparation techniques and applications in food, functional foods, and bioactive delivery: A review, Comprehensive Reviews in Food Science and Food Safety, Vol. 20(2), p. 1280-1306, 2021
29. A. Faham, and M. Fathi, Fabrication and characterization of novel nanofibers from cress seed mucilage for food applications, Journal of Applied Polymer Science, Vol. 135, p. 45811–45816, 2018.
30. C. Blasco, and Y. Picó, Determining nanomaterials in food, TrAC Trends in Analytical Chemistry, Vol. 30(1), p. 84–99, 2011.

About the Editors

Dr. Tejpal Dhewa *is a highly accomplished faculty member in the School of Interdisciplinary and Applied Science at the Central University of Haryana, Mahendergarh. He has diverse experience in teaching, research, administration, and industry, and has made significant contributions to the field of Nutrition, Microbiology, and Food Safety. Dr. Dhewa is instrumental in ensuring food safety and security across the country as the Coordinator of the FoSTaC FSSAI Centre-Central University of Haryana and a Member of the Scientific Panel FSSAI. He has developed and coordinated the highly popular SWAYAM UGC-PG MOOC on "Food Microbiology and Food Safety," which has completed seven cycles with over 29,000+ learners. Dr. Dhewa's research works have been published in national and international journals of high impact, and he has earned recognition and funding from the MoE, SERB, DST, Government of India, and DST-Haryana.*

Prof. (Dr.) Anil Kumar Puniya *(Principal Scientist & Former Head Dairy Microbiology, ICAR-National Dairy Research Institute, Karnal, Haryana) is a renowned dairy microbiology scientist whose research has led to the discovery of two new species of anaerobic fungi. He has significantly improved dairy product production and quality, and his sustainable livestock management strategies have revolutionized the industry. With an impressive portfolio of over 200 published papers and three edited books, Prof. Puniya has received numerous national and international awards and honors for his contributions to the field.*

***Dr. Anil Panghal** has an impressive background in the food industry, with experience working as a Production Manager at Nestle for almost a decade. He is currently an Assistant Scientist in the Department of Processing and Food Engineering at Chaudhary Charan Singh Haryana Agricultural University in India. His areas of expertise include bioprocessing, manufacturing, food chemistry, food science and technology, FSMS, and nutrition. Dr. Panghal has also conducted research on the extraction of bioactive compounds from plants and has developed several products. He holds a Ph.D. in Food Technology, with a focus on the molecular and physicochemical quality aspects of commercial wheat varieties. With an impressive portfolio of over 100 published papers and six edited books, Dr. Anil Panghal has received numerous national and international awards and honors for his contributions to the field.*

Index

Also of Interest

Other Books in the series, "Bioprocessing in Food Science"

HARVESTING FOOD FROM WEEDS, Edited by Prerna Gupta, Navnidhi Chhikara, and Anil Panghal, ISBN: 9781119791973. An exciting new volume in the groundbreaking new series, "Bioprocessing in Food Science," Harvesting Food from Weeds focuses on the latest processes, industrial applications, and leading research on this important, evolving new area, for engineers, scientists, students, and other industry professionals.

MICROBES IN THE FOOD INDUSTRY, Edited by Navnidhi Chhikara, Anil Panghal, and Gaurav Chaudhary, ISBN: 9781119775584. This newest volume in the groundbreaking new series, "Bioprocessing in Food Science," focuses on the latest processes, industrial applications, and leading research on microbes in the food industry, for engineers, scientists, students, and other industry professionals.

NOVEL TECHNOLOGIES IN FOOD SCIENCE, Edited by Navnidhi Chhikara, Anil Panghal, and Gaurav Chaudhary, ISBN: 9781119775584. The third book in the series, "Bioprocessing in Food Science," covers the latest and greatest new technologies in food science.

FUNCTIONAL FOODS, Edited by Navnidhi Chhikara, Anil Panghal, and Gaurav Chaudhary, ISBN: 9781119775560. Presenting cutting-edge information on new and emerging food engineering processes, *Functional Foods*, the second volume in the groundbreaking new series, "Bioprocessing in Food Science," is an essential reference on the modeling, quality, safety, and technologies associated with food processing operations today.

THERMAL FOOD ENGINEERING OPERATIONS, Edited by Nitin Kumar, Anil Panghal, and M. K. Garg, ISBN: 9781119775591. Presenting cutting-edge information on new and emerging food engineering processes, Thermal Food Engineering Operations, the first volume in the new series, "Bioprocessing in Food Science," is an essential reference on the modeling, quality, safety, and technologies associated with food processing operations today.

Other related titles from Scrivener Publishing

FUNCTIONAL POLYMERS IN FOOD SCIENCE, Part 1: Food Packaging, Edited by Giuseppe Cirillo, Umile Gianfranco Spizzirri, and Francesca Iemma, ISBN: 9781118594896. Provides an extensive overview of the link between polymer properties and food quality. *NOW AVAILABLE!*

FUNCTIONAL POLYMERS IN FOOD SCIENCE, Part 2: Food Processing, Edited by Giuseppe Cirillo, Umile Gianfranco Spizzirri, and Francesca Iemma, ISBN: 9781118595183. Provides an extensive overview of the link between polymer properties and food quality. *NOW AVAILABLE!*

ADVANCES IN FOOD SCIENCE AND TECHNOLOGY Volume 1, Edited by Visakh, P.M., Sabu Thomas, Laura B. Iturriaga, and Pablo Daniel Ribotta, ISBN: 9781118121023. Written in a systematic and comprehensive manner the book reports recent advances in the development of food science and food technology areas. *NOW AVAILABLE!*

ADVANCES IN FOOD SCIENCE AND NUTRITION Volume 2, Edited by Visakh P.M., Laura B. Iturriaga, and Pablo Daniel Ribotta, ISBN: 9781118137093. This important book comprehensively reviews research on new developments in all areas of food chemistry/science and nutrition. *NOW AVAILABLE!*

Applied Water Science Volume 1: Fundamentals and Applications, Edited by Inamuddin, Mohd Imran Ahamed, Rajender Boddula and Tauseef Ahmad Rangreez, ISBN 9781119724766. Edited by one of the most well-respected and prolific engineers in the world and his team, this is the first volume in a two-volume set that is the most thorough, up-to-date, and comprehensive volume on applied water science available today. *COMING IN SUMMER 2021*

Printed in the USA/Agawam, MA
September 16, 2026

910661.137